耐火材料实用汉英词典丛书

耐火材料原料
实用术语汉英词典

A Chinese-English Dictionary
of Practical Terms for Refractory Raw Materials

主　编　马北越　罗旭东
副主编　任鑫明　张文国　叶　航　李广明

北　京
冶　金　工　业　出　版　社
2023

内 容 提 要

本书共 10 章，首先介绍了耐火材料的定义及按照不同方法对耐火材料进行分类，接着按化学成分分类分别介绍了氧化硅系耐火材料、铝硅酸盐系耐火材料、氧化镁系耐火材料、尖晶石族耐火材料、碳复合耐火材料、非氧化物系特种耐火材料、隔热耐火材料、耐火材料结合剂等不同体系涉及的原料术语词、术语词组、术语例句、术语例段的汉英对照，最后介绍了耐火材料原料常用词汇术语查询，包括汉英对照、英汉对照。

本书可作为无机非金属材料（耐火材料）和冶金工程等专业的研究生、本科生及工程技术人员学习和查阅用书，也可供相关专业教师参考之用。

图书在版编目（CIP）数据

耐火材料原料实用术语汉英词典/马北越，罗旭东主编 . —北京：冶金工业出版社，2023. 3
（耐火材料实用汉英词典丛书）
ISBN 978-7-5024-9446-9

Ⅰ. ①耐… Ⅱ. ①马… ②罗… Ⅲ. ①耐火材料—术语—词典—汉、英 Ⅳ. ①TQ175-61

中国国家版本馆 CIP 数据核字（2023）第 046446 号

耐火材料原料实用术语汉英词典

出版发行	冶金工业出版社	电　　话	（010）64027926
地　　址	北京市东城区嵩祝院北巷 39 号	邮　　编	100009
网　　址	www. mip1953. com	电子信箱	service@ mip1953. com

责任编辑　于昕蕾　美术编辑　彭子赫　版式设计　郑小利
责任校对　王永欣　李　娜　责任印制　禹　蕊
三河市双峰印刷装订有限公司印刷
2023 年 3 月第 1 版，2023 年 3 月第 1 次印刷
710mm×1000mm　1/16；17.5 印张；352 千字；271 页
定价 98.00 元

投稿电话　（010）64027932　投稿信箱　tougao@ cnmip. com. cn
营销中心电话　（010）64044283
冶金工业出版社天猫旗舰店　yjgycbs. tmall. com
（本书如有印装质量问题，本社营销中心负责退换）

前　言

　　耐火材料是高温工业不可缺少的基础材料，其应用涉及钢铁、有色金属、建材、机械、石油化工、陶瓷和玻璃制造等领域。耐火材料工业的发展已成为国民经济发展的基础条件之一。随着人类社会的发展与冶金等工业技术的进步以及低碳经济时代即将来临，对耐火材料的要求日益提高，除了追求其长寿命外，希冀其具有洁净熔体等功能化作用，也希望其在使用过程中对工业炉衬和高温下容器部件具有显著节能减排作用。

　　国内从事耐火材料研发、生产和应用单位众多，每年发表上千篇耐火材料方面的学术论文、国家/国际专利和科研成果，多项产品远销世界各地。因此，在科研成果撰写和产品销售宣传等各个层面，都需要科研单位和企业界人士精准掌握耐火材料实用术语及对应的英文表述。通过此书编写，旨在让读者掌握耐火材料不同体系原料规范的汉语和英语表述方法，提高读者的英文表述能力，提升耐火材料从业人员的整体水平和汉英互译素质，为耐火材料行业的国际化发展之路做出贡献。

　　本书章节安排中，设置了术语词、术语词组、术语例句和术语例段，方便读者查阅和学习。本书共分为 10 章，由东北大学马北越教授、辽宁科技学院罗旭东教授担任主编，东北大学任鑫明博士、山东国茂冶金材料有限公司张文国董事长、奥镁（中国）有限公司叶航博士和大石桥市美尔镁制品有限公司李广明副总经理担任副主编。编写工作分工如下：马北越负责第 1、5、6、7 章，罗旭东、李广明和马北越负责第 4、8 章，任鑫明和马北越负责第 10 章，张文国和任鑫明负责第 9 章，叶航、任鑫明和马北越负责第 2、3 章。在编写过程中，得到

I

国内多名耐火材料专家教授的指导和帮助，在此一并表示感谢。

限于编者水平和时间仓促，在内容和编排上可能会有欠妥之处，敬请广大读者批评指正。

马北越　罗旭东

2022 年 11 月于沈阳

目　　录

1 绪 论

按照国际标准，耐火材料定义为：化学与物理性质允许其在高温环境下使用的非金属（并不排除含有一定比例的金属）材料与产品（ISO 836，107）。中国标准沿用了 ISO 标准，定义耐火材料为物理与化学性质适宜于在高温下使用的非金属材料，但不排除某些产品可含有一定量的金属材料（GB/T 18930—2002）。

耐火材料的分类方法很多，按照化学性质可将耐火材料分为酸性耐火材料、碱性耐火材料和中性耐火材料。酸性耐火材料是指以二氧化硅为主要成分的耐火材料；碱性耐火材料是指通常以氧化镁、氧化钙或两者共同作为主要成分的耐火材料；中性耐火材料是指在高温下不和酸性耐火材料、碱性耐火材料、酸性或碱性渣或溶剂发生明显化学反应的耐火材料。

按照供给形态，可将耐火材料分为定形耐火材料和不定形耐火材料。定形耐火材料是指具有固定形状的耐火制品与保温制品，其分为致密定形制品与保温定形制品，致密定形制品是总气孔率低于 45% 的制品，保温定形制品为总气孔率高于 45% 的制品。按照形状的复杂程度，定形耐火制品又分为标型砖和异形砖等。不定形耐火材料是指由骨料（颗粒）、细粉、结合剂和添加剂组成的混合料，以散装为交货状态直接使用，或者加入一种或多种不影响耐火材料使用性能的合适液体后使用。在有些不定形耐火材料中还可以加入少量金属、无机或有机纤维材料。不定形耐火材料主要包括浇注料、捣打料、喷射料、可塑料、干式料、接缝料、涂料、挤压料、炮泥和泥浆等。

按照耐火材料中各组分（颗粒、细粉）之间的结合形式，可将耐火材料分为化学结合、水化结合、树脂结合、有机结合和陶瓷结合等。化学结合是指在室温或更高温度下通过化学反应（不是水化反应）产生硬化形成的结合，包括无机或无机-有机复合结合，常见于各种不烧制品中。水化结合是指在常温下，某种细粉与水发生化学反应产生凝固和硬化而形成的结合，常见于浇注料中，如水泥结合浇注料等。树脂结合是指含有树脂的耐火材料在较低温度下加热，由于树脂固化、炭化而产生的结合，常见于含碳耐火材料中。有机结合是指在室温或稍高温度下凭借有机物质产生硬化而形成的结合，常见于不烧制品中。陶瓷结合是指在一定温度下，因烧结或液相形成而产生的结合，常见于烧成耐火制品中，如烧成砖等。

　　按照耐火材料经过高温烧结与否，可将其分为烧成耐火材料和不烧耐火材料。烧成耐火材料是指经过高温烧结的耐火材料。不烧耐火材料是指没有经过高温烧结的耐火材料，主要包括水化结合为主的不定形耐火材料，化学结合、树脂/沥青/焦油结合的耐火材料等。

　　按照生产方式，可将耐火材料分为机压成型耐火材料、手工成型耐火材料、浇注耐火材料和熔铸耐火材料等。需要指出的是，熔铸耐火材料是指先将耐火材料配料熔融后再铸入到模型中使其凝固而制得的耐火材料。

　　按照化学成分，可将耐火材料分为氧化硅系耐火材料、铝硅酸盐系耐火材料、氧化镁系耐火材料、尖晶石族耐火材料、碳复合耐火材料、非氧化物系特种耐火材料等。氧化硅系耐火材料是指以二氧化硅为主要成分的耐火材料，通常二氧化硅的含量不低于93%，常包括硅质耐火材料和半硅质耐火材料。铝硅酸盐系耐火材料是指以三氧化二铝和二氧化硅为主要成分的耐火材料，按照三氧化二铝含量的不同可分为黏土质耐火材料（$30\% \leqslant w(Al_2O_3) < 45\%$）和高铝质耐火材料（$w(Al_2O_3) > 45\%$）等。氧化镁系耐火材料是指氧化镁含量大于80%的耐火材料，包括镁质耐火材料、镁钙质耐火材料和镁硅质耐火材料等。尖晶石族耐火材料包括镁铝质耐火材料、镁铬质耐火材料。碳复合耐火材料是指由氧化物、非氧化物和石墨等炭素材料构成的复合材料，包括碳-氧化物系耐火材料和碳-氧化物-非氧化物系耐火材料。非氧化物系特种耐火材料包括碳化硅/氮化硅质耐火材料和赛隆质耐火材料。

　　按化学成分分类是耐火材料最常见的分类方式，所以本书据此进行章节设置。另外，从耐火材料体系完整度方面考虑，把隔热耐火材料和耐火材料结合剂单列为两章内容，以便读者能更全面了解耐火材料体系脉络。

2 氧化硅系耐火材料原料

2.1 硅质耐火材料原料

2.1.1 术语词

硅石（石英岩）silica,quartzite,ganister

β-方石英 beta-cristobalite

柯石英 corsite

方石英化 cristobalization

电熔石英 electro-quartz

凯石英 keatite

焦石英 lechatelierite

半稳定性方石英 meta-cristobalite

变质石英岩 metaquartzite

石英 silex,quartz

石英砂岩 silicarenite

超石英 stishovite

磷石英 tridymite

鳞石英 tridymite

鳞石英化 tridymization

2.1.2 术语词组

无定形硅石 amorphous quartzite

胶结硅石 cementing quartzite

结晶硅石 crystalline quartzite

致密硅石 dense quartzite

细粒胶结硅石 fine-grained cement quartzite

微晶硅石 microcrystalline quartzite

多孔硅石 porous quartzite

再结晶硅石 recrystallized quartzite,recrystallized quartzite

砂岩硅石 sandy quartzite

半结晶硅石 semi-crystalline quartzite

极慢转化硅石 sluggish inversion quartzite

软硅石 soft quartzite

硬硅石 stiff quartzite

石英砂的风力分级 air classification of quartzite grit

各向异性针状方石英 anisotropic acicular cristobalite

各向异性鳞片状方石英 anisotropic flaky cristobalite

各向异性稳定性方石英 anisotropic stable cristobalite

3

石英砂 arenaceous quartz, quartz sand, ganister sand

带状赤铁矿石英岩 banded hematite quartzite

巴西石英 Brazilian pebble

散石英砂 chip packing of quartz

晶状石英 crystalline quartz

高铁石英 ferruginous quartz

细分散石英 finely dispersed quartz

游离石英 free quartz

熔融石英 fused quartz

熔融石英块 fused quartz lump

石英粉 ground quartz, quartz powder, silica powder

石英质砾石 quartz gravel

低温型方石英 low cristobalite

粗晶石英 macrocrystalline quartz

粉状石英 powdered quartz

石英晶体 quartz crystal

石英纤维 quartz fiber

石英角岩 quartz hornfels

石英板 quartz lamina

石英矿物 quartz mineral

石英页岩 quartz shale

石英转变 quartz transformation

石英长石砂 quartz-feldspathic sand

次生石英 secondary quartz

烧结石英 sintered quartz

慢速转化石英 sluggish type transformation quartzite

脉石英 vein quartz, gangue quartz

2.1.3　术语例句

硅石的化学加工是硅矿开发的一条重要途径。

The chemistry processing of **silica** is an important path of the silicon mineral development.

该组合物含有至少一种环氧树脂，至少一种溶剂和官能化胶态**硅石**填料。

The composition has at least one epoxy resin, at least one solvent and a filler of functionalized colloidal **silica**.

针对硅砖使用过程中易断裂的弊病，在其生产工艺的基础上，以碳化硅代替部分**硅石**克服之。

On account of the disadvantage-ready cracks of **silica** bricks in use and on the basis of their production technology, part of silica is replaced by silicon carbide to conquer the disadvantage.

对较低熔化温度而言，用于耐热浆的材料是石膏作黏合剂和用粉末状**硅石**作耐温材料的混合物。

These materials used for the slurry are a mixture of plaster, a binder and powdered **silica**, a refractory, for low temperature melts.

4

在常温下，硅不和空气反应，但是在高温下它和氧气反应，形成一层**硅石**，不再继续发生反应。

At ordinary temperatures silicon is impervious to air, but at high temperatures it reacts with oxygen, forming a layer of **silica** that does not react further.

在基材表面涂覆一种处理液，该处理液主要由胶体**二氧化硅**组成，同时添加碳酰肼作为醛捕捉剂。

A coat film surface on a substrate is coated with a treatment liquid that is composed of mainly colloidal **silica** added carbohydrazide as aldehyde capture agent.

含有河砾石或**石英岩**骨料的混凝土的切割难度中等。

Concrete with river gravel or **quartzite** aggregate is of medium difficulty to cut.

五强溪水电站**石英岩**人工砂石骨料加工厂是我国最大的人工骨料加工系统。

Wuqiangxi **quartzite** aggregate processing plant is the largest aggregate processing system in China.

测区内发现磁铁**石英岩**残留体、粗面斑岩、大理岩、钠长岩等新岩石类型。

The new type of rocks such as the residuum of magnetite **quartzite**, trachyte porphyry, marble and sodaclase are found in the survey area.

临桂县矿产资源的开发利用主要是以灰岩为主，重晶石、黏土、页岩、**石英岩**等为辅。

The development and utilization of mineral resources in Lingui County give priority to limestone, while barite, clay, shale, **quartzite** as subsidiary.

数百万年前，变质片岩、片麻岩和**石英岩**，在乞力马札罗山附近广阔的平原形成了壮观的平顶岛礁。

Millions of years ago, metamorphic schists, gneisses and **quartzites** formed impressive, flat-topped inselbergs on a vast plain in the shadow of Kilimanjaro.

其岩石类型主要有云母石英片岩、**石英岩**、大理岩、炭质千枚岩等，因富含炭质而呈灰色、灰黑色。

The rocks of this formation, which are black or light black for rich of C, consist mainly of mica quartz schist, **quartzite**, marble, carbonaceous phyllite.

　　长石石英云母片岩、**石英岩**和大理岩的原岩为沉积岩，与副变质的长英质片麻岩和基性火山岩一起构成了古老的表壳岩组合。

The quartz-mica schist, **quartzite** and marble were metamorphosed sedimentary rocks, and together with paragneiss and mafic rock, constituted the old supracrustal series.

　　撞击所释放的强烈高温会在许多类型的岩石中形成一种热冲击**石英**。

The intense heat of the impact would produce heat-shocked **quartz** in many types of rock.

　　到目前为止，在墨西哥尤卡坦半岛已经发现了几个这样的二级陨石坑，而且，在墨西哥和海地都发现了热休克**石英**。

To date, several such secondary craters have been found along Mexico's Yucatan Peninsula, and heat-shocked **quartz** has been found both in Mexico and in Haiti.

　　地质学家准确辨认出花岗岩是**石英**、长石和云母的混合物。

The geologist does well to regard granite as a compound of **quartz**, felspar, and mica.

　　在一般的煤的样板中，**石英**和其他矿物质的含量是相当低的。

In normal coal samples, **quartz** and other minerals are found only in trace amounts.

　　在矿石硬度表上，它的硬度为8，这表明刀子不能割开它而它可在**石英**上划痕。

In the mineral table of hardness, it has a rating of 8, which means that a knife cannot cut it, and that topaz will scratch **quartz**.

　　可选性分析是**石英砂**选矿前需要进行的必要手段。

It is necessary to conduct separation analysis prior to beneficiation of **quartz sand**.

　　此外还有石灰石、铁矿、红柱石、**石英砂**和高岭土等。

In addition, there are limestone, iron ores alusite, **quartz sand** and kaolin etc.

　　广泛分布的花岗岩及**石英砂**、黏土是重要的建筑原料资源。

The widely distributed granite, **quartz sand** and clay are important resource of construction materials.

均质**石英砂**滤料是当前国内水厂过滤技术应用的主要填料之一。

Uniform **quartz sand** media is one of the main media in present filtration technology application.

当陶瓷超滤膜澄清食醋时，先采用经济简便的**石英砂**过滤方法预处理食醋。

Quartz sand filtration was used for pretreatment before vinegar was ultrafiltered with ceramic membrane.

本文简介了湿法生产**石英砂**的工艺过程、主要设备、关键步骤及成品含量分析。

This article introduces briefly the productive process of **quartz sand** in wet method, main equipment key steps and analysis of the product content.

2.1.4　术语例段

引入复合添加剂是改善低碳 MgO-C 耐火材料高温稳定性和抗渣性的一种有效手段。本工作中，首先由电瓷废料经埋碳法在 1500℃、1550℃ 和 1600℃ 保温 4h 合成 Al_2O_3-SiC 复合粉体，然后将合成的 Al_2O_3-SiC 复合粉体作为添加剂添加至低碳 MgO-C 耐火材料中。系统研究了其添加量（质量分数为 0、2.5%、5.0% 和 7.5%）对耐火材料性能的影响。研究发现，增加热处理温度利于电瓷废料中莫来石和**石英**向氧化铝和碳化硅的转化。此外，添加 Al_2O_3-SiC 复合粉体可有效改善低碳 MgO-C 试样的性能；尖晶石致密层和高黏度孤立层的形成是改善低碳 MgO-C 试样抗氧化性和抗渣性的内在原因。本工作提供了电瓷废料再利于的思路，提出了低碳 MgO-C 耐火材料性能优化的一种可选择策略。

Introducing composite additives are an efficient means to improve the high-temperature stability and slag resistance of low-carbon MgO-C refractories. In this work, Al_2O_3-SiC composite powder was firstly synthesized from electroceramics waste by carbon embedded method at 1500℃, 1550℃, and 1600℃ for 4h, and then the as-synthesized Al_2O_3-SiC composite powder was used as an additive to low-carbon MgO-C refractories. The effects of its addition amounts of 0, 2.5wt.%, 5.0wt.%, and 7.5wt.% on the properties of the refractories were investigated in detail. It was found that increasing the heat treatment temperature is beneficial to the phase conversion of mullite and **quartz** to alumina and silicon carbide in the electroceramics waste. Furthermore, the addition of Al_2O_3-SiC composite powder effectively improves the performance of low-carbon MgO-C samples, and the formation of spinel dense layer and high-viscosity isolation layer is the internal reason for the improvement of the oxidation resistance and slag resistance of low-carbon MgO-C samples. This work provides ideas for the reuse of electroceramics waste and presents an alternative strategy for the performance optimization of low-carbon MgO-C refractories[1].

　　以烧结刚玉、电熔刚玉、烧结莫来石和高纯石英砂为原料，经1750℃烧成制得刚玉-莫来石耐火材料。探究其在热风炉工况下的损毁机理。结果表明：有4种形貌的莫来石呈现：使用过程中进一步发育长大的莫来石原料；热风煤气中介稳态的 SiO(g) 扩散迁移进入烧结刚玉颗粒孔隙中反应形成的莫来石；与刚玉相互扩散固溶后于其颗粒表面形成的"钉扎"结构的莫来石；热风介质中的部分杂质元素进入莫来石晶格形成的莫来石固溶体。"钉扎"结构的莫来石对刚玉-莫来石砖物理性能的保持起着重要作用。未观察到 CO 或碱蒸气（主要为 K_2O 和 Na_2O 等）在用后砖体内的富集及其对砖体的侵蚀。在高温和低氧分压等因素作用下，部分杂质以气相挥发，导致用后砖杂质含量降低，气孔率略增加。使用后的刚玉-莫来石耐火材料结构与性能优良，能够实现热风炉的长寿[2]。

　　A corundum-mullite refractory was prepared with sintered alumina, fused alumina, sintered mullite and high-purity **quartz sand** via sintering at 1750℃. The damage mechanism of corundum-mullite refractory for hot stove was investigated. The results show that four types of mullite with different morphologies are mullite in raw materials that is further grown during hot stove operation, mullite formed by reaction between alumina and SiO(g) from hot air gas diffusing and migrating into the pores of sintered alumina particles, mullite with "pinned" structure formed on the surface of the corundum particle, and mullite solid solution due to the solution from some impurity elements in the hot air medium. The "pinning" mullite has a contribution to the properties of the material. Little CO or alkali vapor gas(mainly K_2O and Na_2O, etc.) appears to erode the refractory severely or be enriched in the brick. Some impurities are volatilized in the gas phase at a high temperature and a low oxygen partial pressure, resulting in a decrease in the impurity content and a slight increase in the porosity of the used brick. The corundum-mullite refractory in hot stove operation has superior performance and structure, thus realizing a long campaign for the hot stove[2].

　　设计了一种新型多孔陶瓷支撑膜。这种新型支架以**多晶石英砂**和方解石为原料制成。在这项工作中，已利用挤压方法生产了两种构型的支架（管状和扁平状）。研究发现，烧结支架的开放开孔率、孔径分布、平均孔径（APS）、强度和渗透性主要取决于方解石（CaCO₃）添加剂的质量比。结果表明，方解石添加量（质量分数）在 15%~35%，在 1375℃ 左右烧结 1h，可以获得最佳的支架性能。所研制的管状陶瓷支架的 APS 为 6.3~12μm，开孔率为 42%~55%，渗透性为 16~68m³/(h·m²·bar)❶，抗折强度为 8~18MPa，此研究结果有望为膜技术的广泛应用提供广阔的前景。

A new type of porous ceramic supports for membranes has been designed. The new supports have been fabricated from **polycrystalline quartz sand** and calcite raw materials. In this work, two configurations of support (tubular and flat) have been produced using extrusion method. The open porosity, the pore size distribution, the average pore size (APS), the strength and the permeability of sintered supports have been found to depend mainly on the weight ratio of calcite(CaCO₃) additive. The results showed that with the addition of 15wt.% -35wt.% of calcite and sintering temperature of about 1375℃ for 1h the best characteristics of sintered supports could be obtained. The developed tubular ceramic supports with the APS 6.3-12μm, open porosity 42%-55%, the water permeability 16-68m³/(h·m²·bar) and flexural strength 8-18MPa hopefully offer many perspectives for a wide use in membranes technology[3].

传统的硅砖以其在高温下的尺寸稳定性和耐酸性熔渣而著称。砖由**石英岩**经压制和高温烧结制成。这些材料的主要缺点是低温下的高热膨胀和对更宽温度范围内波动的敏感性。为了克服传统砖的不足，人们开发了熔融石英基材料。它们可能具有不同的黏合剂，包括水泥、硅胶和磷酸盐，并且可以通过压制或铸造来生产。熔融石英耐火材料的主要特点是热膨胀系数低，可以得到大而复杂的预制形状。它们的缺点是二氧化硅在高温下的潜在结晶和缺少实际使用情况的统计。因此，需要对熔融石英耐火材料的特性进行深入研究。多年来，人们对硅砖的性能已经进行了广泛的研究。而可用于熔融石英的有限数据主要是针对处于原始无定形状态的材料。在室温下对硅砖和结晶熔融石英材料试样进行的三点弯曲试验表明，硅砖表现出明显的脆性断裂，而熔融石英出现了应变软化。分析认为，硅砖的疲劳韧性受晶粒互锁、大晶粒开裂和熔体形成的损伤愈合的影响。

Conventional silica bricks are known for their dimensional stability at high temperature and resistance to acidic slag. The bricks are produced from **quartzite** by pressing and high temperature sintering. The main drawback of these materials is high thermal expansion at low temperatures and resulting sensitivity to fluctuations in wider temperature range. To overcome the deficiency of the conventional bricks fused silica-

---

❶ 1bar=0.1MPa。

based materials have been developed. They may feature different binders, including cement, silica gel and phosphates, and can be produced either by pressing or casting. The main features of fused silica refractories are low coefficient of thermal expansion and the possibility to obtain large and complex pre-cast shapes. Their drawback is the potential crystallization of silica at high temperature and the limited-service statistics. The latter necessitates thorough research of the material properties. Properties of silica bricks has been extensively studied over years. Limited amount of data available for fused silica is mainly for the material in its original amorphous condition. Three-point bending tests done at room temperature performed on samples of silica brick and crystallised fused silica material has indicated a rather brittle failure for the former and developed strain softening for the latter. The fatigue in silica brick is affected by grain interlocking, large grain cracking, and the damage healing by the melt formation[4].

　　具有不同几何形状和成分的陶瓷膜的开发扩展了工业应用的可能性，在增加渗透性、膜面积的体积模块和化学、热及机械电阻方面具有优势。在科学研究中，使用低成本的原材料已成为一种趋势。本工作的目的是在1100~1500℃的温度下合成陶瓷块与聚醚砜溶液的混合物，采用浸没沉淀法，从氧化铝和**石英岩残渣**中制备具有中空几何纤维结构的膜。通过化学分析、X射线衍射、粒径分布、扫描电镜、显气孔率、弯曲强度和渗透性对中空纤维膜进行了表征。结果表明：烧结温度对莫来石相的形成、显气孔率和渗流性能有直接的影响。烧结温度越高（1400~1500℃），莫来石相的形成越快，孔隙率越低，膜的渗透性越小，但高温下中空纤维膜的抗弯强度有所增加。

The development of ceramic membranes with different geometries and compositions extends the possibilities of industrial applications, inducing advantages in terms of increased permeability, membrane area by volume module and chemical, thermal and mechanical resistance. The use of low-cost raw materials is a trend that has grown in scientific research. The aim of this work is to prepare membranes with hollow fiber geometry from alumina and **residue of quartzite**, by the technique of immersion precipitation in distilled water from a mixture of ceramic mass with a solution of polyether sulfone and, synthesized in temperatures of 1100℃ to 1500℃. The hollow fiber membranes were characterized by chemical analysis, X-ray diffraction, particle size distribution, scanning electron microscopy, apparent porosity, flexural strength and permeated water flow by the membranes. The results indicated that the sintering temperature has direct influence on the formation of the mullite phase, and the properties of apparent porosity and permeate flow. The higher the sintering temperature (1400-1500℃) increase the formation of the mullite phase, the lower the porosity, as well as the lower the permeate water flow in the membranes. However, there was increase in flexural strength in the hollow fiber membranes with high temperature[5].

在 Si 和 FeSi 生产中，硅的主要来源是 SiO_2，以**石英**的形式存在。与 SiO_2 反应生成 SiO 气体，进一步与 SiC 反应生成 Si。在加热过程中，石英会转化为其他 SiO_2 修饰物，**方石英**为稳定的高温相。向方石英转变是一个缓慢的过程。对几种工业石英源进行了研究，结果表明，在不同类型的石英中，其速率有很大差异。在加热过程中，对这些石英源之间的其他差异，如软化温度和体积膨胀，也进行了研究。石英与方石英的比例会影响 SiO_2 的反应速率。本文还讨论了石英类型之间所观察到的差异的工业后果和其他含义，从而开发出一种新的实验方法，并对几种新的石英源进行了调查，证实了早期观察到的不同源之间的巨大变化。以前期的研究结果作为基础，研究了数据的可重复性，并考察了气氛的影响。

In Si and FeSi production, the main Si source is SiO_2, in the form of **quartz**. Reactions with SiO_2 generate SiO gas that further reacts with SiC to Si. During heating, quartz will transform to other SiO_2 modifications with cristobalite as the stable high-temperature phase. Transformation to **cristobalite** is a slow process. Its rate has been investigated for several industrial quartz sources and has been shown to vary considerably among the different quartz types. Other differences in behavior during heating between these quartz sources, such as softening temperature and volume expansion, have also been studied. The quartz-cristobalite ratio will affect the rate of reactions involving SiO_2. The industrial consequences and other implications of the observed difference between quartz types are discussed. In the current work, a new experimental method has been developed, and an investigation of several new quartz sources has confirmed the earlier observed large variation between different sources. The repeatability of the data has been studied and the effect of gas atmosphere investigated. The results from the earlier work are included as a basis for the discussion[6].

在大多数钢铁行业中，通过高炉（BF）路线生产的铁水仍然是最优选的路线。在高炉操作中，**石英岩**用作助熔剂，以调节所需的炉渣化学性质和碱度（B2 为 1～1.08），以获得最佳的炉渣性能，例如低液相线温度、高硫容和低黏度。本文介绍了在 JSW 钢铁有限公司进行的调查，以确定除石英岩之外的其他二氧化硅来源，以用作高炉中的助熔剂，维持所需的炉渣化学成分，从而实现平稳运行并降低成本。卡纳塔克邦地区可得到的**带状赤铁矿石英岩**（BHQ）矿石包含 28%～33%

（质量分数）的铁，45% ~ 46%（质量分数）的二氧化硅，1.3% ~ 1.5%（质量分数）的氧化铝，可部分替代高炉石英岩。BHQ 的矿物学和相分析表明，赤铁矿和硅石交替分布，而块矿中赤铁矿和硅石分布均匀。BHQ 的这种相分布和还原性指数（约66%）通过形成铁橄榄石（$2FeO \cdot SiO_2$）作为初渣。发现细粉生成量相对较高（质量分数6.3% ~6.4%），但在可接受的范围内。BHQ 中的铁含量也有助于向炉中输入总铁，有利于选择减少炉料中含铁材料的当量。JSW 4 号高炉的工厂试验表明，整体操作性能得到了改善，每次进料的铁输入量增加，由于氧化铝输入量低，炉渣率降低了 8kg/tHM，并且最终炉渣化学成分更容易实现。

Hot metal produced through blast furnace (BF) route is still the most preferred route in majority of the steel industries. In the blast furnace operation, **quartzite** is used as flux to adjust the desired slag chemistry and basicity (B2 of 1-1. 08) to achieve optimum slag properties such as low liquidus temperature, high sulphur carrying capacity and low viscosity. This paper describes the investigations carried out at JSW Steel Ltd. to identify alternate sources of silica other than quartzite for use as flux in BF to maintain the desired slag chemistry for smooth operation and cost reduction. **Banded hematite quartzite** (BHQ) ore available in Karnataka region contains 28wt.% -33wt.% Fe, 45wt.% -46wt.% silica, 1. 3wt.% -1. 5wt.% alumina and is found suitable for use as partial replacement for quartzite in blast furnaces. Mineralogy and phase analysis of BHQ reveals alternate bands of hematite and silica whereas in lump ore, hematite and silica are uniformly distributed. This phase distribution and reducibility index of BHQ (about 66%) affect the softening start temperature by forming fayalite ($2FeO \cdot SiO_2$) as primary slag. Fines generations are found to be comparatively higher (from 6. 3wt.% to 6. 4wt.%) but in acceptable range. The Fe content in the BHQ also contributes to the overall Fe input to the furnace favoring an option to reduce the equivalent amount of iron bearing materials in the charge. Plant trials at JSW steel blast furnace #4, indicate an improvement in overall operational performance resulting in increased Fe input per charge, reduction in slag rate by 8kg/tHM due to low alumina input and easier achievement of final slag chemistry[7].

本文对**石英**和**石英岩**的选材进行了研究，以确定分别在硅铁和纯硅生产中使用它们的优先级。冶金行业中通常使用最高质量的石英和石英岩，但尽管化学成分相似，并非所有类型的硅原料都适合生产硅铁和纯硅。在硅铁和硅的加热和碳热生产的工艺条件下可以观察到行为差异。这些差异尤其取决于杂质的性质和含量，以及单个晶粒的粒度（块度）和微观结构。本研究主要集中在确定硅原料的物理化学和冶金性能，该研究的组成部分还包括创建一种用于确定石英（或石英岩）还原性的新方法，这种方法可用于实际工业过程，并且应该非常可靠。本文介绍了实验室实验的结果以及对单个石英（或石英岩）的理化和冶金性能的评估。根据对测试试样性能的比较，确定了它们的使用优先级。这些原材料在工业条件下的使用有望实现更好的生产参数，例如更高的产量和产品质量以及更低的电耗。

This article deals with material research of selected types of **quartz** and **quartzites** in order to determine the priority of their use in the production of ferrosilicon and pure silicon, respectively. The highest quality quartzes and quartzites are commonly used in metallurgy, but not all types of these silicon raw materials are suitable for the production of ferrosilicon and pure silicon, despite their similar chemical composition. Behavior differences can be observed in the process conditions of heating and carbothermic production of ferrosilicon and silicon. These differences depend, in particular, on the nature and content of impurities, and the granularity (lumpiness) and microstructure of individual grains. The research focused primarily on determining the physicochemical and metallurgical properties of silicon raw materials. An integral part of the research was also the creation of a new methodology for determining the reducibility of quartzes (or quartzites), which could be used for real industrial processes and should be very reliable. The results of the laboratory experiments and evaluation of the physicochemical and metallurgical properties of the individual quartzes (or quartzites) are presented in the discussion. Based on comparison of the tested samples' properties, their priority of use was determined. The use of these raw materials in industrial conditions is expected to result in the achievement of better production parameters, such as higher yield and product quality and lower electricity consumption[8].

二氧化硅纳米颗粒（SiO_2 NPs）已在许多先进领域得到应用。在本研究中，我们报告了一种低成本、简单和安全的制备 SiO_2 NPs 粉末的技术路线。这是一种很有前途的方法，因为它对不同的应用程序均表现出高产率（91.7%）。以商业硅粉为原料，采用超声波法和碱湿化学蚀刻法（一步法）制备 SiO_2 NPs。研究发现，影响从晶体硅生产晶体（**石英、方石英和鳞石英**）和非晶 SiO_2 NPs 的主要因素是蚀刻溶液（KOH）浓度和超声处理时间。SiO_2 NPs 的粒径与3%（质量分数）KOH 的超声处理时间成反比，但与6%（质量分数）KOH 的超声处理时间成正比。同时，通过 X 射线衍射、傅里叶变换红外光谱、拉曼光谱和扫描电镜对合成的纳米颗粒进行了系统的表征。

Silica nanoparticles (SiO$_2$ NPs) have found applications in many advanced areas. In this study, we report a powder technology route as a low cost, simple and safe method for the fabrication of SiO$_2$ NPs. It is a promising method as it has high throughput(91.7%) for different applications. The SiO$_2$ NPs were prepared by an ultra-sonication technique and an alkali wet chemical etching process(one step), starting from commercial silicon powder. The main factors that affect the production of crystalline(**quartz**, **cristobalite**, and **tridymite**)and amorphous SiO$_2$ NPs from crystalline silicon are the etching solution (KOH) concentration and the sonication time. The particle size of the SiO$_2$ NPs is inversely related to the sonication time for 3wt.% KOH but directly proportional to the sonication time for 6wt.% KOH. The synthesized nanoparticles were systematically characterized by X-ray diffraction, Fourier transform infrared spectroscopy, Raman spectroscopy and scanning electron microscopy[9].

研究了不同掺量的天然**石英**（质量分数为 2.5% ~ 20%）与氟磷灰石 （FAp）混合制备的氟磷灰石–硅酸盐复合材料的烧结行为。将复合材料在1000 ~ 1350℃的温度下在空气中进行无压烧结。研究了温度对复合材料致密化、相组成、化学键合和维氏硬度的影响。结果表明，所有试样均为混合相，主要成分为 FAp 和方沸石，次要成分为**方石英**、硅灰石、硅酸钙和（或）白石，并且主要成分取决于石英含量和烧结温度。其中，石英含量（质量分数）为 2.5% 的复合材料烧结性能最佳。在1100℃烧结时，2.5% 石英-FAp 复合材料的体积密度最高，体积密度为3g/cm^3，维氏硬度大于 4.2GPa。研究发现，石英的加入改变了复合材料的微观结构，在1000℃烧结时呈现棒状形貌，在1350℃烧结时呈现规则的圆形晶粒结构。而对于石英含量高的复合材料，在晶粒表面观察到了湿润的现象，这与瞬时液相烧结有关。

The sintering behaviour of fluorapatite(FAp)-silicate composites prepared by mixing variable amounts of natural **quartz**(2.5wt.% to 20wt.%) and FAp was studied. The composites were pressureless sintered in air at temperatures from 1000℃ to 1350℃. The effects of temperatures on the densification, phase formation, chemical bonding and Vickers hardness of the composites were evaluated. All the samples exhibited mixed phase, comprising FAp and francolite as the major constituents along with some minor phases of **cristobalite**, wollastonite, dicalcium silicate and/or whitlockite dependent on the quartz content and sintering temperature. The composite containing 2.5wt.% quartz exhibited the best sintering properties. The highest bulk density of 3g/cm^3 and a Vickers hardness of > 4.2GPa were obtained for the 2.5wt.% quartz-FAp composite when sintered at 1100℃. The addition of quartz was found to alter the microstructure of the composites, where it exhibited a rod-like morphology when sintered at 1000℃ and a regular rounded grain structure when sintered at 1350℃. A wetted grain surface was observed for composites containing high quartz content and was believed to be associated with a transient liquid phase sintering[10].

熔融石英作为熔体单晶生长的坩埚材料具有优良的热学和化学性质，其高纯度和低成本使其对高纯度晶体的生长特别有吸引力。然而，在某些类型的晶体的生长过程中，熔体和石英坩埚之间需要一层热解碳涂层。本文介绍了一种利用真空气相输运技术制备热解碳涂层的方法。该方法被证明可以有效地在各种尺寸和形状的坩埚上产生相对均匀的涂层。所合成的热解碳涂层的特征在于光学衰减测量。在每个包覆过程中，随着热解持续时间的增加，涂层厚度都趋于指数增长，平均涂层厚度随有效己烷蒸气体积与热解涂层表面积之比近似线性增加。用这种工艺涂覆的石英坩埚成功培育出直径为2in❶ 的NaI单晶，并且发现NaI晶体的表面质量随着涂层厚度的增加而提高。

Fused quartz has excellent thermal and chemical properties as crucible material for single crystal growth from melt, and its high purity and low cost makes it especially attractive for the growth of high-purity crystals. However, in the growth of certain types of crystals, a layer of pyrolytic carbon coating is needed between the melt and the quartz crucible. In this article, we describe a method for applying pyrolytic carbon coating by vacuum vapor transport. The method is shown to be effective in yielding relatively uniform coating on a wide range of crucible sizes and shapes. The resultant pyrolytic carbon coating is characterized by optical attenuation measurements. In each coating process, the thickness of coating is shown to approach a terminal value with an exponential tail as the duration of pyrolysis increases, and the average thickness roughly increases linearly with the ratio of the volume of available hexane vapor to the surface area of pyrolytic coating. Quartz crucibles coated by this process have been used to successfully grow up to 2-in-diameter NaI single crystals, and the surface quality of NaI crystal was found to improve as the thickness of coating increases[11].

本研究采用**脉石英碱腐蚀酸浸工艺制备高纯石英**。采用碱腐蚀法，可以去除石英颗粒表面的杂质，使石英表面的裂纹加深和扩大，并伴有腐蚀坑的存在，有利于酸浸进入颗粒，并在进一步酸浸过程中与内部杂质发生反应。经酸溶液浸出后，脉状石英中存在的杂质可浸入溶液中，从而得到高纯度石英。研究了NaOH浓度、反应温度和反应时间对碱腐蚀过程中杂质元素Fe、Al、K、Na和Li去除率的影响，得到了最佳条件。当NaOH浓度为12%、温度为200℃、反应时间为

---

❶ 1in=2.54cm。

15

100min 时 的 去 除 率 最 高： Fe（65.84%）、K（50.26%）、Na（55.96%）、Li（20%）、Al（17.12%）。用 4mol/L HCl、1mol/L HNO$_3$、0.25mol/L HF 在 200℃ 酸浸 5h，最终可得到纯度为 99.994% 的石英。

Prepare of **high-purity quartz** using **vein quartz** by alkali corrosion and acid leaching processes was investigated in this study. Using alkali corrosion method, the impurities on the surface of quartz particles can be removed and the cracks on the quartz surface can be deepened and enlarged, accompanied by the existence of corrosion pits, which is conducive to the acid leaching into the particles and the reaction with the internal impurities during further acid leaching. After leaching by acid solution, the impurity present in the vein quartz can be leached into the solution and high-purity quartz can be obtained. Influence of NaOH concentration, reaction temperature and reaction time affecting on removal rate of impurity element Fe, Al, K, Na and Li and in the alkali corrosion process were studied to obtain the optimum condition. Maximum removal rate of Fe(65.84%), K(50.26%), Na(55.96%), Li(20%), Al(17.12%) can be obtained at the condition: NaOH concentration 12%, 200℃ temperature and 100min of reaction time. After acid leaching using 4mol/L HCl、1mol/L HNO$_3$、0.25mol/L HF at 200℃ for 5h, the high-purity quartz (99.994%) can be obtained eventually[12].

在工业生产和生活过程中会产生大量的含油废水。含油废水对环境造成了严重的污染，对人类的生存环境构成了严重的威胁。深床过滤是常用的含油废水深度处理方法，作为操作单元与其他方法联用。在过滤单元，滤床中油、水和滤料表面三相共存，滤料的表面润湿性对除油效率有很大的影响。对于疏水亲油滤料，当油珠颗粒迁移到滤料表面时，受到滤料表面的亲油作用，将会黏附于滤料表面被去除。显然，当滤料的疏水亲油性越强，油珠越容易黏附于滤料表面，其去除效率越高。而对于亲水疏油滤料，当油珠靠近滤料表面时，受其疏油作用会保持为球形，不容易从比油珠粒径小的孔隙流走，故也可能具有较高的去除效率。另外，就水头损失而言，滤料的疏水性越强，其水头损失越大，反之亦然。因此，提高滤料表面的疏水性可提高滤床的除油效率，但会增大过滤的水头损失；反之，若提高滤料表面的亲水性，则会降低水头损失。近年来，材料领域在特殊润湿性表面方面的研究取得了很大的进展，并在油水分离领域被广泛研究。应用于油水分离的特殊润湿性表面主要包括超疏水超亲油表面和超亲水水下超疏油表面。其油水分离方式主要有两种：一是选择过滤，二是超疏水吸附。而在硬质颗粒滤料过滤方面，特殊润湿性滤料还鲜有研究。为此，本文研究了超疏水超亲油滤料和超亲水水下超疏油石英砂滤料的制备及其除油性能，并研究了两种不同润湿性滤料混合的过滤性能。（1）通过浸涂法将十三烷酸和十八烷基三氯硅烷的有机长链烷烃接枝到了石英砂滤料表面，但没有构造粗糙结构，水的接触角达到了 134.2°，滤料表面成为了高疏水性。通过浸涂法将三（羟甲基）氨基甲烷和十八烷基三氯硅烷复合接枝到**石英砂**滤料表面，同时用纳米氧化锌颗粒构造微纳米粗糙结构，增强了滤料的疏水性，水的表观接触角在 150°以上，油的表观接触角在 0°左右，得到了超疏水超亲油石英砂滤料。所制备的滤料具有很强的耐机械磨损性和较强的耐特殊环境性。（2）将壳聚糖与戊二醛交联，再通过化学反应接枝到石英砂滤料表面，并用二氧化硅纳米颗粒构造微纳米粗糙结构，获得了水下油的接触角为 152.1°±1.2°的超亲水水下超疏油石英砂滤料。制备的壳聚糖包覆的超亲水水下超疏油石英砂滤料具有一定的耐久性、耐碱性、耐机械磨损和抗空气暴露性能。将马铃薯残渣用 NaOH 和超声活化后，再利用水性聚氨酯涂覆于石英砂滤料表面，制备得到水下油的接触角为 151.7°±1.3°的马铃薯残渣包覆的超亲水水下超疏油石英砂滤料。制备的马铃薯残渣包覆的超亲水水下超疏油石英砂滤料具有一定的耐久性、耐碱性、耐机械磨损和抗空气暴露性能。由于马铃薯残渣为废弃物，将其资源化利用，具有很好的实用性，并且成本低廉。（3）制备的超疏水超亲油石英砂滤料对水的吸附量仅为未改性石英砂滤料的 2%，对机油和菜籽油的吸附量比未改性石英砂滤料则分别提高了 54% 和 45%，对不同油的吸附量为未改性石英砂滤料的 5~20 倍，吸水性大大降低，吸油性得到了提高。超疏水超亲油石英砂滤料对机油和菜籽油乳化液废水的去除率分别达到了 99.65% 和 96.46%，对油水混合液的分离效率均大于 92.8%，具有很好的油水分离性能。（4）相对于未改性石英砂滤料，制备的两种超亲水水下超疏油石英砂滤料对乳化油废水的除油效率具有一定程度的提高，对油水混合液的分离效率有大幅提高。对水的渗透系数增大了 11%~19%，对油的渗透系数降低了 6%~8%。优秀的油水分离性能及其较低的水头损失为超亲水水下超疏油石英砂滤料进行油水分离提供了一种新的方案。（5）对于未改性石英砂滤料和超疏水超亲油石英砂滤料按照一定比例混合的混合滤料，其水的表观接触角 θ 与混合滤料中超疏水超亲油石英砂滤料占比 x 的关系为：$\cos\theta = 1/(1.33x+0.53) - 1.41$，清洁混合滤料床的水头损失 H_0 与混合滤料的水的接触角 θ 关系为：$H_0 = [180Lv^2(1-e)^2(0.021\theta^{2.75}+0.96)]/ge^3(\Phi d)^2$，说明滤料的疏水性增大了水头损失。（6）混合滤料床既增大了除油效率，又降低了水头损失。对于过滤处理含油废水的品质因子，未改性石英砂单一滤料远远小于混合滤料。对于混合滤料，当滤料粒径越大时，滤料的疏水性越强，其品质因子越大。总之，本文首次制备了四种具有特殊润湿性的石英砂滤料，研究了其性质和除油性能，首次研究了混合滤料床的润湿性、水头损失和除油性能，具有很强的创新性[13]。

A lot of oily wastewaters will be produced in the process of industrial production and life. Oily wastewater has caused serious pollution to the environment and posed a serious threat to the living environment of human beings. Deep bed filtration is a common method for deep treatment of oily wastewater. In the filtration unit, oil, water, and filter media surface coexist in the filter bed. For hydrophobic and oleophilic filter media, when oil drops migrate to the surface of the filter media, they will adhere to the surface of the filter media and be removed because they are subjected to the affinity of the filter media surface. Obviously, the more hydrophobic and lipophilic the filter media is, the easier the oil drops will adhere to the surface of the filter media and the higher the removal efficiency will be. However, for hydrophilic and oleophobic filter media, when the oil drop is close to the surface of the filter media, it will remain spherical due to its oleophobic effect, and it is not easy to flow away from the pores with smaller particle size than the oil drop, so it may also have a high removal efficiency. In addition, in terms of head loss, the more hydrophobic the filter media is, the greater the head loss will be, and vice versa. Therefore, improving the hydrophobicity of the filter media surface can improve the oil removal efficiency of the filter bed, but will increase the head loss of filtration; On the contrary, if the hydrophilic surface of the filter media is improved, the head loss will be reduced. In recent years, great progress has been made in the research of special wettable surfaces in the field of materials, and it has been widely studied in the field of oil-water separation. The special wettability surfaces used for oil-water separation mainly include superhydrophobic and super oleophilic surface and superhydrophilic and underwater superoleophobic surfaces. However, there is little research on the special wettability filter media in the aspect of hard particle filter media filtration. For this reason, this paper studies the preparation and oil removal performance of superhydrophobic super oleophilic, superhydrophilic and underwater superoleophobic quartz sand filter media, and studies the filtration performance of mixed filter media with two different wettability filter media. (1) Tridecane acid and octadecyl trichlorosilane is grafted to the surface of **quartz sand** filter media by dip-coating method, but no rough structure structured water contact angle of quartz sand filter media is 134.2°, and the quartz sand filter media has become high hydrophobic surface. By the dip-coating method, three (hydroxymethyl) aminomethane and octadecyl trichlorosilane composite were grafted to the surface of quartz sand filter media. At the same time, to enhance the hydrophobicity of filter media, ZnO nano particles structured rough micro/nano structure. The apparent contact angle of water at the prepared quartz sand filter media above 150°, oil apparent contact at 0°, the superhydrophobic and super oleophilic quartz sand filter media obtained. The filter media has strong mechanical wear resistance and hashed environmental resistance. (2) Chitosan crosslinked with glutaraldehyde, again through the chemical reaction they are grafted to the surface of quartz sand filter media, and silica nanoparticles coarse structure, the superhydrophilic and underwater superoleophobic quartz sand filter media prepared with the underwater oil contact angle is $152.1° \pm 1.2°$. The chitosan-coated superhydrophilic and underwater superoleophobic quartz sand filter media has certain durability, alkali resistance, mechanical wear resistance and air exposure resistance. Potato residue with NaOH and ultrasonic activation, coated on the surface of the quartz sand filter media using waterborne polyurethane. The as prepared of the underwater oil contact angle of is $151.7° \pm 1.3°$. The potato residue-coated superhydrophilic and underwater superoleophobic quartz sand filter media has certain durability, alkali resistance, mechanical wear resistance and air exposure resistance. Because potato residue is a waste, it has good practicability and low cost. (3) As prepared superhydrophobic and super oleophilic quartz sand filter media for water adhesive capacity was only about 2% of the pristine quartz sand filter media, the oil and rapeseed oil adhesive capacities are increased by 54% and 45% than the pristine quartz sand filter media, respectively, for the adhesive capacity of different organic reagent for 5-20 times of pristine quartz sand filter media. Superhydrophobic and super oleophilic quartz sand filter media can remove 99.65% and 96.46% of engine oil and rapeseed oil emulsion wastewater, respectively, and the separation efficiency of oil and water mixture is greater than 92.8%, with good oil-water separation performance. (4) Compared with the pristine quartz sand filter, the two kinds of superhydrophilic and underwater superoleophobic quartz sand filter media can improve the oil removal efficiency of emulsified oil wastewater to a certain extent and the separation efficiency of oil-water mixture to a large extent. The permeability coefficient of water increased by 11% -19%, and that of oil decreased by 6% -8%. Excellent oil-water separation performance and low head loss provide a new scheme for oil-water separation of superhydrophilic and underwater superoleophobic quartz sand filter media. (5) As for the mixed quartz sand filter media with pristine and superhydrophobic and super oleophilic quartz sand filter media, the relation between their apparent contact angle of water and proportion x of superhydrophobic and super oleophilic quartz sand filter media is as follows: $\cos\theta = 1/(1.33x + 0.53) - 1.41$. As for head loss H_0 in clean mixed filter bed, its contact angle with water in mixed filter bed is: $H_0 = [180Lv^2(1-e)^2(0.021\theta^{2.75} + 0.96)]/ge^3(\Phi d)^2$. This indicates that the hydrophobicity of the filter media increases the head loss. (6) The mixed filter bed not only increases the oil removal efficiency, but also reduces the head loss. For the quality factors of oily wastewater treatment, the single filter media of pristine quartz sand is far less than the mixed filter media. For mixed filter media, the larger the particle size, the stronger the hydrophobicity and the higher the quality factor. In conclusion, four kinds of quartz sand filter media with special wettability were prepared for the first time, and their properties and oil removal properties were studied. Wettability, head loss and oil removal performance of mixed filter bed were studied for the first time[13].

2.2 半硅质耐火材料原料

2.2.1 术语词

叶蜡石 pyrophyllite,pyrauxite,roseki
叶蜡石基 pyrophyllite-based
明矾石质叶蜡石 alunite-pyrophyllite

红柱石质叶蜡石 andalusite-pyrophyllite
硬水铝石质叶蜡石 diaspore-pyrophyllite
石英质叶蜡石 quartz-pyrophyllite

2.2.2 术语词组

叶蜡石块 pyrophyllite in lumps
叶蜡石粉 pyrophyllite in powder,pyrophyllite powder

叶蜡石矿 pyrophyllite deposit,pyrophyllite ore
叶蜡石选矿 processing of pyrophyllite ores

2.2.3 术语例句

叶蜡石的红外光谱有两种基本振动形式。
There are two basic vibration forms in the infrared spectrum of **pyrophyllite**.

叶蜡石相对热辐射率较低。
The relative heat radiation rate of **pyrophyllite** is fairly low.

叶蜡石是高温高压合成金刚石的关键辅助材料。
Pyrophyllite is a key assistant material in production of synthetic diamond.

在含石英较多的叶蜡石红外谱图中，**叶蜡石**的吸收谱带减弱或消失。
In the spectrum of **pyrophyllite** containing more quartz the absorption band tends

to damp or vanish.

分析讨论了蜡石砖的生产工艺特点及与**叶蜡石**加热过程相变化的关系。

Process trait of **roseki** bricks and its relationship with phase transformation have been studied and discussed.

本发明为一种阻燃聚合物，特别是天然无机矿物**叶蜡石基**阻燃聚合物及其制备方法。

The invention relates to a flame retarding polymer, in particular to a natural inorganic mineral **pyrophyllite-based** flame retarding polymer and a preparation method thereof.

矿粒的扫描电镜照片表明：**叶蜡石**颗粒主要呈薄片状，伊利石和高岭石颗粒呈不规则形状。

The scanning electron photomicrograph of aluminosilicate particle shows that the **pyrophyllite** particles take the shape of thin slice, but illite and kaolinite are particles with irregular appearance.

采用季铵盐 DTAL 作捕收剂，研究了氯化钠对一水硬铝石和**叶蜡石**浮选的影响及其作用机理。

The effects and mechanisms of sodium chloride on the flotation of diaspore and **pyrophyllite** with the cationic quaternary ammonium DTAL as a collector were studied.

高温高压合成超硬材料需要密封介质，**叶蜡石**是目前国内使用最普遍的密封、传压、保温、绝缘材料。

Pyrophyllite is the most common seal, pressure transfer and heat preservation, insulation material in domestic at present.

峨眉**叶蜡石矿**储量大，矿石类型主要有石英叶蜡石型、硬水铝石叶蜡石型、叶蜡石型、明矾石叶蜡石型等。

Emei **pyrophyllite deposit** possesses a large number of reserves with different types of pyrophyllite ore, such as quartz-pyrophyllite, diaspore-pyrophyllite, pyrophyllite and alunite-pyrophyllite etc.

主要研究了油酸钠作捕收剂，草酸钠、水杨酸钠等四种小分子有机调整剂对一水硬铝石、叶蜡石、高岭石三种单矿物的作用。

The effect of four small-molecule organic regulator to three single minerals such as diaspore, **pyrophyllite** and kaolinite was investigated using sodium oleate as collector.

2.2.4　术语例段

为解决以粉煤灰为原料制备轻质隔热砖体积密度过大的问题，利用**叶蜡石**、粉煤灰、苏州土为主要原料，锯末为造孔剂，通过控制**叶蜡石**加入量（质量分数分别为20%、30%和40%）和烧成温度（1250℃、1300℃、1350℃和1400℃）合成了体积密度小于0.89g/cm³的轻质隔热耐火材料，并利用XRD、SEM、导热仪等对试样进行表征。结果表明：当叶蜡石加入量（质量分数）为30%，烧成温度为1400℃时，制备的隔热砖试样的线收缩率为6.6%，显气孔率为57%，体积密度为0.75g/cm³，耐压强度为2.7MPa，在350℃时热导率为0.152～0.216W/(m·K)。由此说明，通过利用**叶蜡石**的高温下分解生成莫来石和非晶石英相所产生的体积膨胀效应抵消坯体高温下产生的收缩，结合废弃物粉煤灰以及黏土可制备轻质隔热耐火材料[14]。

To solve the problem of over high density of lightweight insulation refractory bricks prepared with fly ash, new lightweight insulation refractory materials with density <0.89g/cm³ were synthesized using pyrophyllite, fly ash, and Suzhou clay as the main starting materials and saw dust as the pore forming substance, and controlling the addition of the **pyrophyllite** (20%, 30%, and 40% by mass) and the treating temperature (1250℃, 1300℃, 1350℃, and 1400℃). The synthesized materials were characterized by the XRD, SEM, and the thermal conductivity measuring instrument. The results show at **pyrophyllite** addition of 30% and treat temperature of 1400℃, the material can achieve linear shrinkage of 6.6%, apparent porosity of 57%, bulk density of 0.75g/cm³, compressive strength of 2.7MPa, and thermal conductivity at 350℃ of 0.152-0.216W/(m·K). This indicates that the **pyrophyllite** decomposition at high temperatures forms mullite and amorphous quartz introducing volume expansion, which counteracts some shrinkage at high temperatures. So, it is feasible to use pyrophyllite, fly ash waste and clay to prepare lightweight insulation refractory materials[14].

以**叶蜡石**和天然石墨为原料，通过碳热还原反应在氩气气氛下合成了Al_2O_3-SiC复合粉体。研究了复合微粉对低碳MgO-C耐火材料抗渣侵蚀与渗透及抗氧化性的影响，并讨论了Al_2O_3-SiC复合粉体抗渣性和抗氧化性的机理。研究结果表明，合成温度对Al_2O_3-SiC复合粉体的指标影响显著；以**叶蜡石**和天然石墨为原料在 1600~1700℃氩气气氛下可以成功合成 Al_2O_3-SiC 复合粉体，在 1700℃下合成粉体的粒径为 1~2μm；添加合成的 Al_2O_3-SiC 复合粉体明显改善了低碳 MgO-C 耐火材料的抗渣侵蚀与渗透性能，一定程度上提高了材料的抗氧化性。由于形成镁铝尖晶石相，增加了渣的黏度，有效抑制了渣对耐火材料的侵蚀和渗透。

Al_2O_3-SiC composite powder was synthesized with **pyrophyllite** and natural graphite as raw materials by carbothermal reduction reaction under argon atmosphere. The effect of composite powder addition on the slag penetration and corrosion resistance as well as oxidation resistance of the refractories was investigated, and the slag resistance and oxidation resistance mechanisms of the Al_2O_3-SiC composite powder were also discussed. The research results show that the synthesis temperature has a great influence on preparation of Al_2O_3-SiC composite powder. The Al_2O_3-SiC composite powder can be synthesized at 1600-1700℃ under argon atmosphere, with **pyrophyllite** and natural graphite as raw materials, and particle sizes of the composite power synthesized at 1700℃ are 1-2μm. The slag penetration and corrosion resistance of low-carbon MgO-C refractories can be remarkably improved by adding the synthesized Al_2O_3-SiC composite powder, and the oxidation resistance has an improvement to some extent. The increase of slag viscosity and the formation of $MgAl_2O_4$ can effectively inhibit the slag penetration and corrosion for the refractories[15].

叶蜡石（$Al_2Si_4O_{10}(OH)_2$）是一种层状硅酸盐，通常与石英、云母、高岭石、绿帘石和金红石矿物伴生。在纯净状态下，**叶蜡石**具有低导热性和导电性、高折射率、低膨胀系数、化学惰性以及对熔融金属和气体的耐腐蚀性等独特性能。这些特性使其在耐火材料等不同行业中备受青睐，如陶瓷、玻璃纤维和化妆品行业，造纸、塑料、涂料、农药等行业的填料，作为肥料工业中的土壤改良剂，并在橡胶和屋顶工业中用作除尘剂。在许多工业应用中，**叶蜡石**还可以作为高岭石、滑石和长石等不同矿物的经济替代品。为了增加其市场价值，**叶蜡石**必须具有高铝（Al_2O_3）含量，保持无任何杂质，并具有尽可能高的白度。

Pyrophyllite($Al_2Si_4O_{10}(OH)_2$) is a phyllosilicate often associated with quartz, mica, kaolinite, epidote, and rutile minerals. In its pure state, **pyrophyllite** exhibits unique properties such as low thermal and electrical conductivity, high refractive behavior, low expansion coefficient, chemical inertness, and high resistance to corrosion by molten metals and gases. These properties make it desirable in different industries such as refractory; ceramic, fiberglass, and cosmetic industries; as filler in the paper, plastic, paint, and pesticide industries; as soil conditioner in the fertilizer industry; and as a dusting agent in the rubber and roofing industries. **Pyrophyllite** can also serve as an economical alternative in many industrial applications to different minerals as kaolinite, talc, and feldspar. To increase its market value, **pyrophyllite** must have high alumina (Al_2O_3) content, remain free of any impurities, and possess as much whiteness as possible[16].

叶蜡石是一种硅酸铝矿物，被用作生产铝硼硅酸盐玻璃的原料。为了将叶蜡石作为原料应用到玻璃制造过程中，粉末的粒度分布必须满足以下要求：150～45μm≥60%（质量分数）且低于45μm≤40%（质量分数）。因此，通过了解叶蜡石的破碎特性，尝试开发叶蜡石生产满足这些要求的粉末的技术。叶蜡石在1150℃焙烧时，会导致叶蜡石原料的变化，从而导致破碎特性的差异。在相同的破碎条件下，焙烧叶蜡石粉碎后产生的粉末与未加工的叶蜡石的粉末相比，细颗粒（45μm）的数量减少了。粉体的X射线衍射分析（XRD）和扫描电子显微镜（SEM）结果以及水的吸附量证实了这一差异是由焙烧过程后原始叶蜡石的变化造成的。此外，研究了叶蜡石未烧样和焙烧样的熔融特性，以便在铝硼硅酸盐玻璃生产中应用。玻璃产品的特征是透明且为无定形相。

Pyrophyllite, an aluminum silicate mineral, was used as a raw material for the production of aluminoborosilicate glass. For the application of **pyrophyllite** to the glass manufacturing process as a raw material, the particle size distribution of the powder must the following requirements: 150-45μm≥60wt.% and below 45μm≤40wt.%. Hence, an attempt was made to develop the technology for pyrophyllite to produce a powder that meets these requirements through understanding the crushing characteristics. Roasting of pyrophyllite at 1150℃ resulted in the alternation of raw pyrophyllite and led to differences in the crushing characteristics. Crushing of the roasted pyrophyllite produced a powder with a reduction in the amount of the fine particles (45μm) compared to the powder of the raw pyrophyllite under identical crushing condition. This difference is attributed to the alternation of the raw pyrophyllite after the roasting process, as confirmed by powder X-ray diffraction (XRD) and scanning electron microscopy (SEM) results as well as by the amount of water adsorption. The melting characteristics of the raw and roasted pyrophyllite samples were investigated for application to the aluminum borosilicate glass production. The glass products are characterized as colorless and in an amorphous phase[17].

研究了以**叶蜡石**为原料，采用火花等离子烧结（SPS）法制备莫来石的工艺。**叶蜡石粉末**与α-Al_2O_3粉末混合，化学计量成分为$3Al_2O_3 \cdot 2SiO_2$。所有试样的第一步致密化发生在大约930℃，这是由于施加的压力导致了颗粒的重排。

XRD 结果表明，在 1400℃时，莫来石呈正交相，晶体结构为 $Al_6Si_2O_{13}$，次级相为氧化铝。随着烧结温度的升高，氧化铝衍射峰强度逐渐降低。随着烧结温度的升高，合金的密度增大。热处理后试样的显微组织显示，试样仍为等轴晶组织。XRD、显微组织、密度硬度和断裂韧性值表明，为了充分利用 SPS 技术，叶蜡石-氧化铝应在 1600℃下加工 10min。研究表明，本地开采的廉价叶蜡石粉可以转化为莫来石产品。

A processing technique was developed to synthesize mullite from **pyrophyllite** using spark plasma sintering(SPS). **Pyrophyllite powder** was mixed with α-Al_2O_3 powder in a stoichiometric composition of $3Al_2O_3 \cdot 2SiO_2$. The first densification step was observed at about 930℃ for all the samples and is due to the application of pressure that induced rearrangement of particles. The XRD results show an orthorhombic phase of mullite with a crystal structure $Al_6Si_2O_{13}$ and a minority phase of alumina at 1400℃. The intensities of alumina peaks gradually decrease with increasing sintering temperature. The densities increase as the sintering temperature increases. The microstructure of heat-treated samples revealed that the samples retained their equiaxed grain structure. The XRD, microstructure, densities hardness and fracture toughness values suggest that pyrophyllite-alumina be processed at 1600℃ for 10min in order to take full advantage of the SPS technique. This study shows that an inexpensive locally mined pyrophyllite powder could be converted to mullite products[18].

瓷土（高岭土）在传统的陶瓷混合物中逐渐被**叶蜡石**取代。添加 5%（质量分数）的**叶蜡石**替代瓷土，与 1300℃烧结的传统坯体相比，烧结强度提高了约 24%。当叶蜡石逐渐取代高岭石时，烧成的试样中莫来石的比例增加。当叶蜡石加入量（质量分数）超过 7.5%时，叶蜡石脱羟基释放出的无定形 SiO_2 抑制了莫来石的进一步再结晶。当叶蜡石含量（质量分数）高于 7.5%时，物相组成变化不大。从叶蜡石去羟基化过程中结构重组的角度来分析整个过程。当叶蜡石含量（质量分数）超过 7.5%时，试样显微组织中存在大量较小尺寸的未溶石英和孤立孔隙，可以阻碍裂纹的扩展，从而提高力学性能。莫来石晶体的大小和形状在很大程度上是由它们生长的液体基质的流动性控制的，而这也与温度和成分有关。

China clay(kaolin) was progressively replaced by **pyrophyllite** in a conventional porcelain mix. Addition of 5% **pyrophyllite** as a replacement of China clay improved the fired strength by about 24% compared to that of the conventional body fired at 1300℃. Percentage of mullite was found to increase in the fired specimens when kaolinite was progressively replaced by pyrophyllite. However, beyond 7.5wt.% pyrophyllite addition, amorphous SiO_2 released from pyrophyllite dehydroxylate inhibited further recrystallization of mullite. There was very insignificant change in the phase compositions with mixes having pyrophyllite content higher than 7.5wt.%. Entire phenomenon has been explained on the basis of structural reorganization of pyrophyllite during dehydroxylation. Presence of large amount of undissolved quartz of smaller size as well as isolated pores in the microstructures of specimens containing pyrophyllite more than 7.5wt.% are assumed to hinder the propagation of crack and thereby improving the mechanical properties. The size and shape of mullite crystals is to a large extent controlled by the fluidity of the liquid matrix from which they grow, and this is again a function of temperature and composition[19].

实验的主要原料包括花岗岩废料、长石和黏土，以及作为发泡剂的碳化硅粉末。表所示为 X 射线荧光法确定的这些原材料的化学成分。图所示为花岗岩废料、长石和黏土的 X 射线衍射图谱。花岗岩废料的主要物相是石英、白云石和微晶石；长石的主要物相是白云石和微晶石，黏土中高岭土的特征峰清晰可见，其次是**叶蜡石**。

The primary raw materials for the experiment include granite waste, feldspar, and clay as well as silicon carbide powder as a foaming agent. Table shows the chemical compositions of these raw materials as determined via X-ray fluorescence. Figure shows the X-ray diffraction patterns for granite waste, feldspar, and clay. The primary phases of granite waste are clearlyquartz, albite, and microcline; the primary phases of feldspar are albite and microcline and the characteristic peaks of kaolin in clay are visible, followed by **pyrophyllite**[20].

应该指出的是，考虑到原始黏土的不同，这些参数有一些变化。例如，据报道，在含有石英的绿泥石-蛭石（伊利石）黏土中：$T_v = 1200℃$，$T_d = 1150℃$，$BD = 2.30g/cm^3$。因为与高岭土黏土相比，这些黏土含有大量的熔剂。在生绢云母黏土（绢云母-高岭土和绢云母-**叶蜡石**-高岭土黏土）中，当作为前驱体或莫来石材料加热时，$T_v \approx 1250℃$，BD 最大值在 $2.41 \sim 2.52g/cm^3$ 范围内的结果已经被报道。而在其他原料黏土（石英岩黏土）中，T_v 和 T_d 的值在 $1200 \sim 1250℃$ 之间，BD 在 $2.25 \sim 2.43g/cm^3$ 之间。一般来说，这些特征与含有石英的高岭石和白云母（伊利石/绢云母）-高岭土的热分解行为一致。

It should be noted that there are some variations in these parameters considering raw clays. For instance, in chlorite-muscovite(illite) clays containing quartz, it has been reported $T_v = 1200℃$ and $T_d = 1150℃$ with $BD = 2.30g/cm^3$ because these clays contain a large number of fluxes as compared to kaolinitic clays. In raw sericite clays(sericite-kaolinite and sericite-**pyrophyllite**-kaolinite clays), as precursors or mullite materials by heating, values of $T_v \approx 1250℃$ and *BD* maximum values in the range 2.41-2.52g/cm^3 have been reported. In other raw clays(quartzitic clays), the values of T_v and T_d are in the range 1200-1250℃ with *BD* in the range 2.25-2.43g/cm^3. In general, these features are in accordance with the thermal behaviour of kaolinitic and muscovite(illite/sericite)-kaolinitic clays containing quartz[21].

将处理工业废水的浸没式陶瓷膜反应器与粒状活性炭（GAC）颗粒相结合，以控制膜污损和有机物去除效率。粒状活性炭颗粒沿膜表面悬浮，只在反应器内进行大批量循环，不喷入任何气体。在恒定流量过滤模式下，对涂有 Al$_2$O$_3$ 层（CPM）和无涂层（UPM）的膜支架进行了比较。膜支架由 80%（质量分数）的**叶蜡石**和 20%（质量分数）的氧化铝组成。在上流速度为 0.031m/s 的情况下，仅通过不含 GAC 颗粒的装置再循环，观察到 CPM 和 UPM 的污损率分别为 0.011bar/h 和 0.013bar/h。随着 GAC 颗粒的悬浮，污损的缓解得到了极大的加强，在相同的上流速度下，CPM 的效果比 UPM 更明显（90% 对 57%）。此外，GAC 悬浮液使 CPM 的临界流量比没有炭颗粒的情况下高出 46%。UPM 的有机物去除效率低于 CPM，而污损率则大得多，可能是由于有机染料化合物造成的孔隙堵塞。对于这两种膜，沿膜表面悬浮的 GAC 颗粒提高了有机物去除效率，高于 90%。有机物去除率通过增加渗透量得到提高，但随着上流速度的提高，有机物去除率会变低。

Submerged ceramic membrane reactor treating industrial wastewater was combined with granular activated carbon(GAC) particles to control membrane fouling and organic removal efficiency. The GAC particles were suspended along the membrane surface under bulk recirculation only through the reactor without any gas sparging. Membrane support coated with Al$_2$O$_3$ layer (CPM) and uncoated one (UPM) was compared at constant flux mode of filtration. The membrane support consisted of 80% of **pyrophyllite** and 20% of alumina. Under up-flow velocity of 0.031m/s through bulk recirculation only without GAC particles, the fouling rates were observed as 0.011bar/h and 0.013bar/h for the CPM and UPM, respectively. With suspension of GAC particles, fouling mitigation was enhanced considerably, and this effect was more pronounced with CPM than UPM under the same upflow velocity(90% vs. 57%). In addition, the GAC suspension increased critical flux by 46% higher with CPM than that observed without the carbon particles. The organic removal efficiency of the UPM was lower than that of CPM while the fouling rate was much greater probably due to pore blocking caused by organic dye compounds. For both membranes, suspension of GAC particles along the membrane surface increased organic removal efficiency higher than 90%. The organic removal efficiency was enhanced by increasing permeate flux, but it became lower as upflow velocity was higher[22].

本文探讨了加工工艺对**叶蜡石**和α-氧化铝粉末合成莫来石的影响。研究了烧制温度的影响，以优化制造工艺从而获得更好的力学性能。首先在立式磨机中将具有化学计量组成的原料粉末充分湿磨。然后，使用放电等离子烧结（SPS）在1400～1700℃的温度范围内，在50MPa的压力下以100℃/min的加热速率，在不同的保温时间下烧结干燥后的粉末。研究结果表明，在1400℃和1600℃下烧结的试样的体积密度分别为3.25g/cm^{3}和3.17g/cm^3，保温时间为10min。采用XRD和SEM/EDS表征了试样的烧制转变和微观结构。试样的SEM显微照片在1600℃下烧结，保温时间为10min，显示试样保留了其等轴晶粒结构。XRD结果显示试样烧制后α-氧化铝含量降低。获得的硬度和断裂韧性值分别高达11.73GPa和1.99MPa·m$^{1/2}$。

This paper addresses the effects of processing technique on the synthesis of mullite from **pyrophyllite** and alpha alumina powder. The influence of firing temperatures was examined in order to optimize the fabrication process to achieve a suitable mechanical property. Feedstock powders, with stoichiometric composition, were wet milled in an Attritor mill. The dried powders were consolidated using spark plasma sintering(SPS) in the temperatures range 1400℃ to 1700℃ under pressure of 50MPa, with a heating rate of 100℃/min at different holding times. Densities of 3.25g/cm^3 and 3.17g/cm^3 were obtained for the samples sintered at 1400℃ and 1600℃ respectively, with 10min holding times. The XRD and SEM/EDS were employed to characterize the firing transformation and microstructure of the samples. The SEM micrograph of samples, sintered at 1600℃ with 10min holding times, revealed that the sample retained its equiaxed grain structure. The XRD results show a reduction in alpha-alumina content after the sample was fired. Hardness and fracture toughness values up to 11.73GPa and 1.99MPa·m$^{1/2}$ respectively were obtained[23].

煤气化炉渣（简称CGS）是煤气化过程中不可避免的副产品，是我国排放量较大的固体废物之一，它是由煤炭中的矿物质转变成灰分后，在气化炉高温炉膛中心变成熔融液态渣，在重力作用下流入气化炉底部的激冷室激冷形成的。随着煤化工行业的迅速发展，CGS的排放量也逐年上升。大量堆存不仅占用宝贵的土地资源，重要的是会恶化生态环境，威胁人类健康。本论文采用湿化学分析、X射线荧光和衍射分析、显微结构及岩相分析、灰熔点分析和高温熔体黏度和熔化温度（范围）分析，并借助核磁共振、综合热分析质谱联动仪等手段系统研究了4种气化炉型6个不同产地CGS的本征特征；通过CGS、玻璃、天然矿物（**叶蜡石和黏土**）的碳热还原氮化（简称CRN）过程中相组成与显微结构演变研究揭示CGS的CRN机理；采用CRN工艺，通过添加适量的稀土元素制备Sialon荧光粉；最后，为追求大量利用，提出将 CGS 用于制备发泡陶瓷轻质隔热材料。通过上述研究得到主要结论如下：（1）6 种 CGS 的主要化学组分为 Al_2O_3、SiO_2、CaO、Fe_2O_3 和残余碳；其内部原子结合状态均为 SiO_4 四面体与 AlO_4 四面体相互连接的架状结构；德士古、航天和多喷嘴对喷气化炉产生的 5 种渣均含有较多玻璃体以及残余碳颗粒；然而壳牌（QP）渣残碳很少，且其熔融温度较高，为此前 5 种渣可直接 CRN，而 QP 渣则不然。（2）对比玻璃、气化炉渣、天然矿物 CRN 过程，可知 CGS 的 CRN 机理为：在 CRN 反应初期，O-Sialon 和钙长石相先形成；随着氮化温度的升高，O-Sialon 逐渐转变为富氮的 Sialon，如：β-Sialon、α-Sialon 和 15R 多型体，同时，钙长石相转变为钙铝黄长石相；碳热还原反应末期，β-Sialon，15R 多型体进一步转化为 α-Sialon，如果碳组分在低温下没有被 O-Sialon 的转变消耗完，那么钙铝黄长石相最终会进一步转变为 Sialon 相；除此之外，通过对比实验发现，CRN 过程中高温液相和气相的生成促进了棒状、纤维状形貌的氮化产物的生成。（3）无论是添加稀土氧化物 CeO_2 还是 Eu_2O_3，除杂后均可获得纯度较高的 Ca-α-Sialon 粉体。除碳和除铁极大地提高了荧光粉的发光性能。随着氮化温度的升高，Ca-α-Sialon 粉体聚集体的形貌越来越清晰，并且发展成为针状聚集体 $Ce(Si,Al)_3(O,N)_5$ 和细条状聚集体 $Eu_2(Si,Al)_5(N,O)_8$。Ca-α-Sialon：Ce 在 383nm 紫外光激发下 Ce^{3+} 发生 $^4f^05d1 \rightarrow {}^4f1$ 的能级跃迁呈绿光，Ca-α-Sialon：Eu 在 420nm 紫外光激发下 Eu^{2+} 发生 $^4f^65d \rightarrow {}^4f7$ 的能级跃迁呈黄光。（4）利用碳化硅作为发泡剂以 CGS 为主要原料制备发泡陶瓷是可行的，且 CGS 加入量为 77%，造粒粒度范围为 0.25~1mm，烧成温度为 1160℃条件下制备的发泡陶瓷性能最优，其密度低至 $0.21g/cm^3$，导热系数为 $0.05W/(m \cdot K)$，抗压强度为 1.18MPa。所制材料在新型绿色建筑材料领域展现出巨大的应用潜力[24]。

Coal gasification slag is a by-product from coal gasification process in Integrated Gasification Combined Cycle(IGCC) system. In the gasifier, molten ash originated from the mineral component in coal flows down into a water quenching system, which leading to the formation of glassy CGS. With the rapid development of coal chemical industry, the discharge of CGS is increasing year by year. Massive stockpiling not only occupies valuable land resources, but also, importantly, degrades the ecological environment and threatens human health. In the literature, there are much research on coal gasification technology and downstream extension of polygeneration technology, but few research on CGS, especially on comprehensive utilization of CGS. Based on the above, this thesis, firstly, the intrinsic characteristics of CGSs were studied by means of wet chemical analysis, X-ray fluorescence analysis, X-ray diffraction analysis, petrographic analysis, microstructural analysis, analysis of the ash melting point and high temperature melt viscosity and melting temperature(range) analysis, and nuclear magnetic resonance

(NMR)analysis methods, which provide the foundation for the further comprehensive of the slags; Secondly, by compared experiments, the formation of phase composition and microstructure in the process of carbothermal reduction nitridation(CRN) of CGSs, glass and minerals(**pyrophyllite** and clay), reveals the mechanism of CRN of CGS; Thirdly, Sialon phosphors were prepared by CRN process with adding appropriate rare earth elements; Finally, in order to pursue higher utilization rate, CGS was used to prepare foam ceramic lightweight wall material. The main conclusions are as follows: (1) The main chemical components of the six kinds of CGSs were Al_2O_3, SiO_2, CaO, Fe_2O_3 and residual carbon. An internal atom binding state is the frame structure of SiO_4 tetrahedron and AlO_4 tetrahedron. And the first five kinds of slags all contained more spherical vitreous and residual carbon particles. However, the content of the residual carbon of QP slag was very small, and the melting temperature of it was higher. Based on the intrinsic characteristics, it could be inferred that the first five kinds of slag could be directly carbothermal reduction and nitriding, while QP slag was unrealizable. (2) By comparing carbothermal reduction nitridation (CRN) process of the glass, gasification slag and mineral, the CRN mechanism of CGS was indicated: the CRN process was dominated by nitridation of the Ca-Si-Al-O glass component in slag. With the increase of processing temperature, spherical Ca-Si-Al-O particles were firstly nitride into porous Ca-Si-Al-O-N spheres, plate-like O-Sialon grains, and then nitrogen-rich Sialon (β-Sialon, 15R polytypoid and α-Sialon). At the end of CRN, Ca-α-Sialon was found to be the only Sialon phase presenting elongated prismatic morphology. In addition, it was found that the formation of high temperature liquid and formation of gas phases could promote the formation of bar morphology of Ca-α-Sialon during the process of CRN. (3) Ca-α-Sialon powder with high purity can be obtained after removal of impurities by adding CeO_2 or Eu_2O_3 rare earth oxides. The process of removing carbon and iron greatly improves the luminescence performance of phosphors. With the increase of nitridation temperature, the morphology of Ca-α-Sialon powder aggregates became more and more clear, and developed into acicular aggregate $Ce(Si, Al)_3(O, N)_5$ and fine strip aggregate $Eu_2(Si, Al)_5(N, O)_8$. Ca-α-Sialon: Ce emitted green light under the excitation of 383nm ultraviolet light, and Ca-α-Sialon: Eu emitted yellow light under the excitation of 420nm ultraviolet light. (4) It was feasible to use SiC as the foaming agent and CGS as the main raw material to prepare foaming ceramics. And it can be indicated that: the optimum particle size rang was 0.25-1mm, at the sintering temperature was 1160℃, the foam ceramic wall material with 77% CGS presented relatively volume density of 0.21g/cm^3, heat conductivity coefficient of 0.05W/(m·K), and the corresponding compress strength of 1. 18MPa. Moreover, the total amount of CGS is up to 77% , which greatly reduces the cost of raw materials and has played a very beneficial role in environmental protection. Finally, it will have great potential in the application in the field of new building materials[24].

参考文献

[1] Ma Beiyue, Ren Xinming, Gao Zhi, et al. Synthesis of Al_2O_3-SiC composite powder from electroceramics waste and its application in low-carbon MgO-C refractories [J]. International Journal of Applied Ceramic Technology, 2022, 19 (3): 1265-1273.

[2] 和弦, 李勇, 张秀华, 等. 长寿热风炉用刚玉-莫来石耐火材料损毁机理研究 [J]. 硅酸盐学报, 2019, 47 (12): 1-6.

[3] Kouras N, Harabi A, Bouzerara F, et al. Macro-porous ceramic supports for membranes prepared from quartz sand and calcite mixtures [J]. Journal of the European Ceramic Society, 2017, 37 (9): 3159-3165.

[4] Andreev K, Tadaion V, Zhu Q, et al. Thermal and mechanical cyclic tests and fracture mechanics parameters as indicators of thermal shock resistance—case study on silica refractories [J]. Journal of the European Ceramic Society, 2019, 39 (4): 1650-1659.

[5] S S L Oliveira, S S L Oliveira, R S B Ferreira, et al. Development of hollow fiber membranes with alumina and waste of quartzite [J]. Materials Research, 2019, 22: e20190171.

[6] Ringdalen E. Changes in quartz during heating and the possible effects on Si production [J]. JOM, 2015, 67 (2): 484-492.

[7] Venkatesan J, Ubayadullah M, Mrunmaya K P, et al. Use of banded hematite quartzite (BHQ) in blast furnace for partial replacement of quartzite [J]. Transactions of the Indian Institute of Metals, 2017, 70 (10): 2529-2536.

[8] Legemza J, Findorák R, Bul'ko B, et al. New approach in research of quartzes and quartzites for ferroalloys and silicon production [J]. Metals, 2021, 11 (4): 670.

[9] Nabil M, Mahmoud K R, El-Shaer A, et al. Preparation of crystalline silica (quartz, cristobalite, and tridymite) and amorphous silica powder (one step) [J]. Journal of Physics and Chemistry of Solids, 2018, 121: 22-26.

[10] Kherifi D, Belhouchet H, Ramesh S, et al. Sintering behaviour of fluorapatite-silicate composites produced from natural fluorapatite and quartz [J]. Ceramics International, 2021, 47 (12): 16483-16490.

[11] Suerfu B, Souza M, Calaprice F. Pyrolytic carbon coating of fused quartz by vacuum vapor transport [J]. Journal of Crystal Growth, 2019, 516: 40-44.

[12] Shao Hui, Zang Fangfang, Ji Mengjiao, et al. Prepare and mechanism of high purity quartz by alkali corrosion and acid leaching process using vein quartz [J]. Silicon, 2022, 14: 12475-12483.

[13] 刘建林. 特殊润湿性石英砂滤料的制备及其过滤除油性能研究 [D]. 兰州: 兰州交通大

学，2019.

[14] 陈若愚，李远兵，向若飞，等. 叶蜡石加入量对轻质隔热耐火材料性能的影响 [J]. 耐火材料，2016，50（1）：25-28.

[15] Ma Beiyue, Zhu Qiang, Sun Yong, et al. Synthesis of Al_2O_3-SiC composite and its effect on the properties of low-carbon MgO-C refractories [J]. Journal of Materials Science and Technology, 2010, 26（8）：715-720.

[16] Ali M A, Ahmed H A M, Ahmed H M, et al. Pyrophyllite：An economic mineral for different industrial applications [J]. Applied Sciences, 2021, 11（23）：11357.

[17] Seo J, Kim S, Bae I K, et al. Roasting of pyrophyllite for application in aluminoborosilicate glass production [J]. Geosystem Engineering, 2020, 23（3）：123-130.

[18] Sule R, Sigalas I. Effect of temperature on mullite synthesis from attrition-milled pyrophyllite and α-alumina by spark plasma sintering [J]. Applied Clay Science, 2018, 162：288-296.

[19] Mukhopadhyay T K, Ghatak S, Maiti H S. Effect of pyrophyllite on the mullitization in triaxial porcelain system [J]. Ceramics International, 2009, 35（4）：1493-1500.

[20] Pan Mengbo, Li Xiang, Wu Xiaopeng, et al. Preparation of thermal insulation materials based on granite waste using a high-temperature micro-foaming method [J]. Journal of Asian Ceramic Societies, 2022, 10（1）：223-229.

[21] Sánchez-Soto P J, Eliche-Quesada D, Martínez-Martínez S, et al. Study of a waste kaolin as raw material for mullite ceramics and mullite refractories by reaction sintering [J]. Materials, 2022, 15（2）：583.

[22] Ahmad R, Aslam M, Park E, et al. Submerged low-cost pyrophyllite ceramic membrane filtration combined with GAC as fluidized particles for industrial wastewater treatment [J]. Chemosphere, 2018, 206：784-792.

[23] Sule R, Sigalas I. Effect of temperature on mullite synthesis from attrition-milled pyrophyllite and α-alumina by spark plasma sintering [J]. Applied Clay Science, 2018, 162：288-296.

[24] 袁蝴蝶. 煤气化炉渣本征特征及应用基础研究 [D]. 西安：西安建筑科技大学，2020.

3 铝硅酸盐系耐火材料原料

3.1 黏土质耐火材料原料

3.1.1 术语词

风化黏土 aeroclay
白黏土(陶土)argil
黏土 clay
黏土质的 clayish
硬化黏土 clunch
硬黏土 leck
亚黏土(砂质黏土)loam
多水高岭土 ablykite
黑高岭土 hisingerite
高岭石族矿物 kandite

高岭土 kaolin
高岭石 kaolinite
高岭土化 kaolinization
偏高岭土 metakaolin
偏高岭石 metakaolinite
胶状高岭土 schroetterite
膨润土(膨土岩)amargosite,bentonite
高硅膨润土 distribond
蒙脱石 smectite,askanite

3.1.2 术语词组

酸性黏土 acid clay
活性黏土 active clay
冲击黏土 alluvial clay
球状黏土 ball clay
铝质黏土 bauxite clay
结合黏土 bonding clay
煅烧的黏土 calcined clay
黏土熟料 burned fireclay,chamotte powder

黏土熟料砂 chamotte sand
黏土粗坯 clay blank
黏土结合剂 clay bond
黏土的黏结强度 clay bond strength
黏土砖车间 clay brick workshop
黏土压球 clay briquette
黏土分散性 clay dispersity
黏土基质 clay matrix

33

黏土矿物 clay mineral

黏土矿物结构 clay mineral structure

黏土料球 clay pellet

黏土磷酸盐结合剂 clay phosphate binder

黏土的可塑性 clay plasticity

黏土矿岩 clay rock

黏土砂 clay sand

黏土页岩 clay shale

黏土的烧结性 clay sintering capacity

黏土的天然颗粒组成 clay size distribution

黏土板岩 clay slate

黏土泥浆 clay slip

黏土浆 clay slurry

黏土悬浮液 clay suspension

膨胀黏土 expanded clay

细分散性黏土 finely dispersed clay

黏土熟料细粉 finely ground fire clay

黏土骨料 fireclay aggregates

黏土的煤结性 fusibility of clay

硬质黏土熟料 hard fireclay chamotte

高级黏土 high-grade clay

高岭石质黏土 kaolinite clay

低塑性黏土 lean clay

黏土原料 clay raw material

塑性黏土 plastic clay

塑性耐火黏土 plastic fireclay

多孔黏土 porous clay

原生黏土 primary clay

生黏土 raw clay, unburned clay

黏土含砂量 sand content of clay

含砂黏土 sandy clay

砂岩黏土矿物 sandy clay mineral

次生黏土 secondary clay

精选黏土 selected clay

半酸性黏土 semi-acid clay

半轻质黏土 semi-plastic clay

贫黏土 short clay

硅质黏土 siliceous clay

烧结黏土 sintering clay

软质黏土 soft clay

碱性高岭土 alkaline kaolin

煅烧高岭土 calcined kaolin

煤系高岭土 coal series kaolinite

粗粒高岭土 coarse-grained kaolin

沉积高岭土 deposited kaolin

黏土高岭土熟料 kaolin clay chamotte

高岭土纤维 kaolin fiber

高岭土熟料 kaolin grog

高岭土矿物 kaolin mineral

高岭土原料 kaolin raw material

高岭土泥浆 kaolin slip

再沉淀次生高岭土 secondary sedimentary kaolin

沉淀高岭土 sedimentary kaolin

净高岭土 washed kaolin

3.1.3 术语例句

黏土和砂岩都是不坚固的岩石构造。

Both **clay** and sandstone are unstable rock formations.

基本成分是石灰石和**黏土**，比例为 2∶1。

The basic ingredients are limestone and **clay** in the proportion 2∶1.

壤土是由大约等份的**黏土**、沙和粉砂合成的。

Loam is a soil with roughly equal proportions of **clay**, sand and silt.

黏土是可塑物质。
Clay is a plastic substance.

这个煤层处于一层**黏土**岩之上。
This seam of coal reposes on a layer of **clay**.

将煤粉与**黏土**混合制成的煤球比块煤便宜得多。
Mixing **clay** with coal is much cheaper than burning large blocks of coal.

垃圾掩埋场以伦敦**黏土**为天然地基，其防渗效果极佳。
The landfill sits on a natural foundation of London **clay** which is more or less impermeable.

沉积型高岭土是中国特有的**高岭土**资源。
The sedimentary kaolin is a unique **kaolin** source in China.

骨瓷的基本成分是加入骨粉的**高岭土**和白墩子。
Bone China porcelain is basically made by adding bone ash to **kaolin** and petuntse.

高岭土的工艺矿物学特征直接决定着它的工业用途。
The characteristics of the technological mineralogy of **kaolin** directly decide its industrial utilization.

介绍了该煅烧高岭土的生产工序及其在乳胶漆的应用。
The production process and its application in latex paint of this **calcinated kaolin clay** were described.

致色杂质降低了**高岭土**的亮度并限制了它的应用范围。
Discoloring impurities lowers the brightness of **kaolin** and limits its use.

剥片后**高岭土**粒径变小，但保持了高岭土的基本层状结构。
The exfoliated **kaolin** fragments are tiny particles, but still keep the basic layered structure of kaolin.

通过**高岭土**矿山开采对环境影响的成因分析，提出了治理对策。

By analyzing the cause of environmental influence for mining **kaolin**, the paper puts forward the harnessing countermeasures.

煅烧**高岭石**–石灰–石膏硬化体的强度与钙矾石生成量的多少相关。

The strength for hardening calcined **kaolinite**-portlandite-gypsum system is derived from the quantity of forming ettringite.

长石溶解形成的**高岭石**发育晶间孔。

The **kaolinite** dissolution from feldspar grows much mineral intercrystal porosity.

证明了该黏土矿石中以结晶良好的**高岭石**为主。

The well crystallized **kaolinite** is proved to be a main component in the clay ore.

埋藏成岩作用能使**高岭石类矿物**新生变形，转化或消失。

Burial diagenesis can cause the **kaolinite group of minerals** to be deformed transformed or destroyed.

黏土矿物主要由伊利石、**蒙脱石**、**高岭石**和绿泥石组成。

The main types of clay minerals are illite, **smectite**, **kaolinite** and chlorite.

高岭石的吸氟量还受**高岭石**用量和含氟溶液浓度的影响。

The amount of **kaolinite** and the concentration of fluoride solution also influence the amount of fluorine adsorbed on kaolinite.

随着煤变质程度的升高，煤层夹矸中**高岭石**的结晶度、有序度也随之升高。

With the rise of the coal rank, the crystallinity of **kaolinite** and its order in the anthracite bed go up.

具较低结构有序度、较大的颗粒粒度并充分分散的**高岭石**有利于插层作用。

The results show that the **kaolinite** with low structure order degree and relatively large particle size is more easily intercalated under well dispersed.

如果多用**膨润土**，铅的温度必须降低。

If more **bentonite** clay is used, the lead temperature must be decreased.

这一发现填补了我国**膨润土**系列的一项空白。

This discovery fills up the gap of the **bentonite** series in China.

介绍活化**膨润土**的简易鉴别方法和膨润土的活化技术。

An easy method to differentiate activated bentonite and the **bentonite** activation technique was introduced.

近年来对**膨润土**的不断研究和开发，进一步拓宽了其应用领域。

In recent years, the continued study and the application on **bentonite** has been further extended.

介绍了马钢球团厂通过稳定系统运行来降低**膨润土**消耗的实践。

The practice of reducing **bentonite** consumption through stabilizing system operation at Pelletizing Plant of Ma Steel is introduced.

根据**膨润土**增稠、触变能力的不同建议不同配方有针对性地选用膨润土。

It was suggested that proper **bentonite** can be selected in different formulations according to its thickening and thixotropic ability.

因此对我国所有的**膨润土**进行深度开发是摆在众多科研工作者面前的问题。

Therefore, all of our in-depth development of **bentonite** was placed in front of numerous scientific research workers.

膨润土是一种新型的环境吸附材料，在环境污染控制和修复中有很好的应用前景。

Bentonites, a new kind of environmental adsorption material, have been widely used in the environmental pollution control and remediation.

改性**膨润土**具有吸附脱色性能优良、废水处理效果好等特点，从而扩大了其应用范围。

The modified **bentonite** has better qualities and ability of wastewater treatment, thus enlarge its applying scope.

蒙脱石含量仅次于伊利石，其变化趋势与伊利石相反。

The content of **smectite** is next to that of illite, and its trend is opposite to that of illite.

蒙脱石除源于大陆派生的岩石风化物外，主要来自火山物质的蚀变。

Smectite mainly came from the alteration of volcanic ash or volcanic glass and the weathering materials derived from land rocks.

蒙脱石含量与各种膨胀势判别指标研究表明，蒙脱石含量是各种膨胀势的主要控制因素。

Study on **smectite** content and the indices for discrimination swell potential indicates that **smectite** content is a principal factor controlling the swell potential.

利用岩石薄片和 X 射线衍射资料，分析了准噶尔盆地腹部地区蒙脱石黏土矿物的不正常转化。

The abnormal transformation of **smectite clay minerals** in central Junggar Basin is analyzed according to the petrographic thin section and X-ray diffraction data.

研究表明玄武岩首先风化形成蒙脱石族黏土矿物，在此过程中稀土元素没有明显的分馏作用，强风化的玄武岩重稀土元素强烈亏损。

It is concluded that the half-weathered basalts have not distinct REE fractionation and strong weathered basalts had HREE loss in the process of the basalts weathered to form **smectite family minerals**.

3.1.4 术语例段

为了改善低碳化 MgO-C 耐火材料的性能，引入复合添加剂是一个有效策略。以黏土为原料，采用电磁感应加热和埋碳法制备出 Al_2O_3-SiC 复合粉体。另外，将在 600A 下经电磁感应加热合成的 Al_2O_3-SiC 复合粉体添加到低碳 MgO-C 耐火材料中（质量分数为 4%）以改善耐火材料的性能。结果表明，当 Al_2O_3-SiC 复合粉体的外加量（质量分数）为 2.5%~5.0% 范围内时，低碳 MgO-C 耐火材料的性能明显得以改善。其显气孔率为 7.58%~8.04%，体积密度为 2.98~2.99g/cm³，常温耐压强度为 55.72~57.93MPa，在 1100℃下三次空冷后的残余强度为 74.86%~78.04%，在 1400℃氧化 2h 后脱碳层厚度为 14.03~14.87mm。因此，快速合成的 Al_2O_3-SiC 复合粉体为低碳 MgO-C 耐火材料提供了一个可选择的性能优化策略。

To improve the properties of low-carbonization of MgO-C refractories, the introduction of composite additives is an effective strategy. Al_2O_3-SiC composite powder was prepared from **clay** using electromagnetic induction heating and carbon embedded methods. Further, the Al_2O_3-SiC composite powder synthesized by electromagnetic induction heating at 600A were added into low-carbon MgO-C refractories (4wt.%) to improve their properties. The results showed that when the addition amount of Al_2O_3-SiC composite powder is within the range of 2. 5wt.%-5. 0wt.%, the properties of low-carbon MgO-C samples were significantly improved, e. g. the apparent porosity of 7.58%-8. 04%, the bulk density of 2. 98-2. 99g/cm^3, the cold compressive strength of 55. 72-57. 93MPa, the residual strength after three air quenching at 1100℃ of 74. 86%-78. 04%, and the decarburized layer depth after oxidized at 1400℃ for 2h of 14. 03-14. 87mm. Consequently, the idea for the rapid synthesis of Al_2O_3-SiC composite powder provides an alternative low-carbon MgO-C refractories performance optimization strategy[1].

沸石可以在铝硅酸盐前体的碱活化过程中获得。这种沸石-地聚合物混合散装材料融合了沸石和地聚合物的优势性能。在本研究中，评估了活化剂的类型和浓度对碱活化偏高岭土和偏卤石的结构和性能的影响。通过煅烧高岭土和埃洛石获得这两种不同的高岭石**黏土**，然后用氢氧化钠和水玻璃活化。

Zeolites can be obtained in the process of the alkali-activation of aluminosilicate precursors. Such zeolite-geopolymer hybrid bulk materials merge the advantageous properties of both zeolites and geopolymers. In the present study, the effect of the type and concentration of an activator on the structure and properties of alkali-activated metakaolin, and metahalloysite was assessed. These two different kaolinite **clays** were obtained by the calcination of kaolin and halloysite, and then activated with sodium hydroxide and water glass[2].

为了满足对可持续产品和服务不断增长的需求，**高岭土**煅烧行业正在开发可优化资源利用的方法。主要的挑战包括更有效地利用原材料以及减少煅烧炉所消耗的能源。实现这一目标的关键在于煅烧过程的优化。这可以通过对生成的煅烧高岭土的质量参数实时记录来实现。本研究提出使用红外光谱作为监测技术以确定煅烧高岭土产品的化学性能。

In response to the growing demand for sustainable products and services, the **kaolin** calcination industry is developing practices that optimize the use of resources. The main challenges include more efficient use of raw materials and a reduction in the energy consumed by the calcination furnace. An opportunity to achieve this lies in the optimization of the calcination process. This can be done by giving real-time feedback on the quality parameters of the generated calcined kaolin. This study proposes the use of infrared spectroscopy as a monitoring technique to determine the chemical properties of the calcined kaolin product[3].

对**高岭土**的浓缩悬浮液进行附着力测试。在低浓度下，试样在顶板和底板上形成锥形沉积物，而中心区域在破裂之前变窄为细丝。相比之下，高浓度试样在明显破裂成两块之前变形为圆柱体。随着试样浓度的增加，试样经历了完全不同的滑移形式，这可以从它们各自的力距离曲线中推导出来。对于给定浓度的黏土，滑移行为的类型可以通过改变表面粗糙度、实验前的初始压缩载荷和板的分离速度来改变。

Adhesion tests were performed on concentrated suspensions of **Kaolin clay.** At low concentrations samples formed conical deposits on both the top and bottom plates with the central region narrowing to a filament before undergoing breakup. In contrast high concentration samples deformed as a cylinder before apparently fracturing into two pieces. As the concentration of the samples was increased the samples underwent quite different forms of slip which it is shown can be deduced from their respective force distance curves. The type of slip behaviour for a given concentration of clay could be modified with changes to surface roughness, the initial compressive load prior to an experiment and with the separation velocity of the plates. The different slip characteristics appear to arise from the concentration dependent way in which particles interact with the rough surface topography[4].

在这项研究中，对土耳其的 Balıkesir-Sındırgı 地区的**铝基高岭土**应用了机械擦洗、筛选和研磨等选矿工艺，以去减缩小粒径后的亚矾石和其他杂质。通过化学和矿物学分析，确定了 SiO_2、SO_3 和 Fe_2O_3 随组成的变化。同时，研究了浓缩高岭土在瓷砖生产中的使用情况。为此目的，准备了三种不同的配方，以便使用浓缩的铝基高岭土代替陶瓷工厂中使用的高岭土。然后将生产的瓷砖的性能与工厂自产产品的性能进行了比较。最终发现，添加 20% (质量分数) 的浓缩铝基高岭土改善了工厂瓷砖的物理性能，并且高岭土内的硫含量很低。

In this research, a beneficiation process including mechanical scrubbing, screening and milling was applied to the **alunitic kaolins** of the Balıkesir-Sındırgı region of Turkey to remove alunite and the other impurities, after size reduction. Chemical and mineralogical analysis of the size groups was made to determine the changes with the composition for SiO_2, SO_3 and Fe_2O_3. Afterwards, the use of concentrated kaolins in ceramic tile production was investigated. For this purpose, three different recipes were prepared in order to use the concentrated alunitic kaolins in place of the kaolin being used in a ceramic factory. The properties of the ceramic tiles produced were then compared with those of the factory's own products. It was finally proven that the use of concentrated alunitic kaolin at a ratio of 20wt.% improved the physical properties of ceramic tiles of the factory and that the sulphur within the kaolin was tolerated[5].

高岭石是一种黏土矿物，具有多种环境、工业和农业应用。**高岭石**的晶体学特性（例如结构无序）对这些应用的影响非常令人感兴趣。过去70年来对**高岭石**结构无序的定性和定量分析揭示了层堆叠无序的三个主要来源：（1）对映异构堆叠；（2）地开石状堆积；（3）层的随机移位。这些堆积障碍对**高岭石**的反应性有什么影响？本研究的目的是研究堆积无序对**高岭石**层的插层和溶解的影响。为了尽量减少粒径对反应性的影响，使用了五种1～2μm的地质**高岭石**，并且它们的结构无序程度各不相同。

Kaolinite is a clay mineral with diverse environmental, industrial, and agricultural applications. The influence of the crystallographic properties of **kaolinite**, e. g., structural disorder, on these applications is of great interest. Qualitative and quantitative analyses of **kaolinite** structural disordering over the last 70 years have revealed three main sources of layer-stacking disordering: (1) enantiomorphic stacking; (2) dickite-like stacking; and (3) random shift of layers. What influence do these stacking disorders have on the reactivity of **kaolinite**? The objective of the present study was to investigate the influence of stacking disorder on the intercalation and dissolution of **kaolinite** layers. To minimize the effect of particle size on reactivity, the 1-2μm fractions of five geologic **kaolinites** were used. The 1-2μm fractions varied in the degree of structural disorder[6].

开发纳米级**高岭石**（nano-kaolinite）的可持续合成路线，对于高岭土工业生产高品质的造纸级**高岭石**具有重要意义。通过参考自然界中的化学风化过程，设计并比较了由钾长石合成纳米高岭石的两条工艺路线。本研究获得了厚度约为14nm的均匀片状形态的**高岭石**。这两种合成路线均可综合利用钾长石合成纯**高岭石**，不仅可用于生产高质量的纸张涂料，还可以用于医疗和其他用途。

Development of sustainable routes for the synthesis of **kaolinite** in nanoscale(nano-kaolinite) is very significant for producing high quality **kaolinite** of paper-coating grade in kaolin industry. Duplicating chemical weathering processes in nature, two routes were developed and compared for the synthesis of nano-kaolinite from K-feldspar. **Kaolinite** of uniform plate-like morphology with thickness of around 14nm was obtained in this study. Both synthesis routes may lead to the comprehensive utilization of K-feldspar for the synthesis of pure **kaolinite** for not only high-quality paper-coatings but also medical and other uses[7].

使用电位滴定、电动力学测量和与导数等温线求和（DIS）建模相关的低压气体吸附分析了不同来源的五种**高岭石**的表面特性。数据显示，试样结构在决定表面特性方面很重要。综合分析结果清楚地表明，所有试样都存在一个永久的负电层。对于五种**高岭土**可以观察到的相同特征是：由于电荷层的存在，滴定曲线随着离子强度的增加而向低 pH 值移动；质子量的消耗也受到永久电荷层的影响；虽然低，但永久电荷决定了电动力学行为。形状各向异性以及基底和边缘表面的电荷分布是了解**高岭土**颗粒在水介质中行为的关键参数。喀麦隆试样组表现出明显的永久电荷，以及与颗粒尺寸和形状各向异性有关的大比表面积。

The surface properties of five **kaolinites** of various origins were analyzed using potentiometric titration, electrokinetic measurements and low-pressure gas adsorption associated to derivative isotherm summation (DIS) modeling. The data show that sample structure is important in determining the surface properties. The combined analytical results clearly show the existence of a permanent negative layer charge for all the samples. The general features that could be observed for the five **kaolinites** are: the titration curves are shifted to lower pH with increasing ionic strength due to the layer charge; the amount of consumed proton is influenced by the permanent layer charge; although low, the permanent charge determines the electrokinetic behavior. The shape anisotropy and the charge distribution on the basal and edge surfaces are crucial parameters to understand the behavior of **kaolinite** particles in aqueous media. The Cameroonian sample of the set exhibits a significant permanent charge together with large specific surface area associated to fine particle size and considerable shape anisotropy[8].

了解纳米气泡存在下脉石矿物的夹带行为对于纳米气泡技术在泡沫浮选中的应用是必要的。在这项研究中，研究了用减压产生的纳米气泡水和自来水浮选时**高岭石**颗粒的夹带。与自来水相比，纳米气泡水在浮选中增强了**高岭石**颗粒的夹带。流变学测量与沉降测试被进一步用于检查纳米气泡对黏土颗粒结合的影响及其与**高岭石**颗粒夹带的相关性。纳米气泡的存在似乎可以诱导和稳定**高岭石**片晶的 E-E 接触，从而形成多孔的三维结构。这些具有丰富间隙空隙的开放结构沉降速度较低，因此更容易被夹带回收。

Understanding the entrainment behaviour of gangue minerals in the presence of nanobubbles is necessary for the application of nanobubble technology in froth flotation. In this study, the entrainment of **kaolinite** particles in flotation with nanobubble water produced by decompression, and tap water, was investigated. Compared to tap water, nanobubble water enhanced the entrainment of **kaolinite** particles in flotation. Rheology measurements together with settling tests were further employed to examine the effect of nanobubbles on clay particle association and its correlation with the entrainment of **kaolinite** particles. The presence of nanobubbles appears to induce and stabilize E-E contacts of **kaolinite** platelets, resulting in the formation of porous three-dimensional structures. These open structures with abundant interstitial voids had a low settling velocity and therefore were more readily recovered by entrainment[9].

高岭石是油藏中常见的 1∶1 层状黏土矿物，也是一种天然的固体酸催化剂。一些研究表明，**高岭石**对重油的氧化具有催化作用。其催化效果取决于酸位点的类型和数量，并与其热转化，特别是脱羟基过程密切相关。采用热分析和 Ozawa-Flynn-Wall 等转化法研究了**高岭石**及其热转化对重油氧化的影响。结果表明，**高岭石**对重油氧化的高温氧化（HTO）和燃料沉积（FD）阶段有显著影响，但对低温氧化（LTO）阶段的影响较小。在 HTO 阶段，CO_2 释放速率峰值显著降低，而 H_2O 释放速率峰值温度显著升高。随着**高岭石**添加量的增加，稠油氧化反应 FD 阶段的最大活化能和 HTO 阶段的平均活化能逐渐降低。值得注意的是，**高岭石**在热处理后的转化，特别是 400℃ 以上的脱羟基过程，导致重油 HTO 阶段峰值温度逐渐下降，但对重油氧化 LTO 阶段的影响不大。

Kaolinite is a common 1:1 layered clay mineral in oil reservoirs and also a natural solid acid catalyst. Some studies have shown that **kaolinite** has a catalytic effect on the oxidation of heavy oil. Its catalytic effect depends on the type and amount of acid sites and is closely related to its thermal transformation, particularly the dehydroxylation process. The effects of **kaolinite** and its thermal transformation on the oxidation of heavy oil were investigated with thermal analysis and the Ozawa-Flynn-Wall isoconversional method. The results revealed that **kaolinite** had significant effects on the high-temperature oxidation(HTO) and fuel deposition(FD) stages of heavy oil oxidation, but its effect on the low-temperature oxidation(LTO) stage was small. In the HTO stage, the temperature at which the CO_2 release rate peaked decreased dramatically, whereas the temperature at which the H_2O release rate peaked increased remarkably. With the increase in the content of added **kaolinite**, the maximum activation energy in the FD stage of the oxidation reaction of heavy oil and the average activation energy in the HTO stage gradually decreased. It was noteworthy that the transformation of **kaolinite** upon thermal treatment, especially the dehydroxylation process above 400℃, resulted in a progressive decline in the peak temperature in the HTO stage of heavy oil, but had little influence on that in the LTO stage of heavy oil oxidation[10].

膨润土作为高放废物（HLRW）处置库中的岩土屏障材料被研究。最重要的研究课题之一是膨润土在界面处的稳定性，例如**膨润土**/金属或膨润土/水泥。碱性水泥特别危险，因为它们可能会溶解部分膨润土。因此，该反应被广泛研究。膨润土在不同应用中的特性和性能因矿床而异，甚至在某些矿床内也不同。因此，假设与水泥接触的不同膨润土的稳定性可能不同。基于此，该研究的目的是比较一组显著的特征良好的膨润土与波特兰水泥的反应性，并了解差异的原因。比较膨润土标准反应的参数是固定的：3个月，80℃，25%（或33%）水泥。将所得降解程度与基本膨润土参数进行比较。改变的程度与苏打水溶性硅石（天然膨润土中存在不同数量的活性硅石）的数量有相当好的相关性（$R^2 = 0.7$），这证明苏打水溶性硅石能够缓冲反应。这种具有较高钠溶性二氧化硅含量的膨润土显示出较少的蒙脱石降解，因此在水泥/膨润土界面处的性能优于其他具有较少苏打溶性二氧化硅的膨润土。改变黏土/水泥比例，表明绿土降解需要10%～20%的水泥。低于该值未观察到反应。观察到的线性曲线可以得出如下结论，在选定的溶液/固体比下，1g水泥能够降解约0.6g膨润土（蒙脱石）。1个月后反应基本完成，在60～80℃之间没有观察到差异。

Bentonite is investigated as a geotechnical barrier material in repositories for high-level radioactive waste(HLRW). One of the most important research topics concerns the stability of **bentonite** at interfaces such as bentonite/metal or bentonite/cement. Alkaline cements are particularly hazardous because they may dissolve at least part of the bentonite. This reaction is, therefore, investigated extensively. Bentonite properties and performances in different applications vary from one deposit to another and even within some deposits. The hypothesis, therefore, was that the stability of different bentonites in contact with cement could be different. Hence, the aim of the study was to compare the reactivity of a significant set of well characterized bentonites with Portland cement and to understand the reason for the differences. For bentonite comparison standard reaction parameters were fixed:3 months,80℃,25% (or 33%) cement. The resulting degree of degradation was compared with basic bentonite parameters. A fairly good correlation ($R^2 = 0.7$) of the extent of alteration with the amount of soda soluble silica(reactive silica which to variable amounts is present in natural bentonites) was obtained which proved, that soda soluble silica was able to buffer the reaction. Such bentonites with a higher amount of soda soluble silica showed less smectite degradation and hence will perform better at the cement/bentonite interface than others with less soda soluble silica. Varying the clay/cement ratio showed that smectite degradation requires 10%-20% cement. No reaction was observed below this value. Linear curves were observed which allow to conclude that 1g cement is able to degrade about 0.6g of bentonite (smectite)at the selected solution/solid ratios. The reaction was almost complete after 1 month and no difference was observed between 60℃ and 80℃ [11].

在试验场地测试用于密封高放射性废物储存库的**膨润土**屏障的性能，但在增强的热条件下矿物稳定性程度仍然是一个有争议的话题。本研究重点关注 SKB ABM5 实验，该实验运行了 5 年（2012～2017 年），局部最高温度达到 250℃。使用 XRD 和 Rietveld 精修、SEM-EDX 并通过测量 pH 值、CEC 和 EC 研究了 5 种**膨润土**。从距加热管 0.1cm、1cm、4cm 和 7cm 处的膨润土块中提取的试样显示出与水平热梯度类似的不同阶段的变化。靠近具有较低 CEC 值的接触面的**膨润土**发生以 Si^{4+} 被 Al^{3+} 四面体取代的形式的蒙脱石变化和一些八面体金属取代物，可能与氧化沸腾过程中加热器腐蚀产生的三价铁/亚铁有关，部分膨润土层中发生黄铁矿溶解和酸化。这种改变还与较高量的赤铁矿和少量方解石溶解有关。然而，由于没有任何膨润土显示出任何蒙脱石损失，只是在加热器–膨润土接触处显示出更强的变化，因此密封剂被认为基本保持完整因此密封剂被认为基本完好无损。

Pilot sites are currently used to test the performance of **bentonite** barriers for sealing high-level radioactive waste repositories, but the degree of mineral stability under enhanced thermal conditions remains a topic of debate. This study focuses on the SKB ABM5 experiment, which ran for 5 years(2012 to 2017)and locally reached a maximum temperature of 250℃. Five **bentonites** were investigated using XRD with Rietveld refinement, SEM-EDX and by measuring pH, CEC and EC. Samples extracted from **bentonite** blocks at 0.1cm, 1cm, 4cm and 7cm away from the heating pipe showed various stages of alteration related to the horizontal thermal gradient. **Bentonites** close to the contact with lower CEC values showed smectite alterations in the form of tetrahedral substitution of Si^{4+} by Al^{3+} and some octahedral metal substitutions, probably related to ferric/ferrous iron derived from corrosion of the heater during oxidative boiling, with pyrite dissolution and acidity occurring in some bentonite layers. This alteration was furthermore associated with higher amounts of hematite and minor calcite dissolution. However, as none of the bentonites showed any smectite loss and only displayed stronger alterations at the heater-bentonite contact, the sealants are considered to have remained largely intact[12].

膨润土因其低渗透性、高吸附能力、自密封特性和在自然环境中的耐久性而被认为是处理高放废物（HLW）的候选缓冲材料。地下水流引起的**膨润土**侵蚀可能发生在压实**膨润土**与破碎花岗岩的界面处。**膨润土**絮体的表面侵蚀通常表示为侵蚀阈值。预测**膨润土**絮体的侵蚀阈值需要考虑黏土颗粒之间相互作用产生的内聚力。除了通常对粒度的依赖外，**膨润土**絮凝物的侵蚀阈值和孔隙度测量值之间存在显著相关性。提出了**膨润土**絮体侵蚀阈值的分形模型。黏聚力，即两个黏土颗粒之间的长程范德华相互作用作为侵蚀阈值的来源。通过与文献中发表的实验进行对比，验证了模型的准确性。结果表明，所提出的侵蚀阈值模型与实验数据吻合较好。

Bentonite has been considered as a candidate buffer material for the disposal of high-level radioactive waste (HLW) because of its low permeability, high sorption capacity, self-sealing characteristics and durability in a natural environment. **Bentonite** erosion caused by groundwater flow may take place at the interface of the compacted **bentonite** and fractured granite. Surface erosion of **bentonite** flocs is represented typically as an erosion threshold. Predicting the erosion threshold of **bentonite** flocs requires taking into account cohesion, which results from interactions between clay particles. Beyond the usual dependence on grain size, a significant correlation between erosion threshold and porosity measurements is confirmed for **bentonite** flocs. A fractal model for erosion threshold of **bentonite** flocs is proposed. Cohesion forces, the long-range van der Waals interaction between two clay particles is taken as the resource of the erosion threshold. The model verification is conducted by the comparison with experiments published in the literature. The results show that the proposed model for erosion threshold is in good agreement with the experimental data[13].

风化**膨润土**（日本酸性黏土）是一种在**膨润土**生产过程中作为副产品发现的自然资源，通过环保处理改变了日本酸性黏土与碳酸钠的机械化学混合，无须添加溶剂，得到钠型**膨润土**。产物随后通过常规分散-沉淀法纯化以获得纯化的钠基**膨润土**。改性酸性黏土和纯化产物表现出溶胀性能，能满足常规优质钠基**膨润土**的工业用途，并且可以通过离子交换进一步功能化产物。例如，通过将本产品与阳离子表面活性剂进行离子交换，成功制备了一种对环境中有毒有机化合物有用的吸附剂，亲有机性黏土。所使用的机械化学过程是对自然资源进行实际利用的一种重要的无害环境的方式。

A weathered **bentonite** (Japanese acid clay), a natural resource found as a by-product in the production of **bentonite**, was altered by an environmentally friendly treatment, where the Japanese acid clay was mechanochemically mixed with sodium carbonate without the addition of solvent, to give sodium type **bentonite**. The product was subsequently purified by conventional dispersion-sedimentation method to obtain purified sodium **bentonite**. The altered acid clay and the purified product exhibited useful swelling properties, which are satisfactory for industrial uses of conventional high quality sodium **bentonites**, and further functionalization of the product was possible through ion exchange. As an example, an organophilic clay, which is useful as adsorbent for toxic organic compounds from environments, was successfully prepared by the ion exchange of the present product with a cationic surfactant. The present mechanochemical process is an important environmentally benign option for the modification of the natural resources for the practical uses[14].

由于**膨润土**具有较高的膨胀能力和较低的导水率，因此在垃圾处理场使用压实的膨润土作为衬里材料。垃圾填埋场渗滤液中的盐分可能会导致**膨润土**的扩散双层收缩，进而影响其膨胀和可压缩性。固结是衬里材料的重要特性之一，需要对其进行沉降分析研究。

Due to its high swelling capacity and lower value of hydraulic conductivity, compacted **bentonite** is used as a liner material at the waste disposal site. Salts presents in landfill leachates may cause the diffuse double layer of **bentonite** to shrink which in turn affects its swelling and compressibility behaviour. Consolidation is among important properties of the liner material which need to be studied for settlement analysis[15].

X 射线粉末衍射（XRD）结合 Rietveld 方法是晶体结构分析的常用技术。然而，几乎所有**蒙脱石**组的矿物都表现出湍层无序，从而阻碍了 Rietveld 精修的直接应用。Ufer 等人提出通过在 BGMN 程序中使用单层模型来克服这一限制。该模型现在通常用于含**蒙脱石**试样的定量相分析。然而，一些重要结构特征的精修，如层电荷密度（LCD）和八面体铁含量尚未能实现。这些结构特征通常只能通过分析纯化的**蒙脱石**试样获得。

X-ray powder diffraction (XRD) in combination with the Rietveld method is a common technique for crystal structure analysis. However, almost all minerals of the **smectite** group show turbostratic disorder preventing the direct application of Rietveld refinement. Ufer et al. proposed to overcome this limitation by using a single layer model with the BGMN program. This model is now routinely used for quantitative phase analysis of **smectite**-containing samples. However, the refinement of some important structural features, such as the layer charge density (LCD) and the octahedral iron content has not been achieved successfully yet. Typically, these structural features can be obtained only by analytical methods which require purified **smectite** samples[16].

对各种双八面体**蒙脱石**的实验表明，顺式和反式空层结构的比例、四面体片中的电荷分数和铁含量之间存在弱相关性。这些相关性是由两个独立的样本集证明的，并且在不同的实验室进行了检验，因此被认为具有重要意义。通过量子化学密度泛函对一组模型矿物的相对能和离子交换能进行了精确的计算，验证了这些关系的准确性。因此，人们可以将这些经验相关性主要追溯到能量关系，并通过它们的特殊形成历史合理化一些**蒙脱石**的偏离现象。

The experimental characterization of various dioctahedral **smectites** reveals weak correlations between the ratio of cis-and trans-vacant layer structures, the fraction of charge in the tetrahedral sheet, and the iron content. These correlations are demonstrated by two independent sample sets, examined in different laboratories, and hence are considered to be significant. To rationalize and corroborate these relations, accurate relative energies and ion exchange energies of a set of model minerals were determined by quantum chemical density functional calculations. Accordingly, one may trace these empirical correlations mainly back to energy relations and rationalize some deviating **smectites** by their particular history of formation[17].

自 2013 年以来，双八面体**蒙脱石**形式的膨润土被欧盟批准作为减少黄曲霉毒素对饲料污染的添加剂使用。几项研究表明，**蒙脱石**在隔离黄曲霉毒素方面的有效性存在很大差异。**蒙脱石**的矿物学和物理化学性质与黄曲霉毒素吸附之间的明确相关性尚未确定。为了确定影响**蒙脱石**吸附黄曲霉毒素的最关键的矿物学、化学和物理特性，对来自世界各地不同来源的 29 种膨润土试样进行了评估。

Since 2013, bentonite in the form of dioctahedral **smectite** is an additive authorized in the EU as a substance for the reduction of the contamination of feed by aflatoxins. Several studies indicate a big difference in the effectiveness of **smectites** in sequestering aflatoxins. A clear correlation between mineralogical and physico-chemical properties of **smectites** and aflatoxin adsorption has not been well established. In the effort to identify the most critical mineralogical, chemical, and physical properties that affect aflatoxin adsorption by **smectites**, 29 samples of bentonites obtained from different sources around the world were evaluated[18].

黄曲霉毒素会导致肝脏损伤并抑制免疫力。通过吸附，**蒙脱石**可用于降低黄曲霉毒素的生物利用度。为了进一步降低黄曲霉毒素的毒性并消除对负载黄曲霉毒素的**蒙脱石**的处理，可以将黄曲霉毒素附到无毒或毒性较小的化合物上。本研究的目的是研究温度和交换阳离子对吸附在**蒙脱石**上的黄曲霉毒素 B1 转化的影响。

Aflatoxins cause liver damage and suppress immunity. Through adsorption, smectites can be used to reduce the bioavailability of aflatoxins. To further reduce the toxicity of aflatoxins and to eliminate the treatments of aflatoxin-loaded **smectites**, the ability to degrade the aflatoxin adsorbed to non-toxic or less toxic compounds is desirable. The objective of the present study was to investigate the effects of temperature and the exchange cation on the transformation of adsorbed aflatoxin B1 on **smectite**[19].

我们在 2.5GPa、4.0GPa 和 7.7GPa 的压力和 200~700℃ 的温度下对 NH₄ 掺杂的**蒙脱石**（质量分数约 2% 的 NH₄）进行了高压和高温（HPHT）实验。通过使用 XRD、FTIR、CHN 元素分析和 SEM 进行分析，以确定 NH₄-**蒙脱石**相变及其形态，以及铵的存在形式。研究结果表明，在弱还原环境中，**蒙脱石**可以将氮以铵态（NH₄⁺）形式，通过冷热制度俯冲区输送到地球的更深层。

We performed high pressure and high temperature (HPHT) experiments on NH₄-doped montmorillonite(about 2wt.% of NH₄) under pressures of 2.5GPa,4.0GPa,and 7.7GPa and temperatures from 200℃ to 700℃. Each experiment was analyzed with XRD,FTIR,CHN elemental analysis,and SEM in order to determine the NH₄-**Smectite** phase changes and their morphology,and the presence of ammonium in the runs. Our results show that **smectite** can easily transport nitrogen,speciated as ammonium(NH₄⁺), incorporated into the **smectite** interlayer in mildly reducing environments to deeper levels in the Earth through cold thermal regime subduction zones[20].

本论文针对目前**膨润土**资源开发利用存在的技术落后、资源利用率低、环境污染严重、产业链短、产品性能落后、工艺能耗高等问题，以生态设计思想为指导，按照不同的膨润土原矿特点，研究开发资源利用率高、能源消耗少、对环境友好、产品性能优良的膨润土深加工清洁生产工艺。通过上述工艺得到使用性能优良、环境协调性好、附加值高的新型蒙脱土生态环境材料，并评价所制得的材料性能以及其在费托合成催化过程的应用。本文的研究内容主要包括三部分：（1）不含方英石膨润土深加工工艺研究；（2）富含方英石膨润土深加工工艺研究；（3）有机蒙脱土在费托合成中的应用研究。主要工作如下：（1）针对于不含方英石的膨润土综合利用难题，采用提纯和改型同步进行的湿法钠化工艺制备高纯蒙脱土，通过对湿法改型工艺中的反应条件对产品质量的影响以及它们之间的关系的研究，确定出最佳的单元操作条件，最终得到蒙脱石含量高达 95.6% 的高纯钠基蒙脱土。进而，针对其湿法提纯工艺中"脱水难"的问题，在考虑膨润土的物理化学性质的基础上，依据结构以及性质的不同，选择了五种天然有机高分子絮凝剂羧甲基淀粉钠、羧化壳聚糖、瓜尔胶、羟丙基瓜尔胶、阳离子瓜尔胶，利用分光光度法考察了不同絮凝剂对高纯钠基膨润土矿浆的絮凝性质，并依据其絮凝性质的差别探讨了五种絮凝剂絮凝膨润土的絮凝机理。以膨润土湿法提纯后得到的钠基膨润土浆液为原料，以氧化钙和氯化钙为胶化剂制备了钠基膨润土无机凝胶。比较了胶化剂类型、用量、时间、转速等条件对凝胶性能的影响，并依据其凝胶性能的差别探讨了钠基膨润土无机凝胶的胶化机理。以钙基膨润土原矿为起点，制备出吸水快、吸水量大、白度高、强度高的膨润土猫砂。（2）针对富含方英石膨润土的提纯难题，研究使用碱法提纯方案有效地实现了富含方英石膨润土矿产资源的经济、高效、可持续利用。所制备的高纯蒙脱土其蒙脱石含量高达97%。为了尽可能提高资源的利用率及实现废弃物的资源化利用，研究以上述富含方英石膨润土碱法提取液为原料使用沉淀法制备白炭黑及纳米二氧化硅。考察不同沉淀剂的沉淀效果，研究反应温度、pH 值、浓度、反应时间对白炭黑产品性能及产率的影响。研究不同表面活性剂对所得纳米二氧化硅的粒径的影响，确定了富含方英石膨润土碱法提取液的最佳沉淀方法及条件。该工艺不仅实现了富含方英石膨润土碱法提取废液的回收利用，而且实现了资源循环利用，提高了生产效益，又没有对环境造成污染，体现了绿色工艺的思想。（3）使用有机修饰剂改性后的蒙脱土作为新型载体用于费托合成研究，得益于优良的热稳定性、独特的气体阻隔性和温度开关的性质。有机修饰剂作为催化剂活性组分的有机保护层可以简单有效地实现催化剂反应状态的原位隔离和保护，从而被成功应用于对钴基费托合成反应过程的详细研究。就像将古代生物原始状态保存完好的琥珀一样，聚合物链在原位固化隔离和保护了实际催化剂的结构，以确保离位分析的催化剂状态充分代表了真实催化剂的工作状态。使得费托合成研究中仍然存在争议的一些关键问题得以解决，如还原过程中的钴纳米粒子的相变和团聚、费托反应中活性相态的变化、费托催化剂结构与性能的关系等得以阐明。通过设计不同的有机改性剂，我们可以通过传统的离位表征技术表征各种催化反应的催化剂在真实反应条件下的纳米结构，为将来其他催化体系的研究开辟了一条新途径[21]。

Focusing on the present problems that exist in **bentonite** resource development and utilization such as low-level technology, low resource utilization ratio, serious environmental pollution, short industry chain, poor product performance and high energy consumption; guided by the concept of ecological design, on the basis of the different characteristics of bentonite raw mineral; this thesis develop clean technology of bentonite deep processing with the advantages features including high utilization ratio of resources, less energy-consuming, environmental friendliness and excellent product property. The new montmorillonite eco-materials obtained through the above technique were featured by excellent property, well environmental benignity, and high additional value. Furthermore, the applications of eco-materials performances in Fischer-Tropsch synthesis process were evaluated as well. This thesis mainly includes three parts: （1）deep processing technology of cristobalite-free bentonite; (2) deep processing technology of cristobalite-rich bentonite; (3) application of organic modified montmorillonite in Fischer-Tropsch synthesis. The main work is as follows. (1) For cristobalite-free bentonite, high purity montmorillonite was prepared by wet sodium modification process, which the purification and modification were carried out simultaneously. The optimal unit operation conditions were determined by researching the influence of wet modification process conditions on product quality and the relationship between them. Consequently, the final montmorillonite content of the resulting high purity sodium bentonite is as high as 95.6%. Furthermore, the efficient dewatering of colloidal stable bentonite suspensions presents an intractable challenge for the wet purification process. On the basis of the consideration of the bentonite physical and chemical properties, five natural polymer flocculants, including carboxymethylstach sodium, carboxyl chitosan, guar gum, hydroxypropyl guar gum and cationic hydroxypropyl guar gum, were choose depending on the structure and property difference. The sodium bentonite slurry flocculation conditions of the five flocculants were studied by spectrophotometry. The bentonite flocculation mechanism of the five flocculants was explored based on their different flocculation properties. The sodium bentonite inorganic gel was prepared by bentonite wet purification slurry as raw material, and calcium oxide or calcium chloride as gel agent. The influences of gel agent type, gel agent dosage, dispersion time, dispersion speed and other reaction conditions on gel performance were compared. The mechanism of sodium bentonite inorganic gel was explored based on the difference gel performance. Finally, the bentonite litter with advantages features including fast water uptake, more water uptake volume, high whiteness and high strength was prepared by the calcium bentonite. (2) For cristobalite-rich bentonite mineral resources, alkaline purification scheme was utilized economically, efficiently and sustainably. Consequently, the montmorillonite content of the resulting high purity sodium bentonite through the alkaline purification process is as high as 97%. In order to improve the utilization ratio of resources as far as possible and realize the recycle usage of the wastes, silica and nano SiO_2 were prepared by precipitated the extract wastes of cristobalite-rich bentonite alkaline purification. The effects of different precipitating agents were investigated. The influence of reaction temperature, pH, concentration and reaction time on the properties and yield of silica were compared. The effects of different surfactant on the resulting nano SiO_2 particle size were investigated as well. Then the optimal precipitation method and condition were determined. The alkaline purification process is not only to achieve the cristobalite-rich bentonite alkaline extraction liquid wastes recycle usage, but also to improve production efficiency, and do not pollute the environment. It reflects the thought of green technology. (3) Organic modified montmorillonite was used as a new carrier for Fischer-Tropsch synthesis. Due to the excellent thermal stability, special gas barrier and temperature switch behavior, the organic modifier as the protective coating of catalyst active components were used to in situ isolate and protect the actual catalyst structure, consequently, to study the reaction involved in the cobalt-based Fischer-Tropsch synthesis in detail. Just like the original state of ancient organisms preserved well by amber, the polymer chains were in situ solidified to isolate and protect the actual catalyst structure, which ensured the ex-situ analyzed catalyst state to fully represent the working state of the catalysts. Accordingly, some critical important issues which could not be studied clearly in past days, such as the phase transformation and sintering agglomeration of cobalt nanoparticles in reduction procedure, the identification and phase transformation of the active species during reaction, and the structure-property relationship of catalysts for the FTS, have been thoroughly revealed. By designing different organic modifiers, we can characterize nanostructure evolution of catalysts of various catalytic reactions under operation reaction conditions via ex-situ characterization techniques. Therefore, a new way was opened for the future research of other catalytic systems[21].

3.2 高铝质耐火材料原料

3.2.1 术语词

含铝土的 aluminous
铝矾土(铝土矿) bauxite
铝土化 bauxitization
蓝晶石 kyanite, cyanite, disthene, zianite
红柱石 andalusite, apyre
莫来石 mullite
锰红柱石 viridine
绿硅线石 bamlite

硅线石 sillimanite, fibrolite
莫来石化 mullitization
铝热莫来石 thermit-mullite
人造刚玉 boule
刚玉 corundum
刚玉砂 emery
锆刚玉 zirconia-corundum

3.2.2 术语词组

矾土水泥结合剂 aluminous cement bond
泥质铝矾土 argillaceous bauxite
高铝矾土骨料 bauxite aggregates
矾土水泥(高铝水泥)bauxite cement
高铝矾土熟料 bauxite chamotte
矾土 bauxite ore
低铁铝土矿 bauxite with low iron content
一水软铝石型铝土矿 boehmite bauxite
水硬水铝石型铝土矿 diaspore type bauxite
高铝骨料 high alumina aggregate
高铝粉料 high alumina mass
高铝料 high aluminum material
高铁铝土矿 high iron bauxite
高硅矾土矿(高品位铝土矿)high silicon bauxite
均化铝矾土(矾土均化料)homogenized bauxite,
　homogeneous bauxite
低硅铝土矿 low-silicon bauxite
低品位铝土矿 poor-quality bauxite
重烧矾土熟料 re-sintered bauxite chamotte
高硅铝土矿 siliceous bauxite
白色铝土矿 white bauxite
蓝晶石硅线石精矿 disthene-sillimanite concentrate
蓝晶石硅线石火泥 disthene-sillimanite mortar
蓝晶石精矿 kyanite concentrate
蓝晶石族矿物 kyanite minerals
蓝晶石火泥 kyanite mortar
蓝晶石片岩 kyanite schist facies
红柱石精矿 andalusite concentrate
硅线石精矿 sillimanite concentrate
硅线石片麻岩 sillimanite gneiss
硅线石族 sillimanite group
硅线石页岩 sillimanite shale
合成硅线石 synthetic sillimanite
人工合成莫来石 artificial mullite
刚玉莫来石 corundum mullite
细针状莫来石 fine acicular mullite

细晶莫来石 fine-crystal mullite
电熔莫来石 fused mullite
电熔锆莫来石 fused zirconia mullite
高纯电熔莫来石 high-purity fused mullite
莫来石纤维 mullite fiber
莫来石粉料 mullite mass
针状莫来石 mullite needle
莫来石超细粉 mullite superfine powder
原生莫来石 primary mullite
次生莫来石 secondary mullite
短棱柱状莫来石 short-prismatic mullite
烧结莫来石 sintered mullite
烧结莫来石刚玉熟料 sintered mullite corundum
　chamotte
莫来石的固相合成 solid phase synthesis of mullite
莫来石连晶 twinning mullite
黑褐色的刚玉 adamantine spar
人造刚玉 artificial corundum
高铝矾土基致密电容刚玉 bauxite-based dense
　electro-fused corundum
轻烧刚玉 caustic-burned corundum
铬刚玉 chrome corundum
刚玉熟料 corundum clinker
刚玉晶粒增大 corundum grain growth
致密刚玉 dense corundum
致密电熔刚玉 dense fused corundum
致密白刚玉 dense white fused corundum
电熔棕刚玉 electrically fused brown corundum
粗晶刚玉 macrocrystalline corundum
微晶刚玉 microcrystalline corundum
改性烧结刚玉 modified sintered corundum
烧结刚玉 sintered corundum
电熔亚白刚玉 sub-white electrically fused
　corundum
人造刚玉 synthetic corundum
电熔白刚玉 white fused corundum

3.2.3　术语例句

用**一水硬铝石型铝土矿**在旋流浮选系统中进行了旋流浮选和分级试验对比。

Making use of the **diaspore type bauxite**, the cyclone flotation and classification test had been carried in the system of cyclone flotation.

以某高铁**三水铝石型铝土矿**为原料，研究了铝土矿粒度对烧结矿产质量指标的影响。

The effect of grain size of the bauxite on the sintering performance was studied by using a high iron **trihydrate bauxite** as raw material.

棕刚玉是以**铝矾土**等为原料，经高温熔炼结晶而成。

Brown corundum is fused and crystallized under high temperature from a mixture of **bauxite** and other raw materials.

对比研究了几种**铝矾土**的化学组成、相组成和分散性能。

The chemical composition, petrographic analysis and dispersing property of kinds of **bauxite** are compared.

开发和应用了以**特级优质高铝矾土熟料**作主料的低水泥浇注料。

The exploration and application of low cement castable with **high grade quality super calcined bauxite** as main raw material are introduce in this article.

研究了结合剂种类、石墨和**特级高铝矾土熟料**加入量对镁铝碳制品高温性能的影响。

Effects of variety of binder, additions of graphite and **special grade high-alumina bauxite chamotte** on high temperature properties of magnesia-alumina-carbon products were studied.

探讨了原料配比、粒度以及配料烧结温度、时间等工艺因素对**贫铝矾土**中氧化铝的提取效率的影响。

This article studies the main factors, which affect the extract ratio of aluminum oxide from **lean bauxite**, such as the ratio of raw material, particle size, broiled temperature, and broiled time.

以普通瓷质砖坯料配方为基础配方，引入**熟铝矾土**研制符合超薄瓷质砖性能要求的高强坯料配方。

On the base of ordinary porcelain tile, **sintered bauxite** was introduced to fabricate super-thin porcelain body with high strength.

湖南**铝矾土**试样分散性能好，特别适于原料需要预处理成均匀料浆的工艺过程。

The **bauxite** sample from Hunan has good dispersing property, suitable for the producing process that raw material must be pretreated to well-distributed slurry.

蓝晶石加入量为4%~5%时浇注料的性能较好。

When the adding amount of **kyanite** is 4%-5%, the castable performance is better.

红柱石属多晶型物，含**蓝晶石**和**硅线石**两种不同矿物。

Andalusite is a polymorph with two other minerals, **kyanite** and **sillimanite**.

这主要是由**蓝晶石细粉**向莫来石转化时发生的体积膨胀效应所致。

This is mainly because of the effect of volumetric expansion which occurs when **kyanite powder** is transformed into mullite.

用两个温度下煅烧后的**蓝晶石**为原料，研制烧结蓝晶石制品，对其烧成温度、高温性能、抗热稳定和显微结构作了研究。

Kyanite calcined at two temperatures is used as raw material to prepare calcined **kyanite**, and their firing temperature, high temperature properties, thermal shock and microstructure have been studied.

团麻断裂以西的河南大别山区榴辉岩及其围岩**蓝晶石**石英岩中首次发现柯石英。

Coesite has been first discovered in the eclogites and their country rocks **kyanite** quartzite in the western area of Tuanma Fault from the Dabie Mountains, Henan province.

研究了**蓝晶石**的粒度和加入量对莫来石-刚玉材料性能的影响。

The effect of particle size and additions of **kyanite** on properties of mullite-corundum was studied.

硅线石用途广泛，也是高级耐火材料的重要原料。

Sillimanite is widely used as an important material for high-grade refractory products.

硅线石耐火浇注料适用于流化床锅炉的燃烧部位。

Sillimanite refractory castable is good in use at combustion zone in CFB boilers.

研究了**硅线石**性质、化学成分与耐火材料制品性能的关系。

Also, the relation between ore characteristics and chemical contents of **sillimanite** and its product performance is studied.

概述了**硅线石**作为高铝耐火材料原料的质量要求、应用范围。

The quality requirement and application range of **sillimanite** are introduced for high-aluminium-bearing refractory material.

拟对现有的生产线进行技术改造，达到原年产7000t **硅线石**精矿的设计生产能力。

The project plans to conduct technical reconstruction to the existing production line so as to reach the original design capacity of annually producing 7000 tons of **sillimanite** fine powder.

在综合分析国内外**硅线石**选矿的特点之后，针对河北灵寿硅线石制定了弱磁—浮选—强磁的流程。

A flowsheet of low-intensity magnetic separation-flotation-high-intensity magnetic separation has been developed for **sillimanite** ores from Lingshou Mine in Hebei Province.

简述了**红柱石**的矿产资源分布状况及其特性。

The distribution of mineral resources, properties, and prospect of application about **andalusite** were introduced.

从**红柱石**的结晶程度分析，这种无烟煤是一种低级热接触变质岩。

Having analyzed the degree of crystallization of **andalusite**, it is believed that the anthracite is a low-grade thermal metamorphic rock.

对某**含铁红柱石**进行了选矿试验研究，确定了脱泥—浮选—磁选—酸浸的联合工艺流程。

Beneficiation tests of a **siderous andalusite** were performed, and a combined flowsheet which include desliming-flotation-magnetic separation-acid leaching was developed.

对化学组成和粒度不同的**红柱石**精矿和粗精矿在不同温度下的莫来石化行为进行了 X 射线分析。

Mullitization of concentrated **andalusite** in different chemical composition and different grain size has been determined at different temperatures by means of X-ray analysis.

在详尽分析西峡桑评**红柱石**矿床区域成矿地质背景及矿床基本地质特征的基础上，研究了成矿物质的来源。

On the base of detailed analysis of the regional metallogenic geological background and geological characteristics of Sangping **andalusite**, the source of the metallogenic mass was studied.

采用该流程，取得了精矿总产率为 5.46%，总纯度为 97.11%，**红柱石**总回收率为 65.95% 的试验指标，为该矿的开发利用提供了可行性技术方案。

Then a good performance with concentrate yield of 5.46%, purity of 97.11%, and **andalusite** recovery of 65.95% is obtained, providing a feasible technical scheme for exploration of mine.

晶相组成为圆柱状刚玉、圆粒状锆石和**莫来石**固溶体。

The crystal phase is composed of the cylinder corundum and the circle solid solution of the zirconia and **mullite**.

由**莫来石颗粒**引入的基体拉应力使裂纹倾向于向晶内扩展。

The tensile stress in matrix introduced by **mullite particles** made the cracks incline to extend into the matrix.

氧化锆增韧莫来石陶瓷是优良的高温结构陶瓷材料。

Zirconia toughened mullite ceramics is one of finest performance high temperature structure ceramics.

介绍了**多晶莫来石**纤维在梁、底组合步进式加热炉上的使用效果。

This paper introduces the application effect of **multicrystal mullite** refractory fiber in walking furnace with unity of beam and bottom.

研究了烧结温度和**莫来石含量**对复合材料的介电性能和抗弯强度的影响。

The effect of the sintering temperature and the **mullite content** on the dielectric properties and the bending strength has been studied.

莫来石凝胶是制备**超细莫来石**的优良先驱物，主要分为单相和双相凝胶。

Mullite gels are excellent precursors for preparing **fine mullite**, and mullite gels consist of monophasic gel and diphasic gel.

为了节约能源、改善操作环境，湘钢高线、棒材厂在加热炉试用了**多晶莫来石纤维**。

In order to save energy sources and improve operational condition, the **polycrystalline mullite fiber** was tried using for the heating furnace of wire rod and bar mill in Xianggang.

与**块体莫来石**陶瓷相比，在保证材料强度降低较小的情况下断裂韧性提高了62%。

The fracture toughness of the composite was increased 62% with little decrease of flexure strength when compared with **mullite monolithic** ceramics.

煅烧处理过的高岭土经水热反应后得到纳米级**莫来石**复合晶体，TEM 观测到其晶粒发育良好。

The composite nanocrystalline of **mullite** was prepared via hydrothermal reaction of calcined kaolin, and it was observed that growth of the nanocrystalline grain was good in TEM.

作为一种珍贵的宝石，红宝石是一种红色的矿物**刚玉**。

Rubies, valued as precious gems, are the mineral **corundum** in its red form.

由铝和氧两种元素构成的**刚玉**，会呈现出黄色、灰色或者棕色。

Made up of the aluminum and oxygen elements, **corundum** also can be yellow, gray, or brown.

刚玉晶体结构型氧化铬有高的着色力、高热稳定性及耐化学腐蚀等特点。

Chromium oxide green with **corundum crystal** structure has such characteristics as high thermal stability, high painted capacity, chemical corrosion-resisting capacity, etc.

研究成果可为**刚玉块石**混凝土遮弹层的工程应用和深入理论分析提供依据。

The result can provide reference for technical application and theoretical analysis of the **corundum-rubble** concrete shelter.

在使用**刚玉骨料**的试验中，采用高铁**棕刚玉**的试样的体积密度和强度均最高。

In the experiment using **corundum aggregate**, the specimens prepared with high ferric oxide **brown corundum** aggregate has the highest bulk density and strength.

3.2.4 术语例段

为探讨能否用纯电瓷废料合成莫来石陶瓷，本文对比了以电瓷废料细粉为原料，再添加氧化铝细粉和以纯电瓷废料细粉为原料合成**莫来石**陶瓷。探讨了原料配比和烧结温度对合成的莫来石陶瓷的结构和性能的影响。采用 XRD 和 SEM 分别研究了**莫来石**的物相组成和显微结构。研究表明，随着烧成温度升高，**莫来石**的含量增加，体积密度增大；由于原料采用电瓷废料细粉，烧结活性较大，有利于烧结的进行，因此提高了烧结密度；采用纯电瓷废料合成的**莫来石**陶瓷的体积密度和耐压强度最高，气孔率最小，综合性能最优[22]。

In order to investigate whether pure electroceramics waste could be used to synthesize **mullite** ceramics, the pure electroceramics waste mixed with alumina powders and the pure electroceramics waste as raw materials were compared. The effects of raw materials' composition and sintering temperature on the microstructure and physical properties of **mullite** ceramic were investigated. XRD and SEM were used to study the phase composition and microstructure. The results show that the content of **mullite** is increased with raising sintering temperature, and at the same time the bulk density is escalated. The raw materials are the pure electroceramics waste, thus the sintering activity is greater, and the sintering process can be accelerated, and density is also increased. When the **mullite** is prepared only by the electroceramics waste, the bulk density and compressive strength are largest, the porosity is smallest, and the comprehensive physical properties will be the best[22].

59

引入复合添加剂是改善低碳 MgO-C 耐火材料高温稳定性和抗渣性的一种有效手段。本工作中，首先由电瓷废料经埋碳法在 1500℃、1550℃ 和 1600℃ 保温 4h 合成 Al₂O₃-SiC 复合粉体，然后将合成的 Al₂O₃-SiC 复合粉体作为添加剂添加至低碳 MgO-C 耐火材料中。系统研究了其添加量（质量分数分别为 0、2.5%、5.0% 和 7.5%）对耐火材料性能的影响。研究发现，增加热处理温度利于电瓷废料中**莫来石**和石英向氧化铝和碳化硅的转化。此外，添加 Al₂O₃-SiC 复合粉体可有效改善低碳 MgO-C 试样的性能；尖晶石致密层和高黏度孤立层的形成是改善低碳 MgO-C 试样抗氧化性和抗渣性的内在原因。本工作提供了电瓷废料再利用的思路，提出了低碳 MgO-C 耐火材料性能优化的一种可选择策略[23]。

Introducing composite additives are an efficient means to improve the high-temperature stability and slag resistance of low-carbon MgO-C refractories. In this work, Al₂O₃-SiC composite powder was firstly synthesized from electroceramics waste by carbon embedded method at 1500℃, 1550℃, and 1600℃ for 4h, and then the as-synthesized Al₂O₃-SiC composite powder was used as an additive to low-carbon MgO-C refractories. The effects of its addition amounts of 0, 2.5wt.%, 5.0wt.%, and 7.5wt.% on the properties of the refractories were investigated in detail. It was found that increasing the heat treatment temperature is beneficial to the phase conversion of **mullite** and quartz to alumina and silicon carbide in the electroceramics waste. Furthermore, the addition of Al₂O₃-SiC composite powder effectively improves the performance of low-carbon MgO-C samples, and the formation of spinel dense layer and high-viscosity isolation layer is the internal reason for the improvement of the oxidation resistance and slag resistance of low-carbon MgO-C samples. This work provides ideas for the reuse of electroceramics waste and presents an alternative strategy for the performance optimization of low-carbon MgO-C refractories[23].

以高压电瓷废料为原料，通过气流超细粉碎、圆盘造粒，经 1180~1260℃ 烧结制备油气开采水力压裂用陶粒支撑剂。通过 XRD 和 SEM 分析了不同烧结温度所制陶粒支撑剂试样的物相组成和微观形貌。采用石油天然气行业标准中的方法测试了试样的密度、破碎率及酸溶解度，研究了烧结温度对试样微观结构和性能的影响。结果表明，支撑剂主要物相为**莫来石**和刚玉，随着烧结温度的升高，针状**莫来石**晶粒逐渐长大，并互相交错堆叠形成网格状结构，液相均匀分散并包裹于晶粒，使试样致密化程度提高。但烧结温度过高会导致试样内**二次莫来石化**度变高，部分**莫来石**晶粒异常长大，使支撑剂整体表现出高的空隙率，影响试样强度。1220℃ 制备的试样表现出最高松装密度和最低破碎率，在 69MPa 闭合压力下其破碎率仅为 6.1%，松装密度为 2.74g/cm³，满足 SY/T 5108—2014 中关于低密度陶粒支撑剂的标准要求[24]。

Ceramic proppant for hydraulic fracturing in oil and gas exploitation was prepared from high voltage electric porcelain waste by air flow supper fine grinding, disc granulation and sintering at 1180-1260℃. The phase composition and micro morphology of ceramic proppant samples sintered at different temperatures were analyzed by XRD and SEM. The density, crushing rate and acid solubility of samples were tested by the methods of petroleum and natural gas industry standards, and the effects of sintering temperature on microstructure and properties of samples were studied. The results show that the main phases of proppant are **mullite** and corundum. With the increase of sintering temperature, the acicular **mullite** grains grow up gradually and form a grid structure. The liquid phase is uniformly dispersed and wrapped in the grains, which improves the densification of the sample. However, if the sintering temperature is too high, the degree of **secondary mullite** in the sample will become higher, and some **mullite** grains will grow abnormally, resulting in high porosity of proppant and affecting the strength of the sample. The sample shows the highest apparent density and the lowest crushing rate at 1220℃, while the crushing rate is only 6.1% under 69MPa closing pressure and the apparent density is 2.74g/cm^3, which meets the standard requirements of low-density ceramic proppant in SY/T 5108—2014[24].

粉煤灰是一种典型的严重影响人类健康和生态平衡的工业固体废物。为了实现粉煤灰的循环利用，本文利用粉煤灰和**铝矾土**经反应合成工艺成功制备了多孔**莫来石**陶瓷。系统研究了煅烧温度（1450～1550℃），碳化硅添加量（质量分数为0～15%），钾长石添加量（质量分数为0～16%）对莫来石多孔陶瓷性能的影响。研究发现，增加碳化硅添加量、升高煅烧温度均有利于改善多孔陶瓷的常温耐压强度和抗热震性。因此，添加10%（质量分数）碳化硅、4%～12%（质量分数）钾长石制备的多孔陶瓷具有最佳的总体性能。闭口气孔率、常温耐压强度分别为14.79%～18.57%和217.18～236.67MPa。热震循环次数为7～9次，800℃下的热导率为2.19～2.52W/（m·K）。本工作为粉煤灰的利用提供了一种便利且有前景的方法[25]。

Fly ash is a typical industrial solid waste that seriously affects human health and ecological balance. In order to recycle the fly ash, in this paper, porous **mullite** ceramics were successfully fabricated with fly ash and **bauxite** via reaction synthesis process. Effects of firing temperature(1450-1550℃), silicon carbide addition amount(0-15wt.%), and potash feldspar addition amount(0-16wt.%) on the properties of mullite porous ceramics were systematically investigated. It was found that increasing the silicon carbide addition amount and raising firing temperature was all favorable for improving the cold compressive strength and thermal shock resistance of the porous ceramics. Consequently, the as-prepared porous ceramics added with 10wt. % silicon carbide and 4wt. %-12wt. % potash feldspar had optimal overall performances. The closed porosity and cold compressive strength ranges were 14. 79%-18. 57% and 217. 18-236. 67MPa, respectively. The thermal cycles were 7-9 times, and the thermal conductivity reached 2. 19-2. 52W/(m · K) at 800℃. This work provides a convenient and promising method for the utilization of fly ash[25].

利用粉煤灰、锆英石和氧化铝粉经反应烧结工艺成功制备了氧化锆-莫来石-刚玉复合材料。采用 X 射线衍射仪和扫描电子显微镜分别表征在设定温度1400℃、1500℃和1600℃下保温 4h 合成复合材料物相和显微结构的演化。研究了烧结温度对合成复合材料收缩率、显气孔率和体积密度的影响，并详细讨论了复合材料的形成过程。结果表明，在 1600℃保温 4h 能制备出具有良好烧结性的氧化锆-**莫来石**-刚玉复合材料；氧化锆颗粒均匀分布于**莫来石**基体中，其粒径大约为 5μm；氧化锆-**莫来石**-刚玉复合材料的形成过程由锆英石的分解和**莫来化过程**组成[26]。

Zirconia-**mullite**-corundum composites were successfully prepared from fly ash, zircon and alumina powder by a reaction sintering process. The phase and microstructure evolutions of the composites synthesized at desired temperatures of 1400℃,1500℃ and 1600℃ for 4h were characterized by X-ray diffraction and scanning electronic microscope, respectively. The influences of sintering temperature on shrinkage ratio, apparent porosity and bulk density of the synthesized composites were investigated. The formation process of the composites was discussed in detail. The results show that the zirconia-**mullite**-corundum composites with good sintering properties can be prepared at 1600℃ for 4h. Zirconia particles can be homogeneously distributed in **mullite** matrix, and the zirconia particles are around 5μm. The formation process of zirconia-**mullite**-corundum composites consists of decomposition of zircon and **mullitization process**[26].

以**均化矾土**生料和炭黑为主要原料，制备了轻量**均化矾土**试样，研究了炭黑种类（N990 和 N774）、加入量（质量分数为 0、4% 和 8%）及热处理温度（1450℃、1500℃和1550℃下保温 3h）对试样孔结构的影响。结果表明，随着温度的升高，试样气孔数量逐渐减少，平均孔径增大；炭黑 N990 易于分散，所制得试样的气孔分布均匀且孔径尺寸相对较小；炭黑 N774 粒子难以均匀分散，所制得试样的孔径分布不规律，且其平均孔径随加入量增加而明显增大[27]。

Lightweight **homogenized bauxite** samples were prepared using **homogenized bauxite** raw powder and carbon black as main raw materials. The effects of carbon black type(N990 and N774), addition amounts of carbon black(0,4wt.%, and 8wt.%), and heat-treatment temperatures(1450℃,1500℃, and 1550℃ for 3h) on the pore structure of the samples were investigated. The results reveal that with the heat-treatment temperature increasing, the pore amount in the samples decreases, however, the average pore diameter increases. Pores in the samples with N990 exhibit more uniform distribution and smaller average size since N990 is easy to disperse. While the average pore size distribution of the samples with N774 is less uniform and the average pore size increases with the rising N774 addition since N774 is not easy to disperse[27].

基于"骨料微孔化、基质紧密化"的研究思路，根据颗粒紧密堆积理论，对 Dinger-Funk 方程进行修正，提出了轻量耐火材料基质体积密度预测模型。采用显气孔率低、闭口气孔率高、热导率比普通耐火材料低20% ~50%、抗渣性能与普通原料相当的具有微纳米尺度晶内气孔的微孔刚玉和**矾土**以及镁砂为主要原料，制备了轻量铝镁浇注料。结果表明，此轻量铝镁浇注料不仅抗热震性好，热导率小，而且抗渣性能与普通铝镁浇注料相当。现场应用表明，采用微孔刚玉为骨料制备的钢包工作层用 Al_2O_3-MgO 质预制块使用寿命在同等条件下均优于现行材料[28]。

Based on the proposal of "micro-pored aggregates and dense-packed matrixes" and the particle dense packing theory, the Dinger-Funk equation was modified to establish a prediction model of matrix packing density for lightweight refractories. Lightweight alumina magnesia castables were prepared using micro-pore corundum, **bauxite** and magnesia as starting materials, and the micro-pore corundum used has low apparent porosity, high closed porosity, low thermal conductivity (20%-50% lower than the common corundum), similar slag resistance of the common corundum and micro- or nano-sized intracrystalline pores. The results show that the lightweight alumina magnesia castables not only have good thermal shock resistance and low thermal conductivity, but also have the same slag resistance as the common alumina magnesia castables. The onsite application shows that the service life of the alumina magnesia precast blocks for the ladle working lining prepared with the microporous corundum as aggregates are better than that of the current materials with the same conditions[28].

水铝石–高岭石型中国**铝土矿**的烧结行为分为三个阶段：分解阶段 （400 ~ 1200℃）、二次莫来石化阶段 （1200~1400℃或1500℃） 和再结晶烧结阶段 （高于1400℃或1500℃）。研究发现不同等级铝土矿的可烧结性取决于 Al_2O_3 含量。煅烧铝土矿的成分越接近莫来石，越难烧结。据推测，二次莫来石化和液相作用是影响这些铝土矿烧结的两个主要因素。其中，以二次莫来石形成最多和玻璃含量相对较低为特征的Ⅱ级铝土矿最难烧结[29]。

The sintering behaviour of Chinese **bauxites** of the diaspore-kaolinite type proceeds in three stages, viz: decomposition stage (400-1200℃), secondary mullitization stage (1200-1400℃ or 1500℃) and recrystallization sintering stage (above 1400℃ or 1500℃). The sinterability of different grades of bauxites is dependent on the Al_2O_3 content. The closer is the composition of calcined bauxite to that of mullite, the more difficult is it to sinter. It is postulated that secondary mullitization and liquid phase action are the two principal factors influencing the sintering of these bauxites. Grade Ⅱ bauxites characterized by maximum secondary mullite formation and relatively low glass content are found to be most difficult to sinter[29].

基于**铝矾土**熟料导热系数小，抗滑耐磨性能较好，可用于 HFST（高摩擦表面处理）或沥青混合料的耐磨层，替代或部分替代现有骨料。**铝矾土**熟料按化学成分含量不同主要分为六类。选择**铝矾土**熟料作为骨料不仅是为了经济价值，而且是为了提高集料与沥青的附着力，具有一定的盲目性。采用搅拌静水吸附法和表面自由能理论评估了不同类型**铝矾土**熟料与沥青的附着力。采用灰色关联熵分析法评价了**铝矾土**熟料特征参数对黏附力的影响[30]。

Based on the fact that **bauxite** clinker has minor thermal conductivity and better skid resistance and wear-resisting property, it can be used in HFST(high friction surface treatment) or the abrasion layer of asphalt mixture to replace or partly replace the existing aggregate. **Bauxite** clinker is classified into mainly six types according to different chemical composition contents. The selection of **bauxite** clinker as aggregate is not only for the economic value, but also for improving the adhesion between aggregate and asphalt, which has a certain blindness. This study evaluated the characteristics of different types of **bauxite** clinker. The adhesion of different types of **bauxite** clinker with asphalt was evaluated by means of agitating hydrostatic adsorption method and surface free energy theory. The effect of characteristic parameters of **bauxite** clinker on adhesion was evaluated by grey correlation entropy analysis[30].

为有效利用铝工业废弃物——赤泥和**铝土矿**尾矿泥，减少废弃物对环境的不利影响和对土地资源的占用，在已有研究成果的基础上，研制了赤泥-**铝土矿**尾矿泥泡沫轻质土。通过实验研究了不同比例赤泥和**铝土矿**尾矿泥开发材料的力学性能和微观特征。研究结果表明，随着赤泥含量的增加，合成试样的湿密度和流动性增加。在赤泥含量为 16% 时，合成试样的水稳定性系数达到最大值 0.826，固化 28d 试样的无侧限抗压强度（UCS）达到 1.056MPa[31]。

In order to effectively utilize aluminum industrial waste—red mud and **bauxite** tailings mud—and reduce the adverse impact of waste on the environment and occupation of land resources, a red mud-**bauxite** tailings mud foam lightweight soil was developed based on the existing research results. Experiments were conducted to investigate the mechanical properties and microscopic characteristics of the developed materials with different proportions of red mud and **bauxite** tailings mud. Results show that with the increase in red mud content, the wet density and fluidity of the synthetic sample was increased. With 16% red mud content, the water stability coefficient of the synthetic sample reached its maximum of 0.826, as well as the unconfined compressive strength(UCS) of the sample cured for 28d(1.056MPa)[31].

铝土矿是制造氧化铝的理想原料。除了铝和硅的主要成分外，**铝土矿**还经常与许多有价值的元素结合，例如镓（Ga）、钛（Ti）、钪（Sc）和锂（Li）。氧化铝生产中的**铝土矿**残渣和循环废液通常包含大量有价值的元素，使其成为多金属的潜在来源。这些重要成分的回收可以大大提高氧化铝制造过程的效率，同时减少工业责任和环境影响。本研究对用于从**铝土矿**残渣和循环废液中回收有价值元素的现有技术进行了批判性分析，以深入了解**铝土矿**残渣作为资源而不是废物的更广泛用途。

Bauxite is the ideal raw material for the manufacturing of alumina. Aside from the primary constituents of aluminum and silicon, **bauxite** is frequently coupled with many valuable elements such as gallium(Ga), titanium(Ti), scandium(Sc), and lithium (Li). The **bauxite** residue and circulating spent liquor in alumina production typically include significant amounts of valuable elements, making them a potential source of polymetallic. The recovery of these essential components can greatly increase alumina manufacturing process efficiency while reducing industrial liability and environmental impact. This study gives a critical analysis of existing technology used to recover valuable elements from **bauxite** residue and circulating spent liquor to provide insight into the broader usage of **bauxite** residue as a resource rather than a waste. A comparison of existing process features demonstrates that an integrated process for valuable elements recovery and waste emission reduction is advantageous[32].

本文涉及从**蓝晶石**试样（一种采用微波加热的铝硅酸盐矿物）形成莫来石

陶瓷的研究。在本研究中，蓝晶石在微波烧结炉中与煅烧氧化铝、碳化硅、二氧化锆和黏合剂一起加热以形成莫来石以及制备氧化锆增韧莫来石产品。研究结果揭示了少量的二氧化锆如何有助于在耐火材料和陶瓷应用中增韧莫来石的形成。结果表明，微波加热法是这种增韧莫来石形成的最新颖、最环保的工艺之一，且能在很短的时间内完成。使用 X 射线衍射数据、场效应扫描电子显微镜和能量色散光谱法观察了莫来石形成和氧化锆增韧莫来石形成的结构数据。研究仍在进行中。

This article deals with the investigations on mullite ceramics formation from **kyanite** sample—an alumino-silicate mineral with microwave heating. In the present investigation kyanite was heated in microwave sintering furnace with calcined alumina, silicon carbide, zirconium dioxide and binder for mullite formation as well as for the preparation of zirconia toughened mullite product. The results of these studies reveal that how a small amount of zirconium dioxide helps in toughening the mullite formation for refractory and ceramics applications. The results show that microwave heating method is one of the most novel and eco-friendly process for such toughened mullite formation, well within very short period of time. The observations with mullite formation and zirconia toughened mullite formation were observed with the structural data using X-ray diffraction data, field effect scanning electron microscope and energy dispersive spectroscopy. Future studies are still in process[33].

超低水泥铝矾土-刚玉耐火浇注料因其在性能和成本上的优势越来越受到人们的关注。为提高其体积稳定性和高温性能，将红柱石和**蓝晶石**分别掺入该类耐火材料中，并比较研究了它们对显微组织、热力学性能和抗渣腐蚀性能的影响。这些添加剂的莫来石化引起的膨胀效应不仅抑制了烧结收缩，而且使基体结构更加致密和均匀，这主要是由于产生了许多新的和强的莫来石键。这种改进导致负载下的耐火度和抗蠕变性显著增加。对于添加红柱石的浇注料，这种影响更为明显，其 RUL 增加了 62℃，静止阶段的蠕变率降低了 26%。此外，由于蓝晶石的过度膨胀，加入较高量的**蓝晶石**会对显微组织和性能产生不利影响。此外，讨论了精炼渣对这些耐火浇注料的腐蚀机理，并提出了添加红柱石对耐腐蚀性能的积极作用。

Ultralow-cement bauxite-corundum refractory castables are increasingly attracting attention because of the advantages in performance and cost. To improve its volume stability and high-temperature performance, andalusite and **kyanite** were incorporated into this type of refractory, respectively, and a comparative investigation of their effects on microstructure, thermo-mechanical properties, and slag corrosion resistance was carried out in this work. The expansion effects induced by the mullitization of these additives not only suppressed the sintering shrinkage but also made the matrix structure more compact and uniform, mainly owing to the generation of many new and strong mullite bonds. This improvement resulted in a significant increase in the refractoriness under load and creep resistance. This effect was far more pronounced for the andalusite-added castables, for which the RUL increased by 62℃ and the creep rate at the stationary stage decreased by 26%. Besides, adverse effects on microstructure and properties appeared when adding a higher amount of **kyanite** due to the excessive expansion. Moreover, the corrosion mechanisms of these refractory castables by refining slag were discussed, and the active role of andalusite addition in the corrosion resistance was also proposed[34].

铸造废砂（WFS）在铸造行业中占环境废物的比例最大，并且显示出惊人的持续增长。先前已经对 WFS 的处理进行了研究，并且主要将其作为建筑材料。在这项工作中，研究了浸出有害 WFS 在多孔绝热材料制造中的适用性。通过热力学计算和实验研究评估了晶体组成、微观结构特征、强度和隔热性能以及浸出行为。除 WFS 外，原材料还包括塑料黏土、**蓝晶石**和锯末。WFS 的高添加量促进了致密化，并导致抗压强度的显著增加。试样由 75%（质量分数）的 WFS、15%（质量分数）的塑料黏土、10%（质量分数）的蓝晶石和额外的 20%（质量分数）的锯末组成，在 200℃ 时的最大强度为 0.92MPa，导热系数为 0.11W/（m·K）。根据毒性特征浸出程序分析，基于 WFS 的多孔材料中的目标金属浓度在 US EPA 限制之内，证明了其对环境的友好应用。

Waste foundry sand (WFS) accounts for the largest percentage of environmental wastes in the foundry industry and shows an alarming and continuous increase. The treatment of WFS has been investigated previously and mainly as building materials. In this work, the applicability of leaching hazardous WFS in the manufacture of porous thermal insulation material is investigated. Thermodynamic calculation and experimental investigation are performed to evaluate the crystalline composition, microstructure characteristic, strength, and heat-insulating performance, and leaching behavior. In addition to WFS, raw materials include plastic clay, **kyanite**, and sawdust. High addition of WFS promotes the densification and leads to the remarkable increase in compressive strength. Samples compose of 75wt.% WFS, 15wt.% plastic clay, 10wt.% kyanite, and extra 20wt.% sawdust have a maximum strength of 0.92MPa and thermal conductivity of 0.11W/(m·K) at 200℃. According to toxicity characteristic leaching procedure analysis, the target metal concentrations in porous material based on WFS are within the

US EPA limits, which proves the environmentally friendly application[35].

以刚玉、**蓝晶石**、氧化铝粉、凝胶粉和纳米 ZrO_2 为原料，研究了纳米氧化锆添加量对刚玉浇注料微观结构和性能的影响。结果表明，加入 1.2%（质量分数）的纳米 ZrO_2 可以将浇注料在 110℃ 下的中值孔径从 480nm 降低到 200nm。浇注料经 1100℃ 加热后，常温抗折强度增加 29.95%，达到 24.71MPa，常温耐压强度增加 50.32%，达到 116.37MPa。经 1500℃ 处理后高温抗折强度增加 114.95%，达到 3.59MPa，纳米 ZrO_2 的高表面能可提高相邻颗粒之间的界面能和结合强度。

The effect of nano-zirconia addition on the microstructure and properties of corundum castables was studied using corundum, **kyanite**, alumina powder, gel powder, and nano-ZrO_2 as raw materials. The results showed that adding 1.2wt.% nano-ZrO_2 could reduce the median pore diameter of the castables from 480nm to 200nm at 110℃. After the castables were heated at 1100℃, the cold modulus of rupture increased by 29.95%, to 24.71MPa, and the cold compressive strength increased by 50.32%, to 116.37MPa. The hot modulus of rupture increased by 114.95%, to 3.59MPa after being treatment at 1500℃, the high surface energy of nano-ZrO_2 can increase the interface energy and bonding strength between adjacent particles[36].

红柱石和**蓝晶石**是两种具有相同分子式但晶体结构不同的多态矿物。尽管它们具有很高的经济价值，但通过浮选有选择地回收它们仍然是一个挑战。红柱石的浮选回收率高于**蓝晶石**和油酸钠的浮选回收率。油酸钠主要通过油酸根离子与铝原子的化学相互作用吸附到表面上。油酸钠在两种矿物之间的红柱石表面上的较高吸附与铝原子的较高化学活性有关，这使其具有更好的亲和力，因此油酸钠的吸附能较低。

Andalusite and **kyanite** are two polymorphic minerals with the same formula but different crystal structures. Despite of their high economic values, selectively recovering them by flotation is a challenge. The flotation recovery of andalusite is higher than that of **kyanite** with sodium oleate. Sodium oleate adsorbs onto the surfaces mainly through the chemical interaction of oleate ions with aluminum atoms. The higher adsorption of sodium oleate on andalusite surface between the two minerals is related to the higher chemical activity of the aluminum atoms which leads to the better affinity and therefore the lower adsorption energy of sodium oleate[37].

我们在研究中发现，在还原气氛中烧制的**红柱石骨料**的莫来石化率低于在空气气氛中烧制的红柱石骨料。为了研究气氛对红柱石转化和含红柱石耐火材料性能的影响，分别在空气和埋碳气氛中烧制红柱石粉（≤0.074mm）和含红柱石骨料（1～3mm）的耐火材料。在两种气氛中烧制的红柱石的物相和微观结构分别通过 X 射线衍射和扫描电子显微镜进行表征。从体积稳定性、机械强度和抗热震性方面研究了红柱石耐火材料的性能与烧成气氛的相关性。

We found in our research that **andalusite** aggregate fired in a reducing atmosphere exhibits a lower mullitization rate than that fired in an air atmosphere. For investigating the effect of atmosphere on the transformation of andalusite and the properties of andalusite-containing refractories, andalusite powder (≤ 0.074mm) and refractories containing andalusite aggregate (1-3mm) were fired in air and carbon embedding, respectively. The phases and microstructure of the andalusite fired in both atmospheres were characterized by X-ray diffraction and scanning electron microscopy, respectively. The correlations of the properties of the andalusite-bearing refractories with the firing atmospheres were investigated in terms of volume stability, mechanical strength, and thermal shock resistance. The difference in the properties of the refractories was discussed with respect to the varied transformation rates of andalusite, and in terms of the different viscosities of the silica-rich glass caused by the different atmospheres[38].

为了验证不同莫来石化程度的**红柱石**骨料的热膨胀行为及其对莫来石-刚玉耐火材料性能（抗热震性，尤其是体积稳定性）的影响，对莫来石-刚玉耐火材料的性能进行了未煅烧和不同预煅烧，研究了红柱石骨料（1～3mm）的永久衬里变化、抗热震性、常温抗折强度、弹性模量和显气孔率。实验结果表明，当红柱石的预煅烧温度提高（1300～1500℃）时，试样的体积稳定性增加，因为预煅烧骨料中残留的红柱石较少，高温烧成时耐火试样的体积膨胀较小。同时，红柱石骨料（1～3mm）在较高温度下预煅烧的莫来石-刚玉耐火材料具有更好的常温抗折强度，因为预煅烧温度的提高可以促进莫来石化程度，从而降低微裂纹的长度，也增强了这些制备好的耐火材料的致密化。此外，具有未煅烧和预煅烧红柱石骨料的莫来石-刚玉耐火材料具有良好的抗热震性，含1500℃或1600℃预煅烧红柱石骨料的莫来石-刚玉耐火材料的相应残余强度比仍高于90%。总体而言，通过掺入1500℃预煅烧的红柱石骨料（1～3mm），可以显著提高莫来石-刚玉耐火材料的体积稳定性和破裂的常温抗折强度，并且所有具有未煅烧和预煅烧红柱石骨料（1～3mm）的莫来石-刚玉耐火材料都表现出良好的抗热震性。

In order to verify the thermal expansion behaviour of **andalusite** aggregates with different mullitization degrees and its influence on the properties (thermal shock resistance and especially volume stability) of mullite-corundum refractories, the properties of mullite-corundum refractories with uncalcined and different pre-calcined andalusite aggregates (1-3mm) are investigated in terms of permanent liner change, thermal shock resistance, cold modulus of rupture, modulus of elasticity and apparent porosity. The experimental results show the volume stability of samples increases when the pre-calcining temperature of andalusite is improved(1300-1500℃), as less residual andalusite in the pre-calcined aggregates leads to smaller volume expansion in the refractory samples during firing at high temperatures. Meanwhile, mullite-corundum refractories with andalusite aggregates(1-3mm) pre-calcined at higher temperature have better cold modulus of rupture, because the increase of pre-calcining temperature could promote mullization degree and thereby bring down the length of microcracks and also enhance the densification of these as-prepared refractories. Moreover, mullite-corundum refractories with uncalcined and pre-calcined andalusite aggregates present good thermal shock resistance; the corresponding residual strength ratio of mullite-corundum refractories containing 1500℃ or 1600℃ pre-calcined andalusite aggregates is still higher than 90% . Overall, the volume stability and cold modulus of rupture of the mullite-corundum refractories could be significantly enhanced through the incorporation of 1500℃ pre-calcined andalusite aggregates (1-3mm), and all mullite-corundum refractories with uncalcined and pre-calcined andalusite aggregates(1-3mm)exhibit good thermal shock resistance[39].

这项研究的目的是确定添加碳化硅的煅烧**红柱石**砖的性能。在商业隧道窑中于1300℃混合、成型、干燥和烧结来制备砖块试样。测定了烧结砖试样的体积密度、显孔隙率、透气性和抗碎强度。通过电弧炉（EAF）粉尘回收混合物的静态杯试验研究了熔渣渗透试验。研究了 SiC 添加对试样渗透性能的影响。SiC 的添加会显著影响砖表面电弧炉除尘中金属成分的还原行为。SiC 在砖中表现出自上釉的特性，并作为锌蒸气的防渗屏障。

The aim of this study is to determine the properties of fired **andalusite** brick with silicon carbide addition. Brick samples were prepared by mixing, shaping, drying, and sintering at 1300℃ in a commercial tunnel kiln. The bulk density, apparent porosity, gas permeability, and cold crushing strength of fired brick samples were determined. Slag penetration test was investigated by static cup test with electric-arc furnace (EAF) dust recovery blend. The effects of SiC additions on the penetration properties of the samples have been investigated. The addition of SiC significantly affects the reduction behavior of metallic ingredients in EAF dust recovery on the brick surface. SiC shows the self-glazing properties in the brick and behaves as an antibarrier for zinc vapor[40].

红柱石具有在高温加热时自动分解为莫来石和二氧化硅的独特性能，是生产耐火陶瓷的特殊矿物。从红柱石到莫来石的相变对于红柱石的有效应用起着至关重要的作用。研究了红柱石粉高温分解过程中的微观结构特征和烧结性能。将红柱石粉末与高岭土结合，并在 20MPa 下制备为圆柱体。然后在 1423 ~ 1723K（1150 ~ 1450℃）下烧制。通过压缩试验，X 射线衍射和扫描电子显微镜研究了烧结陶瓷的微观结构和机械强度。

Andalusite has been realized as a special mineral for the production of refractory ceramics due to its unique property to automatically decompose into mullite and silica during heating at high temperature. The phase transformation from andalusite to mullite plays a critical role for the effective applications of andalusite. This study investigated the microstructural characteristics and sinterability of andalusite powder during high-temperature decomposition. The andalusite powder was bonded with kaolin and prepared as a cylinder green body at 20MPa; it was then fired at 1423K to 1723K (1150℃ to 1450℃). The microstructures and mechanical strengths of the sintered ceramics were studied by the compressive test, X-ray diffraction, and scanning electron microscopy[41].

研究并分析了煅烧**红柱石**骨料对莫来石浇注料微裂纹形成和抗热震性能的影响。以红柱石骨料为原料，经高温煅烧制备莫来石浇注料。研究结果表明，红柱石向莫来石转变产生的微裂纹可有效缓解热应力，提高莫来石浇注料的抗热震性。随着红柱石骨料煅烧温度的升高，产生的微裂纹减少。当煅烧温度从 1300℃提高到 1500℃时，莫来石浇注料的抗热震性不断提高。然而，在 1600℃ 煅烧红柱石骨料的莫来石浇注料的抗热震性低于在 1500℃ 煅烧的红柱石骨料的抗热震性。

The effect of calcined **andalusite** aggregates on the micro-crack formation and thermal shock resistance of mullite castables was investigated and analyzed. The mullite castables were prepared from andalusite aggregates calcined at high temperatures. The results show that the micro-cracks from the transformation of andalusite to mullite can effectively relieve thermal stress, improving the thermal shock resistance of mullite castables. The micro-cracks generated decreased with increasing calcine temperatures of the andalusite aggregates. When the calcine temperature was increased from 1300℃ to 1500℃, the thermal shock resistance of mullite castables was found to continuously increase. However, the thermal shock resistance of mullite castables with the andalusite aggregates calcined at 1600℃ is lower than those with the andalusite aggregates calcined at 1500℃[42].

印度海岸的沙滩上富含钛铁矿、独居石、金红石、**硅线石**、石榴石和锆石等战略性矿物。为了突出这些重矿物的重要性，我们使用高级表征技术（例如拉曼光谱，X 射线光电子能谱（XPS）和紫外可见近红外光谱）对结构、表面化学和吸收光谱进行了详细的研究。该研究揭示了锆石和独居石的成矿程度，并区分了石榴石中的同晶系列、透明和不透明矿物、多晶体以及由于其物理和化学过程而引起的晶体对称性的各向异性变化。X 射线光电子能谱显示了表面元素的化学组成、精确的氧化态和化学键。紫外可见近红外光谱解释了元素的电子跃迁和电荷转移谱。这个扩展的数据集为科学界和采矿业提供了有价值的信息，以帮助确定滩涂砂的等级和潜在用途。

The beach sands of the Indian coast are bestowed with strategic minerals like ilmenite, monazite, rutile, **sillimanite**, garnet, and zircon. In order to bring out the importance of these heavy minerals, we present detailed studies on structure, surface chemistry, and absorption spectroscopy using advanced characterization techniques such as Raman spectroscopy, X-ray photoelectron spectroscopy(XPS), and UV-Visible-NIR spectroscopy. The study reveals the degree of metamictization in zircon and monazite and differentiates the isomorphous series in garnet, discrimination of opaque and non-opaque minerals, polymorphs, and the anisotropic changes in crystal symmetry due to their physical and chemical process. X-ray photoelectron spectroscopy illustrates the chemical composition, precise oxidation state, and chemical bonding of surface elements. UV-Visible-NIR spectroscopy reveals the electronic transition of elements and charge transfer spectra. This expansive dataset provides valuable information to the scientific community and mining industries to assist with determining the grade and potential uses of beach sands[43].

测定了天然**硅线石**、红柱石和蓝晶石在 296~1273K 的温度范围内在环境压力下的高温拉曼光谱。在本研究的温度范围内没有观察到相变。三个试样的拉曼模式随温度呈线性变化。确定了 Al_2SiO_5 多晶型物中 Si—O 伸缩振动力常数的温度和压力依赖性。硅线石、红柱石和蓝晶石的等压模式 Grüneisen 参数是根据当前高温拉曼光谱的温度相关性和热膨胀系数的结果确定的。本征非谐性波模态参数为非零，表明硅线石、红柱石和蓝晶石存在固有非谐。最后，基于 Kieffer 模型，计算了等梯度热容和熵等热力学参数。

High-temperature Raman spectra of natural **sillimanite**, andalusite and kyanite were measured in the temperature range of 296-1273K at ambient pressure. No phase transition was observed over the temperature range in this study. Raman modes for the three samples vary with temperature linearly. The temperature and pressure dependence of the force constants for Si—O stretching vibrations in Al_2SiO_5 polymorphs were determined. The isobaric mode Grüneisen parameters of sillimanite, andalusite, and kyanite were determined from the temperature dependent of present high-temperature Raman spectra and previous results of thermal expansion coefficients. The intrinsic anharmonic mode parameters were estimated and nonzero, indicating the existence of intrinsic anharmonicity for sillimanite, andalusite and kyanite. Based on Kieffer model, the thermodynamic parameters, including isochoric heat capacity and entropy, were calculated[44].

蓝晶石和**硅线石**是两种具有相同 Al_2SiO_5 分子式的多晶型矿物，但晶体结构不同。尽管它们具有很高的经济价值，但通过浮选有选择地回收它们仍然是一个挑战。在这项研究中，以油酸钠为捕收剂，研究了两种矿物在不同 pH 值条件下的浮选行为。通过 Zeta 电位测量、红外光谱测量、化学形态测量和 X 射线光电子能谱测量来确定其作用机理。结果表明，在相同的浮选条件下，两种矿物的浮选行为不同。在捕收剂油酸钠存在下，硅线石的浮选回收率远高于蓝晶石的浮选回收率。油酸钠主要通过离子分子二聚体与 pH 值为 8.0 的铝原子的化学相互作用吸附到蓝晶石和硅线石的表面上。两种矿物硅线石浮选回收率较高与铝原子6倍配位的高静电荷密度有关，这导致了较高的捕收剂吸附。

Kyanite and **sillimanite** are two polymorphic minerals with the same formula of Al_2SiO_5, but different crystal structures. Despite their high economic values, selectively recovering them by flotation is a challenge. In this study, the flotation behaviors of the two minerals with sodium oleate as the collector were examined at different pH conditions. Zeta potential measurement, infrared spectroscopic measurement, chemical speciation and X-ray photoelectron spectroscopy measurement were conducted to identify the underpinning mechanisms. It is found that the flotation behavior of both minerals is different under the same flotation condition. The flotation recovery of sillimanite is much higher than that of kyanite in the presence of the collector sodium oleate. Sodium oleate adsorbs onto the surfaces of kyanite and sillimanite mainly through the chemical interaction of the ionic-molecular dimers with aluminum atoms at pH 8.0. The higher sillimanite flotation recovery between the two minerals is related to the higher electrostatic charge densities of the aluminum atoms in six-fold coordination, which leads to the higher collector adsorption[45].

针对废建筑陶瓷（以下简称废建陶）堆积、填埋和在普通混凝土中少量替代砂石等处置方式存在的环境影响和利用价值不高的突出问题和迫切需求，充分挖掘和利用废建陶块状可颗粒化、硬度高、致密有较高强度且耐一定高温等优势，开展废建陶规模化高值利用技术研究具有重要意义。本论文基于废建陶的组分波动大、色料/釉料低熔点组分影响高温性能的瓶颈问题，研究通过组分设计、微结构"核-壳"屏蔽和低熔点组分物相优化调控技术、减弱性能有害组分劣化高温性能的影响，以不同颗粒尺寸废建陶为主要原料，研究可在1000℃使用的耐热材料制备技术，用于高温热工设备建筑基础和烟道/烟囱等部位，研究工作将为废建陶制备高值耐高温材料提供重要依据。主要研究成果如下：（1）以不同粒径的废建陶细粉、颗粒为骨料材料，研究颗粒级配以及热处理条件对全体量废建陶基耐高温材料的影响规律；对废建陶元素组成以及分布进行分析，研究在1100℃热处理后废建陶组分迁移及包覆料对废建陶大小颗粒色料/釉料组分的屏蔽效果。结果表明：不同粒径的废建陶颗粒质量比为粗颗粒：中颗粒：细粉=45：25：30，1100℃保温3h后，全体量废建陶基耐高温材料综合性能较佳；1100℃热处理后，废建陶颗粒元素分布以及形态基本保持稳定，包覆料包裹废建陶颗粒表层形成的壳对废建陶中的低熔点组分有较好的屏蔽效果，有效降低废建陶中有害组分对高温性能劣化程度。（2）分别以黏土/**铝矾土细粉**和偏高岭石/铝矾土细粉为包覆料包裹废建陶颗粒，探究热处理温度、黏土/偏高岭石添加量对废建陶基耐高温材料各项性能的影响规律。结果表明，当热处理温度为1100℃，黏土/偏高岭石添加量（质量分数）为10%，即废建陶的含量（质量分数）为85%时，制备的材料综合性能较好，在废建陶高掺量情况下材料也具有较高强度。（3）探讨废建陶基耐高温材料在不同温度下强度获得机制，黏土细粉和铝矾土生料细粉的塑性有助于提高试样成型时的致密性，其在不同温度下热处理后的常温强度要大于全体量废建陶基耐高温材料；偏高岭石在1000℃左右会再结晶生成莫来石晶核，有助于莫来石晶体的生长，添加偏高岭石有助于提高废建陶基耐高温材料在1000℃的强度。（4）研究废建陶基耐高温材料抗热震性影响因素及高温强度的结果表明：适量气孔有助于提高试样的抗热震性，在常温至1000℃试样的高温强度随着温度升高而增大；抗垃圾焚烧炉飞灰侵蚀实验表明，制备的废建陶基耐高温材料抗飞灰侵蚀性能良好，分析认为废建陶基耐高温材料的垃圾焚烧炉飞灰侵蚀机理为"熔融-渗透"[46]。

In order to solve the severe problem and urgent demand of the environmental influence and low utilization value caused by the disposal of waste architectural ceramics like accumulating,land-filling and a small amount of replacement of sand and gravel in ordinary concrete, the advantages of granulation, high hardness, compactness, high strength, and high temperature resistance of architectural ceramics need to be fully excavated and utilized, and scientific research of high-value utilization of waste architectural ceramics have a significant value. This paper is based on the bottleneck problem of large fluctuations in the composition of waste architectural ceramics, and low melting point components of pigments/glazes affecting high-temperature performance, by component design, microstructure "core-shell" shielding and low-melting component phase optimization and control technology, the impact of harmful components deteriorating high-temperature performance has been reduced. Using waste architectural ceramics with different particle sizes as the main raw materials, the preparation technology of high temperature-resistant materials used at 1000℃ was studied, which can be applied in the construction foundation of high temperature thermal equipment and flue/chimney. The research work will provide an important basis for the preparation of high-value high-temperature-resistant materials from waste architectural ceramics. The main research results are as follows: (1) Using waste architecture ceramics in different diameter as aggregate and fine powder, the influence of parameters of particle grading and heat treatment on the total amount of waste architectural ceramic-based high-temperature materials was investigated. The element composition and distribution of two types of waste architectural ceramics, colored glaze and colorless glaze, were analyzed respectively, and the migration of waste architectural ceramic components after heat treatment at 1100℃ and the shielding effect of coating materials on the pigments/glazes components of waste architectural ceramics were studied. The results show that: when the mass ratio of waste architectural ceramic particles with different particle sizes——coarse particles : medium particles : fine powder = 45 : 25 : 30, after heat preservation at 1100℃ for 3 hours, the comprehensive performance of the total amount of waste architectural ceramic-based high temperature resistant materials is the better; in addition, after heat treatment at 1100℃, the element distribution and morphology of waste architectural ceramic particles remain basically stable. The shell formed by wrapping the surface layer of the waste architectural ceramic particles with the coating material has a good shielding effect on the low melting point components in the waste architectural ceramics, which proves that it can effectively reduce the deterioration of the high temperature performance of the harmful components in the waste architectural ceramics. (2) Clay /**bauxite fine powder** and metakaolin/bauxite fine powder are used as coating materials to wrap waste architectural ceramic particles and explore the influence of heat treatment temperature and clay/metakaolin addition on the properties of waste architectural ceramic-based high-temperature materials. The results show that: when the heat treatment temperature is 1100℃ and the clay/metakaolin additive amount is 10wt.%, that is, the content of waste architectural ceramics is 85wt.%, the overall performance of the prepared materials is the better, that is, the strength of the prepared material is higher when the amount of waste architecture ceramics is high. (3) The strength acquisition mechanism of waste architectural ceramic-based high temperature resistant materials at different temperatures was studied. The results show that the plasticity of fine clay powder and raw bauxite powder can help improve the compactness of the sample, and the normal temperature strength after heat treatment at different temperatures is slightly greater than the total amount. Since metakaolin will recrystallize at about 1000℃ to form mullite crystal nuclei, which will help the growth of mullite crystals, adding metakaolin will help improve the strength of the waste architecture ceramics based high-temperature-resistance materials at 1000℃. (4) The factors affecting the thermal shock resistance and the high temperature strength of the waste architectural ceramics based high-temperature-resistance materials were studied. The results show that a proper number of pores help to improve the thermal shock resistance of the sample. The high temperature strength of the sample at room temperature to 1000℃ increases with the increase of temperature; the anti-fly ash erosion experiment of waste incinerator showed that the prepared waste architectural ceramic-based high temperature resistant material has good fly ash erosion resistance, and the corrosion mechanism of fly ash from waste incinerators based on waste architectural ceramics-based high-temperature materials is "melting-infiltration"[46].

参考文献

[1] Ma Beiyue, Ren Xinming, Gao Zhi, et al. Influence of pre-synthesized Al_2O_3-SiC composite powder from clay on properties of low-carbon MgO-C refractories [J]. Journal of Iron and Steel Research International, 2022, 29: 1080-1088.

[2] Biel O, Rożek P, Florek P, et al. Alkaline activation of kaolin group minerals [J]. Crystals, 2020, 10 (4): 268.

[3] Guatame-Garcia A, Buxton M. Framework for monitoring and control of the production of calcined kaolin [J]. Minerals, 2020, 10 (5): 403.

[4] Hayes M J, Smith M I. Slip in adhesion tests of a kaolin clay [J]. The European Physical Journal E, 2021, 44 (8): 1-11.

[5] Ediz N, Tatar I, Aydın A. The use of alunitic kaolin in ceramic tile production [J]. Journal of the Australian Ceramic Society, 2017, 53 (2): 271-281.

[6] Fashina B, Deng Youjun. Stacking disorder and reactivity of kaolinites [J]. Clays and Clay Minerals, 2021, 69 (3): 354-365.

[7] Yuan Jiangyan, Yang Jing, Ma Hongwen, et al. Hydrothermal synthesis of nano-kaolinite from K-feldspar [J]. Ceramics International, 2018, 44 (13): 15611-15617.

[8] Mbey J A, Thomas F, Razafitianamaharavo A, et al. A comparative study of some kaolinites surface properties [J]. Applied Clay Science, 2019, 172: 135-145.

［9］ Lei Wang, Zhang Ming, Zhang Zhenxin, et al. Effect of bulk nanobubbles on the entrainment of kaolinite particles in flotation ［J］. Powder Technology, 2020, 362: 84-89.

［10］ Zhang Xianghui, Wang Junjie, Wang Ling, et al. Effects of kaolinite and its thermal transformation on oxidation of heavy oil ［J］. Applied Clay Science, 2022, 223: 106507.

［11］ Kaufhold S, Dohrmann R, Ufer K. Determining the extent of bentonite alteration at the bentonite/cement interface ［J］. Applied Clay Science, 2020, 186: 105446.

［12］ Sudheer Kumar R, Podlech C, Grathoff G, et al. Thermally induced bentonite alterations in the SKB ABM5 hot bentonite experiment ［J］. Minerals, 2021, 11 (9): 1017.

［13］ Xu Y F, Li X Y. Fractal approach to erosion threshold of bentonites ［J］. Fractals, 2018, 26 (2): 1840012.

［14］ Hayakawa T, Oya M, Minase M, et al. Preparation of sodium-type bentonite with useful swelling property by a mechanochemical reaction from a weathered bentonite ［J］. Applied Clay Science, 2019, 175: 124-129.

［15］ Dutta J, Mishra A K. Consolidation behaviour of bentonites in the presence of salt solutions ［J］. Applied Clay Science, 2016, 120: 61-69.

［16］ Wang Xiaoli, Ufer K, Kleeberg R. Routine investigation of structural parameters of dioctahedral smectites by the Rietveld method ［J］. Applied Clay Science, 2018, 163: 257-264.

［17］ Kaufhold S, Kremleva A, Krüger S, et al. Crystal-chemical composition of dicoctahedral smectites: an energy-based assessment of empirical relations ［J］. ACS Earth and Space Chemistry, 2017, 1 (10): 629-636.

［18］ D'Ascanio V, Greco D, Menicagli E, et al. The role of geological origin of smectites and of their physico-chemical properties on aflatoxin adsorption ［J］. Applied Clay Science, 2019, 181: 105209.

［19］ Sadia A, Dykes L, Deng Youjun. Transformation of adsorbed aflatoxin B1 on smectite at elevated temperatures ［J］. Clays and Clay Minerals, 2016, 64 (3): 220-229.

［20］ Cedeño D G, Conceicao R V, de Souza M R W, et al. An experimental study on smectites as nitrogen conveyors in subduction zones ［J］. Applied Clay Science, 2019, 168: 409-420.

［21］ 张兵兵. 基于蒙脱土矿物的几种生态环境材料的制备、性能及应用研究 ［D］. 呼和浩特: 内蒙古大学, 2014.

［22］ 贺嘉伟, 马爱琼. 利用电瓷废料原位合成莫来石陶瓷 ［J］. 材料科学与工程学报, 2016, 34 (1): 123-126.

［23］ Ma Beiyue, Ren Xinming, Gao Zhi, et al. Synthesis of Al_2O_3-SiC composite powder from electroceramics waste and its application in low-carbon MgO-C refractories ［J］. International Journal of Applied Ceramic Technology, 2022, 19 (3): 1265-1273.

［24］ 付鹏程, 肖国庆, 丁冬海, 等. 高压电瓷废料制备低密度高强度陶粒支撑剂及其性能 ［J］. 材料导报, 2022, 36 (4): 66-70.

［25］ Ma Beiyue, Su Chang, Ren Xinming, et al. Preparation and properties of porous mullite ceramics with high-closed porosity and high strength from fly ash via reaction synthesis process ［J］. Journal of Alloys and Compounds, 2019, 803: 981-991.

［26］Ma Beiyue, Li Ying, Cui Shaogang, et al. Preparation and sintering properties of zirconia-mullite-corundum composites using fly ash and zircon［J］. Transactions of Nonferrous Metals Society of China, 2010, 20（12）: 2331-2335.

［27］朱惠良, 刘燕, 任博, 等. 炭黑对轻量均化矾土孔结构的影响［J］. 耐火材料, 2021, 55（1）: 30-34.

［28］顾华志, 付绿平, 黄奥, 等. 轻量化耐火材料的研制与应用［J］. 耐火材料, 2021, 55（4）: 309-315.

［29］Zhong Xiangchong, Li Guangping, Chung H S, et al. Sintering characteristics of Chinese bauxites［J］. Ceramics International, 1981, 7（2）: 65-68.

［30］Wu Xirong, Zhen Nanxiang, Kong Fansheng. Effect of characteristics of different types of bauxite clinker on adhesion［J］. Applied Sciences, 2019, 9（22）: 4746.

［31］Ou Xiaoduo, Chen Shengjin, Jiang Jie, et al. Analysis of engineering characteristics and microscopic mechanism of red mud-bauxite tailings mud foam light soil［J］. Materials, 2022, 15（5）: 1782.

［32］Chen Yang, Zhang Tingan, Lv Guozhi, et al. Extraction and utilization of valuable elements from bauxite and bauxite residue: a review［J］. Bulletin of Environmental Contamination and Toxicology, 2022: 1-10.

［33］Srikant S S, Rao R B. Microwave heat treatment on kyanite for mullite formation and the preparation of zirconia toughened mullite［J］. Journal of Microwave Power and Electromagnetic Energy, 2022, 56（1）: 37-44.

［34］Liu Yang, Xu Lei, Chen Min, et al. Effects of andalusite and kyanite addition on the microstructure, thermo-mechanical properties and corrosion resistance of ultralow-cement bauxite-corundum castables［J］. Ceramics International, 2022, 48（10）: 14141-14150.

［35］Xiang Ruofei, Li Yuanbing, Li Shujing, et al. New insight into treatment of foundry waste: porous insulating refractory based on waste foundry sand via a sacrificial fugitive route［J］. Journal of the Australian Ceramic Society, 2021, 57（2）: 427-433.

［36］Qiao Mengke, Li Xiangcheng, Chen Pingan, et al. Pore evolution and slag resistance of corundum castables with nano zirconia addition［J］. Ceramics International, 2021, 47（5）: 6947-6954.

［37］Jin Junxun, Long Yuyang, Gao Huimin, et al. Flotation behavior and mechanism of andalusite and kyanite in the presence of sodium oleate［J］. Separation Science and Technology, 2019, 54（11）: 1803-1814.

［38］Ding Dafei, Ye Guotian, Li Na, et al. Andalusite transformation and properties of andalusite-bearing refractories fired in different atmospheres［J］. Ceramics International, 2019, 45（3）: 3186-3191.

［39］Tang Wei, Zhu Lingling, Mu Yuandong, et al. Effects of pre-calcining temperature of andalusite on the properties of mullite-corundum refractories［J］. Journal of the Australian Ceramic Society, 2021, 57（5）: 1343-1349.

［40］Aslanoglu Z, Baglan C, Soykan H S. Influence of SiC addition on the properties of andalusite brick used in EAF dust recovery kiln［J］. Journal of the Australian Ceramic Society, 2017, 53（2）: 517-522.

［41］Li Bowen, He Mengsheng, Wang Huaguang. Phase transformation of andalusite-mullite and its roles in the microstructure and sinterability of refractory ceramic［J］. Metallurgical and Materials Transactions A, 2017, 48（7）: 3188-3192.

［42］Ding Dafei, Zhao Zhenhua, Huang Danwu, et al. Effect of the calcined andalusite aggregates on the micro-crack formation and thermal shock resistance of mullite castables［J］. Ceramics International, 2022, 48（15）: 21556-21560.

［43］Sundararajan M, Rejith R G, Renjith R A, et al. Raman-XPS spectroscopic investigation of heavy mineral sands along Indian coast［J］. Geo-Marine Letters, 2021, 41（2）: 1-18.

［44］Zhai Kuan, Xue Weihong, Wang Hu, et al. Raman spectra of sillimanite, andalusite, and kyanite at various temperatures［J］. Physics and Chemistry of Minerals, 2020, 47（5）: 1-11.

［45］Jin Junxun, Gao Huimin, Ren Zijie, et al. The flotation of kyanite and sillimanite with sodium oleate as the collector［J］. Minerals, 2016, 6（3）: 90.

［46］史腾腾. 用废建筑陶瓷制备免烧成耐高温材料及其性能研究［D］. 北京: 中国地质大学, 2021.

4 氧化镁系耐火材料原料

4.1 镁质耐火材料原料

4.1.1 术语词

水纤菱镁矿 artinite
铁菱镁矿 breunerite
水镁石 brucite
铜菱镁矿 cupromagnesite
低铁水镁石 ferrobrucite
水菱镁矿 hydromagnesite

多水菱镁矿 lansfordite
菱镁矿 magnesite，giobertite
镁砂 magnesia
三水菱镁矿 nesquehonite
钛菱镁矿 titanomignesite

4.1.2 术语词组

盐湖镁砂（卤水镁砂）brine magnesia
烧成镁砂 burned magnesia
轻烧镁砂（苛性镁砂）light burned magnesia，caustic calcined magnesia
重烧镁砂（死烧镁砂）dead burned magnesia
致密烧结镁砂 dense sintered magnesia
冶金镁砂（补炉镁砂）fettling magnesia
细晶镁砂 fine-grained magnesia
细分散镁砂（高分散镁砂）finely dispersed magnesia
电熔镁砂 fused magnesia
高纯烧结镁砂 high-purity dead-burned magnesia

电熔镁砂高纯晶体 high-purity crystals of fused magnesia
高纯度高密度镁砂 high-purity high-dense magnesite
高纯镁砂 high-purity magnesia
高钙镁砂 high-calcium magnesia
大结晶镁砂 large crystal magnesia
大结晶电熔镁砂 large crystalline fused magnesia
低硅高钙电熔镁砂 low-silica and high-calcia fused magnesia
低铁镁砂 low-iron magnesia
低温煅烧镁砂 low-temperature sintered magnesia

80

竖炉镁砂 magnesia from shaft kiln

镁砂标号 magnesia sort, magnesia mark

菱镁矿床 magnesite deposit

冶金镁砂 metallurgical magnesia

中档镁砂 medium grade sintered magnesia

非晶质菱镁矿 non-crystalline magnesite

沥青涂附镁砂 pitch-coated magnesia

菱镁矿原矿 raw magnesite

海水镁砂 sea-water magnesia

烧结镁砂 sintered magnesia

4.1.3 术语例句

水镁石纤维是天然产出的、罕见的高镁实心纤维。

Brucite fiber is a natural, rare and high-Mg medicine fiber.

水镁石纤维是一种安全的非石棉天然矿物纤维。

Brucite fiber is a kind of natural mineral fiber which is safe and non-asbestos.

大同盆地东大沟湖成化学沉积物试样的实验室分析结果表明，该化学物质是**水菱镁矿**。

The result of the laboratory analysis of the lacustrine chemical sediment samples from Dongdagou in the Datong Basin shows that the chemical matter is a **hydromagnesite**.

铬铁矿及**菱镁矿**矿化通常是与蛇绿岩有关。

Chromite and **magnesite** mineralization is usually related to the ophiolites.

利用低品位的**菱镁矿**为原料，采用碳酸法生产工艺合成轻质碳酸镁。

The synthesis process of light magnesium carbonate from low-grade **magnesite** by carbonation method is described.

以**菱镁矿**煅烧所得氧化镁为原料，水化制备超细氢氧化镁。

Magnesia calcined from magnesite as raw material, and superfine magnesium hydroxide has been produced by hydration.

以**菱镁矿**为原料，经煅烧、盐酸酸化后与氢氧化钠反应合成了纳米级、片状、粒度均匀的氢氧化镁。

Magnesium hydroxide that was of nanometer level and flake shape and had uniform particle size was synthesized by calcination and hydrochloric acid-acidification and with **magnesite** as raw material.

由于**镁砂**为主要成分，使该镁碳砖具有良好的抗渣侵蚀性能。

The magnesite graphite brick has a good slag resistance because the **magnesia** is the primary component.

本文从碳源、**镁砂原料**和添加剂三个角度综述了低碳镁碳材料抗热震性的研究进展[1]。

The research progress of thermal shock resistance of low-carbon magnesia carbon materials was summarized from the aspects of carbon source, **magnesia raw material** and additives[1].

主要原料包括不同粒度的**镁砂**（质量分数>98%）、鳞片石墨（质量分数>98%）、金属铝粉（分析纯）和酚醛树脂（液态、固定碳为48%）。

The main raw materials used were different sizes of **magnesia**（>98wt.%）, flake graphite（>98wt.%）, aluminium powder（analytical grade）and phenolic resin（liquid, 48% fixed carbon）[2].

本发明涉及一种冶炼设备，即一种清洁节能**镁砂**电熔炉。

The invention relates to smelting equipment, in particular to a clean and energy-saving **magnesia** electrical furnace.

镁砂行业的污染主要以烟尘、烟气为主。

Dust and smoke are the key pollutants in **magnesia** industry.

详述了卤水镁砂与**海水镁砂**的特点。

The characteristics of brine and **sea-water magnesia** are described.

文章论证了我国卤水和**海水镁砂**的研究与开发。

The research and development of brine and **sea-water magnesia** in China are demonstrated.

这是一种由**重烧镁砂**制成的混合型低铬镁砖。

This is a hybrid magnesia low chromite brick made from **dead burned magnesia**.

这是一种由重烧镁砂制成的以**烧结镁砂**和铬铁矿为成分配制的标准砖。

This is our standard quality brick based on **burnt magnesia** and chromite, made from dead burned magnesia.

到 2012 年夏天，奥镁公司将在挪威生产出 8 万吨的优质**电熔镁砂**。

Starting in the summer of 2012, RHI will produce up to 80000 tons of high-graded **fused magnesia** in Norway.

以优质**电熔镁砂**、烧结镁铝尖晶石为主要原料，采用纸浆废液作为结合剂研究镁质复合滑板材料。

By using **fused magnesia** and sintered spinel as main materials and sulphite liquor as binders, magnesium compound side gate was researched.

辽宁省的铁、金刚石、硼、**菱镁矿矿床**储量位列中国各大省市之首，它同时也是天然气和石油的重要产地之一。

Liaoning Province has the most iron, boron, diamond, and **magnesite deposits** among all province-level subdivisions of China. Liaoning Province is also an important source of petroleum and natural gas.

中档**烧结镁砂粉**的添加使浇注料高温处理后的线变化率增大和气孔率增加。

Adding middle grade **sintered magnesia powder** can enlarge the line ratio and porosity after high temperature treatment.

电热管由 U 型钢管、**烧结镁砂**和电热丝制成，直接浸在水里热能损耗少。

The electrical heating tube is made of U-shaped steel tube, **sintered magnesia** and heating wire. It is immersed directly in the water for preventing heat loss.

4.1.4 术语例段

考虑到碳捕获及其利用技术快速发展的需要，在混凝土配方中由 Mg-Si 矿物生产的 RMCs 的主要优势是在来源于低碳原料的 MgO 被碳化过程中，它们能够以稳定的碳酸盐形式吸收和永久储存 CO_2。在这些过程中，MgO 与水（H_2O）反应形成具有低强度的多孔**水镁石**（$Mg(OH)_2$）。然而，水合 MgO 具有很强的吸收 CO_2 的能力，并通常可用来生产建筑用的碳酸化产品。换言之，MgO 通过水合溶解形成 $Mg(OH)_2$，然后根据以下反应碳酸化并产生一系列水合碳酸镁（HMCs）：$Mg(OH)_2 + CO_2 + 2H_2O \rightarrow MgCO_3 \cdot 3H_2O$。其中，**三水菱镁矿**（$MgCO_3 \cdot 3H_2O$）是最常见的水和碳酸镁类型。此外，如**水菱镁矿**（$4MgCO_3 \cdot Mg(OH)_2 \cdot 4H_2O$）、**球碳镁矿**（$4MgCO_3 \cdot Mg(OH)_2 \cdot 5H_2O$）和**纤菱镁矿**（$MgCO_3 \cdot Mg(OH)_2 \cdot 3H_2O$）这些亚类型也经常被发现。

Considering the need for the rapid development of carbon capture and utilization technology, the main advantage of RMCs produced from Mg-Si minerals in concrete formulations is their ability to absorb and permanently store CO_2 in the form of stable carbonates during the carbonation process when MgO is sourced from low-CO_2 feedstocks. In such processes, MgO reacts with water (H_2O) to form **brucite** ($Mg(OH)_2$), which generally has a weak and porous structure. However, hydrated MgO has a strong ability to absorb CO_2 and produce carbonated products at a strength useful for construction purposes. In other words, the dissolution of MgO through hydration results in the formation of $Mg(OH)_2$, which is then carbonated according to the following reaction and produces a range of hydrated magnesium carbonates (HMCs): $Mg(OH)_2 + CO_2 + 2H_2O \rightarrow MgCO_3 \cdot 3H_2O$. **Nesquehonite** ($MgCO_3 \cdot 3H_2O$) is the most commonly obtained HMC, yet other phases such as **hydromagnesite** ($4MgCO_3 \cdot Mg(OH)_2 \cdot 4H_2O$), dypingite ($4MgCO_3 \cdot Mg(OH)_2 \cdot 5H_2O$), and **artinite** ($MgCO_3 \cdot Mg(OH)_2 \cdot 3H_2O$) can also be present[3].

通过多种物理和化学技术研究了在水菱镁矿（$Mg_5(CO_3)_4(OH)_2 \cdot 4H_2O$）存在下活性方镁石（MgO）的水化反应。纯 MgO-水混合物水化后得到了结构疏松的水镁石（$Mg(OH)_2$）糊状物，但 MgO-水菱镁矿混合物水化后得到的糊状物能够快速凝固，凝固后的强度能满足应用在建筑领域的要求。在观察的 28 天内，共混物的强度随着水化时间增加而增加，并且即使菱镁矿含量增加至 30% 也没有明显降低。拉曼光谱表明，可能形成了组成介于水镁石、水菱镁矿和水之间的无定形相。由于 MgO 原料中存在少量 CaO 杂质，因此也会形成少量的方解石。热力学计算表明亚晶石（$MgCO_3 \cdot Mg(OH)_2 \cdot 3H_2O$）是该体系的稳定相，但 XRD 或 FTIR 检测均未观察到，分析认为可能是生长动力学不足。

The hydration of **reactive periclase** (MgO) in the presence of hydromagnesite ($Mg_5(CO_3)_4(OH)_2 \cdot 4H_2O$) was investigated by a variety of physical and chemical techniques. Hydration of pure MgO-water mixtures gave very weak pastes of **brucite** ($Mg(OH)_2$), but hydration of MgO-hydromagnesite blends gave pastes which set quickly and gave compressive strengths of potential interest for construction applications. The strengths of the blends increased with hydration time at least up to 28 days and were not significantly decreased by increasing the hydromagnesite content up to 30%. Raman spectroscopy suggests that an amorphous phase, of composition between that of brucite, hydromagnesite and water, may form. Small amounts of calcite also form due to CaO in the MgO source. Thermodynamic calculations imply that the crystalline phase artinite ($MgCO_3 \cdot Mg(OH)_2 \cdot 3H_2O$) should be the stable product in this system, but it is not observed by either XRD or FTIR techniques, which suggests that its growth may be kinetically hindered[4].

致密氧化镁材料和多孔氧化镁材料是氧化镁材料的重要组成部分，分别在不同的领域有关键作用，新型致密或多孔氧化镁材料是当下的研究热点。为此，主要综述了制备致密氧化镁材料的研究进展，不同催化剂对致密氧化镁材料的影响和制备多孔氧化镁材料的研究进展。最后对氧化镁材料的发展提出几点建议[5]。

Dense magnesia materials and porous magnesia materials are important branches of magnesia materials and have played important roles in many important fields. The novel dense or porous magnesia materials are current research hotspots. The preparation progresses of dense magnesia materials and porous magnesia materials, the influence of different catalysts on dense magnesia materials were reviewed. Finally, some suggestions on the development of magnesia materials were put forward[5].

简述了国内外菱镁矿的利用现状，总结了利用**菱镁矿**制备高致密烧结镁砂、镁质晶须、纳米氧化镁材料的研究进展，并展望了高效利用菱镁矿资源的发展方向[6]。

The utilization status of **magnesite** in China and overseas was briefly described, the research progress of preparing high density sintered magnesia, magnesia whiskers and nano magnesia materials from magnesite was summarized, and the development tendency of high efficient utilization of magnesite resources was prospected[6].

本研究通过水化和球磨处理活化**菱镁矿**，以得到的高活性 MgO 和 H_3BO_3 为原料，在 KCl 盐的辅助下合成 $Mg_2B_2O_5$ 晶须。研究了 B/Mg 比和反应温度对晶须相变和形貌演变的影响，并分析了晶须的生长机理。以制备的不同直径的 $Mg_2B_2O_5$ 晶须为原料，采用传统无压烧结法制备了 $Mg_2B_2O_5$ 陶瓷。研究了它们的物理性质，如体积密度、显气孔率和维氏硬度（HV），尤其是 $Mg_2B_2O_5$ 陶瓷的微观结构和微波介电性能（ε，$Q \times f$ 和 τ_f）之间的关系。研究结果表明，较高的 B/Mg 比和较低的反应温度更有利于液固高纵横比 $Mg_2B_2O_5$ 晶须的生长。特别是平均直径（AD）= 198nm 并在 1250℃下烧结的 $Mg_2B_2O_5$ 陶瓷表现出良好的微波介电性能（$\varepsilon = 6.18$，$Q \times f = 18597GHz$，$\tau_f = -82ppm$❶/℃）。本研究为菱镁矿的高附加值利用提供了思路。

In this study, **magnesite** was activated through hydration and balling treatment, and

---

❶ 1ppm = 10^{-6}。

the received MgO with high activity and H_3BO_3 were used as raw materials for synthesizing $Mg_2B_2O_5$ whiskers with the aid of KCl salt. Attention was paid to the effects of B/Mg ratio and reaction temperature on the phase transformation and morphology evolution of the produced whiskers, and their growth mechanism was proposed as well. The $Mg_2B_2O_5$ ceramics were prepared based on the whiskers with different diameters via traditional pressing-sintering. Their physical properties, such as bulk density, apparent porosity and Vickers hardness (HV), were studied. Particular emphasis was placed on the relationship between microstructure and microwave dielectric properties (ε, $Q \times f$, and τ_f) of the $Mg_2B_2O_5$ ceramics. According to the results, the higher B/Mg ratio and the lower reaction temperature were more suitable for the growth of $Mg_2B_2O_5$ whiskers with high aspect ratio conforming to the liquid-solid (LS) mechanism. In particular, the $Mg_2B_2O_5$ ceramics based on the whiskers with average diameter (AD) = 198nm and sintered at 1250℃ exhibited good microwave dielectric properties (ε = 6.18, $Q \times f$ = 18597GHz, and τ_f = -82ppm/℃). This new design principle provides a guidance for fabricating high value-added magnesia-based ceramics from magnesite[7].

以**晶质菱镁矿**为原料、CeO_2 为添加剂，煅烧不同粒径的 MgO 粉末，制备了烧结 MgO。此外，研究了 CeO_2 对烧结 MgO 性能的影响。研究结果表明：（1）添加少量 CeO_2 有益于烧结 MgO 的致密化；（2）最大体积密度为 3.48g/cm³；（3）显孔隙率为 0.35%。在烧结过程中，Ca^{2+} 可以进入 CeO_2 晶格形成 O^{2-} 空位，从而促进 CeO_2 晶粒的生长；同时可填充 MgO 晶粒间的孔隙，提高所制 MgO 的致密度。此外，固溶反应通过限制硅酸盐基质的形成，提高了 MgO 晶粒之间的结合程度。

Sintered MgO was prepared by using MgO powder with different particle sizes by calcination of by **crystal magnesite** as the raw material and CeO_2 powder as an additive. In addition, the effect of the CeO_2 addition on the performance of the sintered MgO was investigated. The results showed that (1) the densification of the sintered MgO with a small amount of CeO_2 addition appreciably promoted, (2) the maximum bulk density reached 3.48g/cm³, and (3) the apparent porosity was 0.35%. In the sintering process, the Ca^{2+} could approach into CeO_2 lattices, forming O^{2-} vacancy at CeO_2 grains, which resulted in promoting the growth of CeO_2 grains filling the pores between the MgO grains, and enhanced the densification of sintered MgO. In addition, the solid solution reaction limited the formation of silicate phase in MgO, improving the bonding degree between the MgO grains[8].

　　烧结镁砂的体积密度在压制过程中受成型方法和生坯坯料尺寸的显著影响。生坯中夹带的空气含量是影响镁砂试样的体积密度的关键因素。本研究采用真空压制成型和常规压制成型两种方法制备了不同尺寸的高密度镁砂试样。使用阿基米德法、水银孔隙率测定法和扫描电子显微镜（SEM）表征了所制烧结镁砂的体积密度、孔径分布等物理性质和显微形貌。研究结果表明，常规压实法制备的镁砂试样的体积密度随着厚度的增加而降低至 $3.40g/cm^3$ 以下，这是由于夹带的空气而引起的大孔和晶间裂纹缺陷。此外，通过 SEM 照片证实了常规压实试样中存在大孔隙和晶间裂纹。然而，真空压实法所制镁砂试样的体积密度均大于 $3.40g/cm^3$。此外，真空压实法所制镁砂试样中的缺陷主要以小圆形孔存在。

The bulk density of **sintered magnesia** is significantly influenced by molding methodology and blank size of the green body during dry pressing. The entrapped air in the green body plays a critical role in determining the bulk density of magnesia samples. Herein, high-density magnesia samples, with different sizes, are prepared by using vacuum compaction molding and conventional compaction molding. The physical properties, such as bulk density and pore size distribution, and morphology or as-sintered magnesia samples were characterized by using Archimedes method, mercury porosimetry, and scanning electron microscopy (SEM). The results indicate that the bulk density of conventional compaction magnesia samples decreased below $3.40g/cm^3$ with the increase of thickness due to the presence of entrapped-air induced large pores and intergranular cracks. In addition, the large pores and intergranular cracks in conventionally compacted samples are observed by SEM images. However, vacuum compaction of magnesia samples resulted in a bulk density of higher than $3.40g/cm^3$ for all thicknesses. Moreover, the defects in vacuum-compacted magnesia samples are mainly in the form of small circular pores[9].

　　采用真空压制成型法制备了**高密度镁砂**，并研究了成型压力和烧结温度对镁砂试样的体积密度、显气孔率、直径收缩率、体积收缩率、孔径分布、冷压强度和抗热震性的影响。通过常规压实成型法制备的试样中存在两个范围的孔分布：$350\sim2058nm$ 和 $6037\sim60527nm$。分析认为大尺寸的孔隙是影响镁砂致密化的主要因素。研究发现，采用真空压制成型有利于大尺寸气孔的消除；在大于 200MPa 成型压力、高于 1600℃ 烧结温度条件下，可制备出理想的高密度镁砂（密度大于 $3.40g/cm^3$）。与使用传统压制成型制备的试样相比，通过真空压制成型制备的镁砂试样表现出更好的烧结性能。

High-density magnesia was fabricated using vacuum compaction molding, and effects of forming pressure and sintering temperature on bulk density, apparent porosity, diameter shrinkage ratio, volume shrinkage ratio, pore size distribution, cold compressive strength, and thermal shock resistance of the magnesia samples were investigated. There were two ranges of pore distribution in samples that were formed via conventional compaction molding, and these ranges were about 350-2058nm and 6037-60527nm. It was considered that the range of larger pores mainly influenced the densification of magnesia. Using vacuum compaction molding, large size pores were removed, and high-density magnesia (with a density greater than $3.40g/cm^3$) was easily prepared when forming pressure was higher than 200MPa and sintering temperature was higher than 1600℃. Magnesia samples prepared via vacuum compaction molding showed better performance compared to that of samples prepared via conventional compaction molding[10].

中国以**粗晶菱镁矿**为原料生产的烧结**电熔镁砂**（FM）作为各种耐火材料的生产受到了全世界的关注。本文研究了具有不同成分的重烧氧化镁（DBM）。研究结果表明，试样 DBM92 的方镁石晶体呈自形，而试样 DBM95 的方镁石晶体呈自形和亚自形。此外，在试样 DBM97 中观察到了大量具有直边界的方镁石-方镁石键合。没有清晰边界的试样 DBM98 的方镁石晶体表现出所有试样中最致密的堆积结构。试样 DBM92 中的方镁石颗粒周围的硅酸盐基质含有镁橄榄石和钙镁橄榄石，而试样 DBM95 中的方镁石颗粒周围的硅酸盐基质主要为钙镁橄榄石和镁硅钙石。另外，在试样 DBM97 中观察到少量硅酸二钙和水镁石晶间相。相比之下，在试样 DBM98 的方镁石晶粒中仅观察到集中在晶间的硅酸二钙。高温抗折强度测试表明，耐碱 DBM95 基砖可以在 <0.5 的钙硅比下制备。最优质的 DBM97 基砖可以用大约 2.2 的钙硅比原料制备，以确保硅酸二钙是唯一的间隙相。两步法 FM 的方镁石晶体尺寸明显大于一步法 FM 的方镁石晶体，并且硅酸盐边界明显减少。采用结构致密、大晶粒尺寸的两步法 FM 制备的镁碳砖表现出优异的抗渣性能。

Sintered and **fused magnesia**(FM) produced from the **macrocrystalline magnesite** in China have attracted attention worldwide for the production of various refractories. Herein, dead burnt magnesia (DBM) with varying compositions was investigated. The results revealed that the periclase crystals of the DBM92 sample were subrounded to rounded euhedral, whereas the periclase crystals of the DBM95 sample were subhedral and idiomorphic. In addition, a significant amount of periclase-periclase bonding with straight boundaries was observed in the DBM97 sample. The periclase crystals of the DBM98 sample with no clear boundaries exhibited the densest packing among the samples. The silicate matrix around the periclase grains of the DBM92 sample contained forsterite and monticellite, whereas that around the periclase grains of the DBM95 sample was mainly composed of monticellite and merwinite. Dicalcium silicate and merwinite were observed in small amounts as interstitial phases in the DBM97 sample. In contrast, only dicalcium silicate, which was concentrated in small triangular pockets, was observed in the densely packed periclase grains of the DBM98 sample. The hot modulus of rupture tests revealed that an alkali-resistant DBM95-based brick can be prepared at a lime-silica ratio of < 0.5. The best quality DBM97-based brick can be prepared at a lime-silica ratio of approximately 2.2, which ensures that dicalcium silicate is the only interstitial phase. The periclase crystals of the two-step FM were significantly larger than those of the one-step FM and exhibited remarkably fewer silicate boundaries. The superior structure and large crystal size of the two-step FM can endow magnesia-carbon bricks with slag corrosion resistance[11].

以**轻烧镁砂**为原料，以**菱镁矿细粉**为致孔剂，制备了一种保温性能良好、微孔结构均匀的新型镁质耐火材料。重要结论如下：（1）**菱镁矿**在 700℃ 热分解过程中形成的 $MgCO_3$ 晶型的（原始母盐假象）多孔方镁石微晶聚集体促进了材料本身的微孔化；（2）经 XRD 和 SEM 表征分析，1600℃ 烧结的试样均由发育良好、孔径为 1.5~4.2μm 的方镁石构成；（3）造孔剂的最佳含量（质量分数）为 20%。在此条件下，1600℃ 的烧结试样具有较低的热导率和热膨胀系数，并且烧结强度较高。以上研究成果，有望应用于冶金工业基础隔热耐火材料领域。

A novel magnesia-based refractory with good thermal insulation performance and uniform microporous structure was prepared by using **caustic calcined magnesia** as raw material and **magnesite fine powder** as a porogenic-agent. The important conclusions were as follows: (1) The formation of porous periclase microcrystal aggregation with $MgCO_3$ crystal form (original salt pseudomorph) promoted the microporation of the material itself during the thermal decomposition of **magnesite** at 700℃; (2) After XRD and SEM characterization analysis, all the samples sintered at 1600℃ consisted of well-developed periclase, with a uniform microporous structure and an average pore size in the range of 1.5-4.2μm; (3) The optimal content of the pore-forming agent is 20wt.%. Under this condition, the sintered sample at 1600℃ has a lower thermal conductivity and thermal expansion coefficient, and higher sintering strength. Based on the above research results, it was expected to be applied in the fields of basic heat-insulation refractory for the metallurgical industry[12].

活性氧化镁是一种重要的镁质化工材料，其用途广泛，是生产其他高纯镁化合物的原料。它主要作为氯丁橡胶及氟橡胶的促进剂与活化剂，用于黏合剂、塑料、油漆和纸张的填料；医药上用作抗酸剂和缓泻剂，用于胃酸过多和十二指肠溃疡病；可作为陶瓷、玻璃、高级保温材料及氧化镁水泥等原料；还可用于冶炼脱硫及燃气和燃料脱硫，等等。烧结镁砂具有高熔点（2800℃）、高电阻和优良的抗渣侵性能，因而作为生产镁质耐火材料的基本原料，被制成各种镁砖在冶金、水泥等行业的高温炉上广泛使用。镁质耐火材料的使用寿命主要取决于原料烧结镁砂的体积密度。尤其是近年来，随着钢铁产业的发展，对镁砂质量的要求越来越高，尤其是对体积密度高于 $3.40g/cm^3$ 的优质烧结镁砂的需求越来越大。辽宁省大石桥市被誉为"世界镁都"，储有占世界总储量20%的品位达40%以上的优质菱镁矿。本研究就以该菱镁矿为原料，利用水化法，制备出了高活性氧化镁和高密度烧结镁砂。本工艺流程短，易操作，成本低，无环境污染，所得产品氧化镁活性很高、烧结镁砂密度大于 $3.40g/cm^3$，具有实际生产意义。此外，还考察了制备条件对氧化镁活性及镁砂体积密度的影响，研究了氧化镁微观结构与其活性之间的关系，最后分析了水化对氧化镁性质的影响，进而对其高温烧结性能的影响。(1) 活性氧化镁的制备：1) 由菱镁矿在850℃下煅烧2h制得氧化镁后，将其水化后得到前驱体氢氧化镁。将氢氧化镁在400℃下煅烧1h得到吸碘值为 $278.82mgI_2/g$ 的高活性氧化镁，其比表面积为 $202.41m^2/g$。2) 随保温时间的延长和煅烧温度的升高，氧化镁的晶型越来越完整，氧化镁由亚稳态非晶体转变为结构紧密、晶格完整的氧化镁晶体，活性逐渐降低。保温时间和煅烧温度都是影响氧化镁活性的主要因素。3) 为得到活性氧化镁，起始加热温度应高于200℃。控制好煅烧温度及保温时间保证氢氧化镁的完全分解并抑制氧化镁晶体的生长，是得到高活性氧化镁的关键。4) 氧化镁的微观结构对其宏观性质有很大的影响。"假晶"形态以及氧化镁的非晶亚稳态结构是令其具有高活性的重要因素。另外，晶格畸变也是一个不可忽视的因素。(2) 高密度烧结镁砂的制备：1) 将菱镁矿在850℃下煅烧2h得到的氧化镁水化成氢氧化镁，并以此为原料，在烧结温度1600℃下制得了体积密度高达 $3.47g/cm^3$ 的烧结镁砂。2) 将该水化工艺应用到低档菱镁矿在1600℃也取得了体积密度为 $3.41g/cm^3$ 的烧结镁砂；同样应用在高钙高铁菱镁矿，使得镁砂的体积密度由未经过水化的 $3.38g/cm^3$ 提高到水化后的 $3.54g/cm^3$。从而进一步验证了该水化方法的应用可靠性。3) 与菱镁矿制得的轻烧氧化镁相比，经过水化处理工艺后得到的轻烧氧化镁较容易破碎，大大降低了细磨的强度和时间，减少了粉尘污染，节约了能源。4) 该研究制备的轻烧氧化镁晶格内部不存在官能团 CO_3^{2-} 的残留物的制约，并且 $Mg(OH)_2$ 在分解过程中释放出的水部分固溶到 MgO 晶格中有效地促进了氧化镁的烧结[13]。

The **active MgO** is one of the most important chemical industry materials. It is widely used in producing other high purity magnesia compounds. The active MgO is mainly used in following aspects: accelerator and activator of chlorobutyl, butyl, acrylonitrile-butadiene and fluorine rubber; fillings of sizing agent, plastic, paint and paper; antacid and laxative in medical field; materials to produce ceramic, enamel, fireproof materials and MgO cement; smelting and flue gas desulfurization. Dead-burned magnesia has been widely used in metallurgy and cement industry as a major ingredient of refractory products, because of its many advantages such as high melting point (2800℃), high electrical resistance and excellent slag corrosion resistance. In the application of the magnesia products in the metallurgical industry, the service life of the magnesia products is mainly affected by the densification of the raw magnesia materials. With the further developing of the metallurgical industry, more dense magnesia materials with excellent properties are requested, especially the bulk density is over $3.40g/cm^3$. Dashiqiao of Liaoning province, called "the capital of magnesium in the world", is rich in high-quantity magnesite. In this study, the magnesite from Dashiqiao is used as raw material to produce high-activity MgO and high-density sintered MgO with hydration technique. In this technic, the process is short and operative, while the cost is low. What is important is that the product MgO has a high activity and the sintering MgO has high density. In addition, we also studied how the producing condition affected the characters and the bulk density of MgO and the relationship between activity of MgO and its microstructure. Finally, the effects of hydration process on the characteristic and sinterability of MgO were analyzed. (1) Studies on high-activity MgO: 1) The magnesite was firstly calcined at 850℃ for 2h to become light burned MgO and then hydrated into $Mg(OH)_2$. High active MgO was obtained by calcined $Mg(OH)_2$ at 400℃ for 1h. The iodine absorption value reached $278.82mgI_2/g$ and the specific surface area was $202.41m^2/g$. 2) Heating temperature and soaking time influenced the activity of MgO greatly. 3) To produce high active MgO, the initial temperature must be above 200℃. In order to obtain high active MgO, the heating temperature and soaking time must be controlled properly to allowed $Mg(OH)_2$ decomposed completely and prevent the MgO grain from growing. 4) It was confirmed that the microstructure has strong influence on the activity. MgO powders, with pseudocrystal structure, unstable amorphous structure and high strain energy, have high activity. (2) Studies on high-density MgO: 1) Sintered MgO with bulk density up to $3.47g/cm^3$ (at 1600℃) was obtained using $Mg(OH)_2$ as raw material, which was made by hydrating the light-calcined magnesia from magnesite ore(at 850℃ for 2h). 2) Using low-grade magnesite as raw material, sintered MgO(at 1600℃) with bulk density of $3.41g/cm^3$ was obtained under the hydration treatment. In addition, compared with MgO clinker that without hydration process, the bulk density of

MgO clinker (high CaO and Fe_2O_3) increased from 3.38g/cm^3 to 3.54g/cm^3. Both of the two above experimental results further verified the stability and reliability of the hydration method in producing high-density MgO clinker from magnesite. 3) The grinding tensity and time of light-calcined MgO decreased remarkably due to the light-calcined MgO easily crack down after hydration treatment, which could save a lot of energy and reduce dust pollution as well. 4) CO_3^{2-} group relic, which has bad effect on MgO sinterability, is still in MgO (magnesite) but for MgO that obtained after-hydration (Mg(OH)$_2$). Moreover, some H_2O from the decomposition of Mg(OH)$_2$ dissolved into MgO, which catalyzed the sintering of MgO[13].

4.2 镁钙质耐火材料原料

4.2.1 术语词

铁白云石 ankerite, ferrodolomite

方解石（冰洲石）calcite, calcspar

白云石（白云岩）dolomite, pearl spar, magnesium limestone

白云石化 dolomitization

水白云母 hydro-dolomite

水白云石 hydromuscovite

镁白云石 konite, magnesiodolomite, magnesite-dolomite

二次煅烧白云石 magdolite

高锰白云石 mangan-dolomite

4.2.2 术语词组

藻白云岩 algal dolomite

泥质白云岩 argillaceous dolomite

人造白云石 artificial dolomite

纯晶白云石 bitter spar

黑白云石粉 black dolomite powder

烧结细粒白云石 burn fine-grained dolomite

石灰质白云石 calcareous dolomite

煅烧白云石 calcined dolomite, dolime, burnt dolomite

方解石晶体 calcite crystal

苛性白云石 caustic dolomite

轻烧白云石 caustic-burned dolomite, soft-burned dolomite

碎屑白云岩 clast dolomite

结晶白云石 crystalline dolomite, bitter spar

死烧白云石 dead-burned dolomite

致密镁钙砂 dense periclase-lime clinker

难烧结白云石 difficulty sintering dolomite

白云石结合剂 dolomite bonding agent

白云石崩解 dolomite disintegration

焦油浸渍白云石颗粒 dolomite grain coated with tar

白云石料 dolomite mass

白云石乳 dolomite milk

白云石粉 dolomite powder

白云石原料 dolomite raw material

白云石料浆 dolomite slurry

白云石质石灰 dolomitic lime

白云石灰石 dolomitic limestone

含白云石的泥灰岩 dolomitic marl

白云石砂 dolomitic sand

白云石质石灰石（白云石化灰岩）dolomitization limestone

二步煅烧白云石 double burned dolomite

易烧结白云石 easily sintered dolomite

富镁白云石 magnesia enriched dolomite

镁钙砂 magnesia-calcium clinker

镁质白云石 magnesite dolomite, magnesitic dolomite, magnesia dolomite

中等结晶白云石 medium-crystalline dolomite

冶金白云石 metallurgical dolomite

油浸白云石砂 oil-immersed dolomite

生白云石 raw dolomite, crude dolomite, green dolomite

烧结白云石 sintered dolomite, calcined dolomite, fired dolomite

烧结粗颗粒白云石 sintering coarse particle dolomite

稳定白云石 stabilized dolomite

合成白云石砂（合成白云石）synthetic dolomite

合成镁钙砂 synthetic magnesia-calcium clinker

合成镁白云石砂 synthetic magnesia-dolomite

焦油结合白云石 tar-bonded magnesitic dolomite

焦油白云石 tarred dolomite

抗水化白云石砂 water resisting dolomite clinker

4.2.3 术语例句

碳酸盐岩中自生碳酸盐矿物主要为文石、高镁方解石，少量**白云石**、**铁白云石**和菱铁矿。

X-ray studies show that carbonate minerals are aragonite, high-Mg calcite, and lesser **dolomite**, **ankerite** and siderite.

铁白云石是八卦庙超大型金矿床的主要脉石矿物，与金的矿化关系十分密切。

Ankerite is one of the main gangue minerals genetically related to gold mineralization in the Baguamiao giant gold deposit.

在数百万年的时间里，地下水在魔鬼洞内的岩石上留下了一种叫做**方解石**的矿物质沉积物。

Over millions of years, groundwater left deposits of a mineral called **calcite**, on the rock within Devil's Hole.

贝壳中的碳酸钙是以文石和**方解石**形式存在。

In the shell, mineral form of calcium carbonate is **calcite** and aragonite.

以**方解石**为研究对象，以表面粗糙度为切入点，探讨了表面粗糙度对水滴在药剂作用前后**方解石**表面黏附的影响，并比较了测量黏附力和计算黏附力[14]。

Calcite was selected as the research object, and surface roughness was taken as the key point, the effect of surface roughness on the adhesion of water droplets on **calcite** surface before and after conditioning with sodium oleate was investigated[14].

多数**白云石**显然是原来**方解石**交代的结果，但肯定不是全部。

Most **dolomite** is obviously the result of replacement of precursor **calcite** but certainly not all.

通常是用菱镁矿、**白云石**和卤水为原料生产的。

Usually use magnesite, **dolomite** and brine as raw material to produce.

声波测井用于区分泥岩和**白云岩**。

The sonic log is used to differentiate shale from **dolomite**.

据估计，北美的碳酸盐岩油气藏有一半是在**白云岩**内。

It is estimated that about half of the carbonate oil and gas reservoirs in North America are in **dolomite**.

利用固体介质活塞圆筒式高压容器测量了**白云岩**与石灰岩的电导率。

The electrical conductivity of **dolomite** and limestone has been studied by means of a high-pressure apparatus of solid medium piston cylinder type.

由**白云岩**加工而成的钙镁粉，用作橡胶填料的实际生产，并不多见。

The calcium-magnesium powder produced from **dolomite** is not commonly used as a filler in rubber production.

认为不同类型的**白云岩**形成于不同的沉积环境，其**白云石化**模式也不相同。

It is concluded that different **dolomites** formed in different environments have different models of **dolomitization**.

泥岩和**白云岩**导致中子孔隙度的读数偏高，密度孔隙度的读数偏低。

Shale and **dolomite** will cause the neutron porosity high and the density porosity low.

花梨山矿区**白云岩**是以白云石为主的碳酸盐岩，是碳酸钙和碳酸镁的复盐岩石。

The dolomite of Hualishan ore area is carbonate that composed by **dolomite**, it is the composite salt rock of calcium carbonate and magnesium carbonate.

通常，粗分散的 MgO 可以通过煅烧菱镁矿和**白云石**，以及通过处理 MgS 或其他类似化合物来生产[15]。

Conventionally, roughly dispersed MgO can be produced through calcination of magnesite or **dolomite**, or by treatment of MgS or other similar compounds[15].

还探讨了**白云质石灰岩**-水玻璃灌浆材料的反应机理。

The reaction mechanism of **dolomite limestone** and water glass was discussed.

用**焦油白云石**做包衬，在德国及欧洲其他国家很普及。

The ladle lining with **tar dolomite** is common in German and in Europe.

矿化表现为"硅化、**白云石化**、黄铁矿化"组合。

The gold mineralization is represented by the association of silicification, **dolomitization** and pyritization.

矿石矿物主要有高岭石、埃洛石、**水白云母**、石英及少量长石。

The minerals of kaolin ores consist mainly of kaolinite as well as halloysite, **hydromuscovite**, quartz associated with small amounts of feldspar.

4.2.4　术语例段

介绍了煅烧各种物料的回转窑选用耐火材料及砌筑、维护、修补内衬的情况。一种具有工作层、保温层和隔热层的预制块与浇注料复合砌筑，而且预制块与浇注料内部嵌有金属锚固件，直接焊接在筒体钢壳上。烧水泥、**白云石**、石灰、氧化铝、废物等的窑，要挂好，并保护好窑皮；烧金属球团、炭素等的窑要防止结圈，选用耐磨、与低熔物液相润湿角大的耐火材料。同时都要做好定期喷补，提高内衬的使用寿命。

The selection, masonry, maintain and repair of refractory materials for rotary kiln of calcining various kinds of materials were introduced. A prefabricate working layer with heat insulation layer and thermal insulating layer was compositely masonry with castings, and there was metal anchor in the prefabricate block and castings, which was directly welded on the shell. For kilns burning cement, **dolomite**, lime, aluminum oxide and waste, etc. , the coating needed to be hung seriously and maintain in good condition. For kilns burning metal pellets, carbon, etc. , it should avoid ring and choose refractory materials with wearing resistance, and big low melting liquid wetting angle. At the same time, all of them needed to do regular gunning in order to improve the service life of lining[16].

冶金生产过程涉及多种固体化学反应，**方解石**分解作为最基础的固体反应之一被广泛研究。目前，对**方解石**分解过程的认识源于实验过程的表观动力学数据，但由于实验过程易受外界环境影响，导致表观活化能偏差很大。不仅如此，基于化学反应计量关系式得到的表观动力学难以表征分解过程的结构信息，对于多步反应也无法合理地分配每一步的贡献。建立**方解石**表面亚稳态结构的热稳定性与表面电流之间的唯象关系。基于此唯象关系，解析了**方解石**表面二氧化碳的吸附动力学行为。**方解石**、**白云石**和菱镁矿具有相似的晶体构型，但因镁含量差异导致它们的表面性质不同。利用电化学方法探究这三种矿物的气体吸附差异，首次强调了有效吸附距离在揭示吸附量差异时的重要性。基于二氧化碳在方解石表面吸附引起表面电学性质的改变，控制表面电学性质即可实现对气体吸附的定向调控[17]。

The metallurgical process involves many solid chemical reactions, and **calcite** decomposition as one of the most basic solid reactions has been widely studied. At present, there are limitations in studying **calcite** decomposition through macro kinetics. This is because the experimental process is affected by the environment, resulting in differences in apparent activation energy. The macro kinetics based on the stoichiometric relation of chemical reaction is difficult to represent the structural information of the decomposition process. It is also impossible to reasonably allocate the contribution of each step for multi-step reaction. The phenomenological relationship between thermal stability of **calcite** surface metastable structure and surface current was established. Based on the phenomenological relationship, the adsorption kinetics of carbon dioxide on **calcite** surface was analyzed. The **calcite**, **dolomite**, and magnesite have similar crystal configurations, but they have different surface properties due to differences in magnesium content. Using electrochemical method to investigate the gas adsorption difference of three minerals, the importance of effective adsorption distance is emphasized for the first time to reveal the difference of adsorption capacity. Based on the change of surface electrical properties caused by carbon dioxide adsorption on calcite surface, the directional regulation of gas adsorption can be realized by controlling the surface electrical properties[17].

本研究中使用的原菱镁矿来自中国海城市。添加剂为 La_2O_3（纯度 ≥ 99.99%，国药化学试剂有限公司）。将原材料研磨至 30 目，然后进行化学成分分析。分析标准基于湿化学方法进行。菱镁矿的化学成分证实存在大量作为杂质存在的 CaO（质量分数为 1.82%）和 SiO_2（质量分数为 4.59%）。图 1 为菱镁矿和 La_2O_3 的 XRD 图谱。可以看到，除菱镁矿外，还检测到**白云石**和石英，从而也证实了 CaO 和 SiO_2 相的存在。

The raw magnesite used in the study was from Haicheng region of China. La_2O_3 (purity ≥ 99.99%, Sinopharm Chemical Reagent Co., Ltd.) was used as an additive. The raw material was ground to 30 mesh, followed by the characterization of the chemical composition. The chemical analysis was carried out by using the standard wet chemical methods. The chemical composition of raw magnesite confirmed the presence of a significant amount of CaO(1. 82wt. %) and SiO_2(4. 59wt. %) as major impurities. Figure 1 shows the XRD patterns of raw magnesite and La_2O_3. Except magnesite, **dolomite** and quartz are also detected, thus, confirming the presence of CaO and SiO_2[18].

镁砖的腐蚀机理已经有了一些研究。Chen 等在 1650℃ 和 1750℃ 和低氧分压（ $0.53×10^{-12}$MPa 和 $3.0×10^{-12}$MPa）条件下，在真空感应炉中采用静态坩埚法研究了镁铬和**镁白云石**耐火材料的腐蚀行为。研究结果表明，MgO 基耐火材料的腐蚀原因主要是 MgO 的溶解。Goto 等采用静态坩埚法在 1400~1450℃ 研究了铝硅酸钙熔渣对 $MgO\text{-}MgAl_2O_4$ 耐火砖的腐蚀机理。研究结果表明，富镁熔渣中二次尖晶石的析出是由 MgO 颗粒的溶解导致的。与以往的类似研究相比，本试验充分考虑了钢水冲刷、熔渣腐蚀、温度波动、交变应力等因素，阐明了镁质耐火材料的降解机理。

Some studies have been carried out on the corrosion mechanism of magnesia bricks. Chen et al. investigated the corrosion behavior of magnesia-chrome and **magnesia-dolomite** refractories through crucible tests in a vacuum induction furnace at elevated temperatures(1650℃ and 1750℃) and low oxygen partial pressures($0.53×10^{-12}$MPa and $3.0×10^{-12}$MPa). The results reveal that MgO-based refractories are corroded due to the dissolution of MgO. Goto et al. investigated the corrosion of MgO-$MgAl_2O_4$ spinel refractory bricks by calcium aluminosilicate slags through crucible tests at 1400-1450℃. The results show that the corrosion mechanism is believed to involve initial dissolution of fine matrix MgO in the slag leading to secondary spinel precipitation from the MgO-enriched slag. However, previous studies mostly focused on laboratory simulation of refining conditions, which could not take the hot molten steel scouring, slag corrosion, temperature fluctuations, alternating stress and other factors into account, affecting understanding on degradation mechanism of magnesia refractories[19].

为了研究耐火材料和钢液中夹杂物之间的相互作用，在铸钢模拟器中进行了浸入试验。有关铸钢模拟器设置的详细信息，可参考 Dudczig 等人的作品。具体实验步骤为，先将大约 38kg 的 42CrMo4 钢在氩气（Ar 4.6，纯度为 99.996%）气氛中在一个特殊的预烧结坩埚（该坩埚由氧化铝/氧化铝-氧化镁-尖晶石和可水合氧化铝组成，不含任何**氧化钙**、二氧化硅或其他添加物以消除浸泡测试期间的不良反应）中熔化。其中，钢液温度和氧含量在脱氧前后由高温传感器间歇测定。然后等钢水温度调整到目标温度 1600℃后，在钢厂脱氧的基础上，应用预期生成的非金属夹杂物。因此，使用 FeO/Fe₂O₃ 粉末的混合物（钢质量的 0.5%）对钢液进行氧化。氧含量从 8~10ppm 上升到 30~40ppm。接着，将铝粉（钢质量的 0.05%）添加到钢液中，使钢液脱氧（氧含量降低到 3~5ppm）。最后，在脱氧步骤之后，将 MgO-C 试样浸入坩埚中心，大约 60mm 深的钢液中 30min。钢液随炉冷却至室温，并继续在氩气气氛下保持 12h。使用单独的坩埚以及来自同一批次的全新钢试样对不同试样进行浸入试验。

To investigate interactions between the refractory material and the inclusion population in liquid steel, immersion tests in a steel casting simulator were performed. Detailed information about the setup of the steel casting simulator can be found in Dudczig et al. Approximately 38kg of 42CrMo4 steel was melted under argon atmosphere(Ar 4.6, purity = 99.996%) in a special pre-sintered crucible consisting of alumina/alumina-magnesia-spinel with a hydratable alumina without any **calcia**, silica, or further additions to eliminate undesired reactions during the immersion test. The temperature and oxygen content of the steel melt were intermittently determined by high-temperature sensor before and after the deoxidation. After adjustment of the temperature of the steel melt to the target temperature of 1600℃, an intended generation of non-metallic inclusions was applied based on deoxidation of the steel plant. Therefore, the steel melt was oxidized using a mixture of FeO/Fe_2O_3 powder(0.5wt.% related to the steel mass). Consequently, the oxygen content rose from 8-10ppm up to 30-40ppm. Afterward, the steel melt was deoxidized by adding aluminum shavings (0.05wt.% related to the steel mass)into the steel melt leading to an oxygen content of 3-5ppm. After the deoxidation procedure, the MgO-C prism was immersed in the center of the crucible, roughly 60mm deep into the steel melt for 30min. After the removing of the immersed prism the inductive heating was switched off and the steel melt was unforced cooled to room temperature still remaining under argon atmosphere for the next 12h. For each immersion trial a separate crucible as well as a fresh steel sample from the same batch was used[20].

以菱镁矿和**白云石**为主体的定形和非定形碱性耐火陶瓷材料在世界范围内生产用于工业炉炉衬，特别是初级和精炼钢炉。实际上，以**白云石**和菱镁矿为原料生产耐火材料有两种方法：一种是在回转窑或竖窑中烧制到1500~1800℃的死烧温度；另一种是用2500℃以上的电熔炉生产。例如烧成的菱镁矿（即电熔镁砂）和电铸脊柱是由电熔炉生产的。通过使用互补的表征方法可以研究这些精细材料在高温下的行为：结构（X射线衍射）、微观结构（扫描电子显微镜（SEM））、宏观（光学和偏光显微镜）、技术（孔隙率、吸水率、密度、弯曲强度和收缩率）、热学（DTA、膨胀、冲击和导电性）和化学性能（耐酸侵蚀）。

Shaped and unshaped basic ceramics-refractories, based on magnesite and **dolomite** are produced worldwide for lining industrial furnaces, especially primary and secondary steel furnaces. Actually, there are two methods to produce refractory by using **dolomite** and magnesite as materials, one is fired in rotary or shaft kilns up to dead burning temperatures of 1500-1800℃, the other is produced by electric smelting furnace with a temperature over 2500℃, for example burned magnesite(i. e. , fused magnesia) and electrocast spine are produced by electric smelting furnace. The behavior of these elaborated materials in high temperature has been investigated through the use of complementary methods of characterization:structural(X-ray diffraction), microstructural (scanning electron microscopy (SEM)), macroscopic (optical and polarized microscope), technological (porosity, water absorption, density, flexural strength, and shrinkage), thermal(DTA, expansion, shock, and conductivity), and chemical(resistance toward acid attack)[21].

水泥生产的总需求量逐年增长。为了提高竞争力，水泥生产商实施了基于新型环保原材料的新运营实践。最近，**镁质白云石**耐火材料被认为是无铬耐火材料，可替代镁质尖晶石和镁质铬铁矿衬砖用于水泥行业，对人类健康和环境有益。然而，由于它们的抗水合性低，它们的使用受到限制。在目前的工作中，研究了海铁矿对**镁质白云石**性质（晶体学、微观结构、物理和力学）的影响。耐火试样通过单轴压力成型，然后进行干燥和烧制过程，达到1600℃的均热温度。根据XRD和SEM分析，水铁矿提高了耐火材料的抗水化性和机械强度。这些改进主要是由于褐煤的形成。铁铝酸钙相通过晶界和三相点扩散，在氧化镁和游离

99

石灰颗粒周围形成项链状微观结构（凝固的液体网络）。此外，观察到这种结构避免了硅酸盐颗粒中膨胀水合过程的进展。与传统的镁质白云石耐火材料相比，含海西石的耐火材料具有更优越的物理和力学性能。通过添加海西石，发现机械强度提高了约35%，硬度提高了约50%。

Year by year, the total demand for cement production has been growing. To be more competitive, cement producers have implemented new operating practices based on new eco-friendly raw materials. Lately, **magnesia-dolomite** refractories have been considered chrome-free refractories that might substitute magnesia-spinel and magnesia-chromite lining bricks for the cement industry with benefits to human health and the environment. However, their use is limited due to their low hydration resistance. In the present work, the hercynite's effect on the **magnesia-dolomite** properties (crystallographic, microstructural, physical, and mechanical) was studied. Refractory specimens were formed by uniaxial pressure, followed by a drying and firing process, reaching a soaking temperature of 1600℃. According to the XRD and SEM analysis, hercynite promotes hydration resistance and mechanical strengthening in the refractory body. These improvements are mainly due to brownmillerite formation. Brownmillerite diffuses through the grain boundary and triple points, forming a necklace-like microstructure (solidified liquid network) surrounding the magnesia and free-lime particles. Furthermore, it was observed that this structure avoids the progress of the expansive hydration process in the portlandite particles. Hercynite-containing refractory specimens exhibit superior physical and mechanical properties than conventional magnesia-dolomite refractories. By hercynite addition, an improvement of about 35% in mechanical strength and about 50% in hardness was found[22].

氧化镁（MgO）与石灰和**白云石**相比，具有熔点高、对碱性熔渣具有优异的耐腐蚀性和水化倾向小等显著特性，使其成为黑色、有色、水泥行业不同领域的耐火材料应用的理想材料。耐火级氧化镁要么通过煅烧菱镁矿矿物获得，要么从海水或内陆盐水中合成获得。与海水氧化镁相比，天然菱镁矿的主要优势在于后者含有较少量的 B_2O_3 作为杂质，它作为助熔剂并大大降低了衍生产品的耐火度。但天然菱镁矿含有不同种类的杂质，主要是 CaO、Fe_2O_3、Al_2O_3、SiO_2 等，在较高温度下与 MgO 反应生成各种低熔点化合物，从而降低烧结试样的高温强度。在更纯的菱镁矿中，方镁石晶粒之间的直接结合程度越大，耐火度越高，但杂质含量越高，方镁石与方镁石晶粒接触不良，从而导致高温强度降低。

Magnesia (MgO) owns various remarkable properties like high melting point, outstanding corrosion resistance against basic slag, and less hydration tendency compared to lime and **dolomite** which makes it suitable material for refractory applications in different areas of ferrous, nonferrous, cement industries. Refractory grades magnesia is either obtained by calcinations of magnesite minerals or synthetically derived from seawater or inland brines. Major advantage of natural magnesite over seawater magnesia is the presence of lower amount of B_2O_3 as impurity in latter, which acts as fluxing agent and drastically degrades the refractoriness of the derived products. However, natural magnesites contain different types of impurities, mainly CaO, Fe_2O_3, Al_2O_3, SiO_2, etc., which react with MgO at higher temperatures and form various low melting compounds and consequently reduces the high-temperature strength of the sintered samples. In purer-grade magnesite, greater degree of direct bonding among periclase grains provides superior refractoriness, but higher number of impurities resulted in poor periclase-to-periclase grain contact, which deteriorates the high-temperature strength[23].

工业固态废物的回收利用一直是废物管理前景中的优先解决方案。这项工作展示了一种通过添加**白云石矿物**的高温烧结将粉煤灰（CFA）业废物转化为钙长石-堇青石基多孔陶瓷膜载体（ACCMS）的新方法。所制备的 ACCMS 的表征包括相组成、微观结构、孔结构、N_2 和水渗透性以及力学性能。根据膨胀测量，添加 28.43%（质量分数）**白云石**的试样的致密化温度扩展至 1100℃，而没有任何**白云石**的试样的致密化温度为 1060℃。对于在 1150℃烧结的试样，当**白云石**添加量（质量分数）从 0 增加到 28.43% 时，其开孔率从 25.2% ±0.5% 增加到 46.5% ±0.2%。表明这种**白云石**添加水平显著抑制了 CFA 的烧结行为。

Recycling of industrial solid-state waste is always the priority solution in the prospect of waste management. This work demonstrated a new method converting coal fly ash (CFA) industrial waste to anorthite-cordierite-based porous ceramic membrane supports (ACCMS), through high-temperature sintering with additions of **dolomite mineral**. Characterization of as-prepared ACCMS included phase composition, microstructure, pore structure, N_2 and water permeability, and mechanical properties. From dilatometric measurements, the densification temperature of a sample with 28.43wt.% **dolomite** addition was extended to 1100℃ compared to 1060℃ for that of a sample without any **dolomite**. For samples sintered at 1150℃, their open porosity increased from 25.2% ±0.5% to 46.5% ±0.2% when **dolomite** addition increased from 0 to 28.43wt.%. It is suggested that this level of **dolomite** addition significantly inhibited the sintering behavior of CFA[24].

另一种建议是通过 ZrO_2 和**白云石**（$CaMg(CO_3)_2$）混合物的反应烧结制造

的 $CaZrO_3$ 基复合材料，这可以生产 $CaZrO_3$-YFSZ（完全稳定的氧化锆）-MgO 材料。白云石在高于 $800℃$ 的温度下分解产生高反应性的 CaO 纳米颗粒，它很容易与 ZrO_2 反应。因此，将形成不含游离石灰并因此对大气水分稳定的最终材料。

An alternative proposal is that of $CaZrO_3$-based composites fabricated by reaction sintering of ZrO_2 and **dolomite** ($CaMg(CO_3)_2$) mixtures, that will produce $CaZrO_3$-YFSZ (fully stabilized zirconia)-MgO materials. Decomposition of dolomite at temperatures higher than $800℃$ originates highly reactive nanometric particles of CaO, which readily react with ZrO_2. Thus, final materials without free lime and, consequently, stable against atmospheric moisture would be formed[25].

为解决白云石质材料极易水化的问题，同时提高耐火材料资源的综合利用率，以**白云石**、**高硅菱镁矿**、硅石、镁橄榄石和锆英石等为原料，采用二步煅烧和消化工艺，制备出以方镁石（MgO）、硅酸二钙（$2CaO \cdot SiO_2$）和硅酸三钙（$3CaO \cdot SiO_2$）等为主要物相的稳定性（锆）镁白云石合成料，并利用合成原料制备烧成稳定性镁白云石材料以及 Al/Si 复合不烧稳定性镁白云石材料。研究了：原料配比、杂质含量和烧成温度等工艺因素对合成料的烧结性能、物相组成和显微结构的影响；烧成稳定性（锆）镁白云石材料常温、高温物理性能和作为水泥回转窑烧成带材料使用时的挂窑皮性能；Al/Si 复合不烧稳定性镁白云石材料的常温、高温物理性能，加热后物理性能的变化以及材料的抗氧化性能，并分析了材料的性能与物相组成和显微结构之间的关系。结果表明：（1）以白云石、高硅菱镁矿和硅石（镁橄榄石）为主要原料，在制备 C/S 比（分子比）为 2.2、2.5、2.8 的稳定性镁白云石合成料的加热过程中，$1000℃$ 时 CaO 和 SiO_2 开始反应生成 $2CaO \cdot SiO_2$，$1100℃$ 时镁橄榄石开始分解为 MgO 和 SiO_2，$1200℃$ 后 CaO 和 $2CaO \cdot SiO_2$ 反应生成 $3CaO \cdot SiO_2$。在制备稳定性锆镁白云石合成料时，$1000℃$ 时锆英石开始分解为 SiO_2 和 ZrO_2，并分别与 CaO 反应生成 $2CaO \cdot SiO_2$ 和 $CaZrO_3$，$1100℃$ 时 ZrO_2 与 CaO 完全反应。（2）随着烧成温度的提高，合成料的体积密度增大。经 $1600℃$ 烧后，稳定性镁白云石合成料与稳定性锆镁白云石合成料的体积密度分别为 $3.07 \sim 3.13g/cm^3$ 与 $3.04 \sim 3.11g/cm^3$。当烧成温度提高至 $1650℃$ 后，试样的密度基本不变。经高压蒸沸法（$125℃ \times 2h$）测试，$1500 \sim 1650℃$ 烧后稳定性镁白云石合成料的水化增重率小于 0.5%，稳定性锆镁白云石合成料具有更好的抗水化性，水化增重率小于 0.3%。（3）随着材料中 C/S 的增大，CaO 的存在形式由以 $2CaO \cdot SiO_2$ 为主过渡到以 $3CaO \cdot SiO_2$ 为主。反应生成的低熔点相 $3CaO \cdot Al_2O_3/2CaO \cdot Fe_2O_3$ 和 $4CaO \cdot Al_2O_3 \cdot Fe_2O_3$ 填充在方镁石和硅酸钙的晶界和孔隙中，总含量为 4.3% ~ 8.8%。随烧成温度的提高，合成料中的气孔数量减少，方镁石均匀分布在 $2CaO \cdot SiO_2$ 和 $3CaO \cdot SiO_2$ 的基体中。在稳定性锆镁白云石合成料中，$CaZrO_3$ 以颗粒状或聚集体形式分布在孔隙中以及方镁石和硅酸钙的晶界间。(4) 1550℃和1600℃制备的烧成稳定性镁白云石材料，显气孔率为16.0%～16.3%，体积密度为3.01～3.04g/cm³；常温耐压强度在117.0～123.2MPa之间。烧成试样的性能与国标 GB/T 2775—2007 中 M-95B 和 M-93 烧成镁砖的要求相当。加入锆英石后，生成的 $CaZrO_3$ 减小了试样中的孔隙，促进了晶体之间的直接结合，提高了试样的体积密度和强度。(5) 烧成稳定性镁白云石试样的挂窑皮性能优于某商品镁铁铝尖晶石砖和直接结合镁铬砖。试样与水泥熟料之间的黏结强度在2.4～3.6MPa之间；未发生熟料层的脱落、粉化现象。加入锆英石（ZrO_2 含量相当于0.5%～2.5%）后，试样与水泥熟料的黏结强度提高，ZrO_2 含量为0.5%时，黏结强度达到7.1MPa。锆英石含量继续增大后，黏结强度逐渐降低。(6) 以稳定性锆镁白云石合成料（ZrO_2 含量为2%～8%）制备的烧成稳定性锆镁白云石材料，随 ZrO_2 含量增大至8%，试样的体积密度提高至3.12～3.14g/cm³，常温耐压强度和抗折强度分别达到128.2～131.8MPa和48.1～56.2MPa，高温抗折强度达到5.6～7.9MPa。随着试样中 ZrO_2 含量的增大，试样与水泥熟料的黏结强度逐渐降低，ZrO_2 含量增大至8%后，黏结强度减小至1.3MPa。(7) Al/Si 粉的加入有利于不烧稳定性镁白云石材料体积密度、强度和抗热震性能的提高。加入 Al 粉后试样的高温抗折强度和抗热震性显著提高，加入4% Al 粉后试样的高温抗折强度（1400℃×0.5h，埋炭）达到14.9MPa，1100℃风冷3次后试样的残余强度保持率达到73.5%。(8) 在400～1400℃热处理后，加入 Al/Si 粉后不烧稳定性镁白云石试样的强度表现为先降低后增大。经400～800℃处理后，试样的常温抗折和常温耐压强度分别降低至3.1～6.5MPa和18.3～26.3MPa，1000～1400℃处理后，耐压强度和抗折强度分别增大至29.6～72.7MPa和5.7～12.2MPa。(9) 加入 Al 粉的试样，1000℃时，Al 粉发生碳化反应生成粒状 Al_4C_3；1200～1400℃后，Al_4C_3 转化为纤维状、针状 AlN 晶须；加入 Si 粉的试样，1400℃后，逐渐生成纤维状 SiC 晶须。原位生成的 AlN 和 SiC 晶须填充在试样的孔隙和颗粒间，形成牢固的非氧化物结合，起到了增强增韧的作用。(10) Al/Si 的加入有利于提高不烧稳定性镁白云石材料的抗氧化性能。加入 Al 粉可以有效提高试样的抗氧化性，Al 粉的适宜加入量为4%。Al/Si 粉氧化后分别生成的 Al_2O_3 和 SiO_2，与试样中的方镁石和硅酸钙反应生成镁铝尖晶石、镁硅钙石（镁蔷薇辉石）和钙镁橄榄石等物相，使试样的结构更加致密，提高了试样的抗氧化性能。Al/Si 加入量过多（10%）时，试样表面致密层的形成阻碍了内部气体的逸出，试样内部出现因气体高温膨胀引起的分层、开裂，导致试样膨胀、变形[26]。

In order to improve the hydration resistance of dolime refractory and make full use of the natural resources, stabilized magnesia dolime materials composed of periclase and calcium silicate were prepared from **dolomite**, **high silica magnesite**, silica, forsterite and zircon by two stage calcination and slaking process. The burned and Al/Si composite unburned stabilized magnesia dolime materials were prepared by the pre synthesized starting materials. The influence of raw material formulation, impurity and firing temperature on the physical properties, phase composition and microstructure of the synthesized raw materials, the physical properties at both room temperature and high temperature as well as coat ability adherence of the fired materials, the physical properties, the evolution of the physical properties during the heating process and the oxidation resistance of the Al/Si composite materials were studied. The results show: (1) In the firing process of synthesizing stabilized magnesia dolime materials in the C/S ratio of 2.2, 2.5 and 2.8 from dolomite, magnesite, forsterite and silica, dicalcium silicate and tricalcium silicate appear at 1000℃ and 1200℃, well forsterite decomposes at 1100℃, respectively. Zircon decomposes at 1000℃ and the decomposed SiO_2 react with CaO to form calcium zirconium below 1100℃. (2) With the rising of firing temperature, the compactness of the materials increases, and they perform superior hydration resistance. After firing at 1600℃, the bulk density of stabilized magnesia dolime materials and stabilized zirconium magnesia dolime materials are 3.07-3.13g/cm^3 and 3.04-3.11g/cm^3 respectively. After firing at above 1650℃, the compactness remains steady. The weight gained for the specimens fired at 1500-1650℃ after testing at high pressure steam (125℃ × 2h) is less than 0.5%, respectively. The stabilized zirconium magnesia dolime materials show higher hydration resistance. (3) With the increasing of the C/S ratio, the major phase of CaO in the material change from dicalcium silicate to tricalcium silicate. C_3A/C_2F and C_4AF generated from the impurities of Fe_2O_3 and Al_2O_3 with CaO locate at the grain boundaries and the pores. The amount is between 4.3%-8.8%, and the specimens synthesized from forsterite have higher content. After firing at higher temperature, pores in the specimens decreases and MgO distributes uniformly in the calcium silicate matrix in the form of island or nonindividual body. In the stabilized zirconium magnesia dolime materials, granular or aggregate $CaZrO_3$ grains locate at the matrix boundaries forming direct bonding between MgO and calcium silicate. (4) The burn ed stabilized magnesia dolime materials prepared at 1550℃ and 1600℃ have the apparent porosity of 16.0%-16.3% the bulk density of 3.01-3.04g/cm^3 and the cold crushing strength of 117.0-

123. 2MPa. The physical properties of specimens meet the requirements of M-95B and M-93 magnesia bricks in GB/T 2275—2007. After incorporated with zircon, $CaZrO_3$ crystals are formed and fill in the pores in the materials, as well as in crease the direct bonding between of the matrix. The compactness and mechanical properties are also strengthened. (5) The stabilized magnesia dolime materials show superior coat ability adherence to the commercial magnesium hercynite brick and directed bonded magnesium chrome brick. The adherence strength between specimens and cement clinker is between 2. 4-3. 6MPa. After incorporated with zircon (equivalent 0. 5% -2. 5% ZnO_2), the adherence strength increases to 7. 1MPa for specimen with ZrO_2 content of 0. 5% but then decrease gradually at high concentration (ZrO_2 content>0. 5%). (6) The burned stabilized zirconium magnesia dolime materials (2% -8% ZrO_2) show strengthened physical properties with increasing of ZrO_2 content. The bulk density increases to 3. 12-3. 14g/cm^3. The crushing strength and the modulus of rupture at room temperature is 128. 2-131. 8MPa and 48. 1-56. 2MPa respectively. The hot modulus of rupture is 5. 6-7. 9MPa. The adherence strength decreases to 1. 3MPa with the ZrO_2 content increases to 8% . (7) The Al/Si composite unburned specimens show increased properties in compactness, mechanical properties and thermal shock resistance. The specimens with 4% Al perform significant increase in the hot modulus of rupture (1400℃ ×5h, carbon embedded;14. 9MPa) and thermal shock resistance, the residual strength ratio is 73. 5% after testing in air cooling for 3 cycles. (8) In the firing process between 400-1400℃ , the strength of the unburned stabilized magnesia dolime specimens first decreases and then increases. After treated between 400-800℃ , the cold modulus of rupture and cold crushing strength decrease to 3. 1-6. 5MPa and 18. 3-26. 3MPa, respectively, and then rise to 29. 6-72. 7MPa and 5. 7-12. 2MPa after fired at 1000-1400℃ . (9) In the specimens with Al added, granular Al_4C_3 is in situ formed at 1000℃ and then change to fibrous or acicular AlN between 1200-1400℃ . For the specimens with Si incorporated, fibrous SiC is formed after treated at 1400℃ . The in situ formed crystals fill between the pores or granules and form tight bonding in the matrix. (10) The Al/Si composite materials show promoted oxidation resistance. The addition of Al powder can effectively improve the oxidation resis tance of the specimens and the optimum addition is 4% Al. The Al_2O_3 and SiO_2 after oxidation react with MgO and calcium silicates and form magnesium aluminate spinel, merwinite or monticellite. The dense oxidized structure is formed and the oxidation resistance increase. The specimens with more than 10% Al/Si addition occur volume expansion and crack between the oxidation layer and original matrix[26].

4.3 镁硅质耐火材料原料

4.3.1 术语词

镁橄榄石 boltonite, forsterite

钙橄榄石 calcio-olivine

纯橄榄岩 dunite

铁橄榄石 fayalite

绿粒橄榄石 glaucochroite

镁铁橄榄石 hortonolite

锰铁橄榄石 knebelite

橄榄石 olivine, peridotite

苦闪橄榄石 olivinite

钙镁橄榄石 shannonite

锰橄榄石 tephroite

菱镁矿 magnesite, giobertite

镁砂 magnesia

琥珀蛇纹石 amber

叶蛇纹石 antigorite

异纤蛇纹石 asbophite

纤维蛇纹石 chrysotile

水蛇纹石 deweylite

镍纤蛇纹石 garnierite

片状蛇纹石 plate-serpentine

蛇纹石 serpentine

纤滑石 agalite, talcum, talc

滑石（寿山石）agalmatolite

滑石棉 asbecasite

水滑石 hudrotalcite

皂石（块滑石）soapstone

块滑石 steatite

滑石质镁 talc-magnesite

4.3.2 术语词组

纯橄榄岩骨料 dunite filler, dumite aggregate

纯橄榄岩泥料 dunite mass

纯橄榄岩火泥 dunite mortar

纯橄榄岩粉 dunite powder

电熔纯橄榄岩 electrically fused dunite

镁橄榄石砂 forsterite sand

橄榄石原料 olivine raw material, peridot raw

material

橄榄石矿砂 olivine sand

黑色蛇纹石 black serpentine

高铁蛇纹石 ferruginous serpentinite

滑石粉 talc flour, talcum powder

含滑石镁石 talcose magnesite

4.3.3 术语例句

以**镁橄榄石**和镁砂为主要原料，黏土为结合剂，利用泡沫法制备了微孔、高强度的镁橄榄石制品。

Using natural **forsterite** and magnesite clinker as raw materials, the forsterite refractories with micropore and high strength has been prepared by foaming method.

有些钻石含有硅酸盐矿物的微小包裹体，通常是**橄榄石**、辉石和石榴石。
Some diamonds contain minute inclusions of silicate minerals, commonly **olivine**, pyroxene, and garnet.

镁质橄榄石几乎总是代表了原生沉淀物。
Forsterite olivine almost invariably represents primary precipitate material.

仅有少数的矿石，像可以看到**橄榄石**和钻石，都是在很久远的年代形成的。
Only a handful, like diamonds and **olivine**, could be found, having been formed far away.

其主要造岩矿物为**橄榄石**、斜方辉石、单斜辉石、斜长石和角闪石。
The major rock forming minerals include **olivine**, orthopyroxene, clinopyroxene, plagioclase and amphibole.

而阿波罗计划也从月球上收集回了各种**橄榄石**。
Various kinds of **olivine** were also collected on the moon throughout the Apollo program.

只有少量金刚石和**橄榄石**产生于遥远的恒星爆炸中。
Only a handful, like diamonds and **olivine**, could be found, having been formed far away in exploding stars.

稳定的带状排列是这些含**橄榄石**岩类杂岩的固定特点。
A consistent zonal arrangement is a constant feature of those complexes that contain **olivine**-bearing rocks.

随着岩浆冷却，首先结晶出的矿物是**橄榄石**和辉石，富含铁和镁。
As magma cools, the first minerals to crystallize are **olivine** and pyroxene, which are rich in iron and magnesium.

钻芯检查的结果显示，**橄榄石**全部蚀变为**蛇纹石**、**滑石**及角闪石。

Olivine has been completely altered to **serpentine**, **talc**, and amphibole in the bore hole examined in detail.

研究了以**蛇纹石**提镁残渣为原料直接合成有机硅化合物的方法。

The method is presented for synthesis of organ silicon compounds directly from residue after extracting magnesium from **serpentine** ore.

蛇纹石玉是我国主要的玉石种类之一，在玉石工艺上具有重要的地位。

Serpentine jade is one main kind of the jade in our country, and it plays an important role in jade craft.

石棉尾矿是一种以**蛇纹石**为主要矿物成分的尾矿，对矿区环境危害严重。

The asbestos tailing is a kind of environment-destruction mineral tailing, which main mineral components are **serpentine**.

介绍了**蛇纹石尾矿**对环境的危害，分析了蛇纹石尾矿的物理化学性质，总结出资源化利用的途径。

Harm to the environment has been introduced. Physical chemical properties of **serpentine tailings** have been analyzed. The main recovery pathways for serpentine tailings have also been presented.

由于废液中含有少量的锰，如何除去其中的锰，使制得的镁制品达到国家标准成为**蛇纹石**综合利用的关键。

Due to small amount of Mn contained in the waste liquor, so, the key point of producing the qualified magnesium products from **serpentine** is the Mn removal.

蛇纹石是一种用于装饰和制作珠宝的深绿色矿物。在自然界中，它是在海水渗入上层地幔时形成的，在俯冲区的深度可以达到200km。

Serpentine is a dark green mineral used in decoration and jewelry. In nature, it is formed when sea water infiltrates into Earth's upper mantle, at depths that can reach 200 km in subduction zones.

本文主要探讨了**蛇纹石**资源化利用的可行性技术经济方案。

This paper examines the feasibility techno-economy project of the recovery of **serpentine**.

这个石头实际上相当稀罕和难于切割，因此在许多情形里其他类型岩石被当作代用品使用，例如绿长石，玄武岩和**蛇纹石**。

This stone is actually quite rare and difficult to cut, so in many cases other types of rock were used as substitutes, for example green feldspar, basalt, and **serpentine**.

滑石是一种天然黏土矿物，由镁元素和硅元素组成。

Talc is a naturally occurring clay mineral composed of magnesium and silicon.

经**滑石-蛇纹石化**，进而碳酸盐化，微量金解离。

After the **talc serpentinization** and carbonation, micro gold has been separated.

本文介绍了**滑石粉**的特殊性质及其在造纸填料、功能性材料和涂布颜料方面的应用。

This article introduces **talc** special properties and its applications in the filler of paper making, pitch absorbent and coating pigment.

爽身粉是一种常用的护肤品，一般用**滑石粉**、淀粉及其他辅料制成。

Talcum powder is a kind of common skin care products, general use of **talcum powder**, starch and other supplementary materials.

对该矿床形成的条件及**滑石**和绿泥石矿石的含量已较清楚。

The conditions of formation and concentration for **talc** and chlorite ores are now well known.

在舱里对**滑石粉**进行分标志将花相当多的时间，最好的办法是在岸上进行分标志。

It'll take considerable time to sort out the marks of **talc powder** in the holds, and the best way is to sort them out ashore.

4.3.4　术语例段

菱镁矿作为我国优势矿产资源之一，广泛应用于冶金、建材、化工等行业。高硅菱镁矿（SiO_2 质量分数>3%）是一种低品质菱镁矿，为难选菱镁矿矿石。采用浮选工艺可以制备出高质量的镁矿产品。将选后高硅菱镁矿合理配料后直接煅烧，可以制备镁硅砂、**镁橄榄石**及复合粉体材料等高附加值耐火材料，还可以制备建材、化工等原料，具有重要的经济意义和科学研究价值。本文对国内外高硅菱镁矿提纯工艺进行归纳，阐述其技术指标和提纯原理，最后介绍高硅菱镁矿的综合利用方法[27]。

Magnesite is one of superior mineral resources in China. It has been widely used in metallurgy, building industry, chemical industry etc. High-silicon magnesite (SiO2 select magnesite with high content of SiO2 >3wt.%) is one of low-grade and difficult-to. High-quality magnesia products are prepared from high-silicon magnesite by flotation. By direct calcination of high-silicon magnesite with reasonable ingredients, high value-added refractory materials such as magnesia-silica, **forsterite** and composite powder materials are prepared. Important raw materials such as building materials and chemical materials are also prepared from high-silicon magnesite. Hence high-silicon magnesite has high economic and scientific research value. In the paper, the purification processes of high-silicon magnesite at home and abroad were reviewed. The purification principles and technical indicators were expounded. Finally, the comprehensive utilization of high-silicon magnesite was introduced[27].

固体废物的回收利用是人类与自然双赢的解决方案。为此，本文以菱镁尾矿和硅切割废料为原料，采用固态反应合成制备了 $MgO-Mg_2SiO_4$ 复合陶瓷。同时系统地研究了烧结温度和原料配比对所制陶瓷性能的影响。研究发现，随着烧结温度从 1300℃ 提高到 1600℃，试样的致密度（从 47.55%～68.12% 提高到 90.96%～95.25%）和常温耐压强度（从 7.34～118.66MPa 提高到 303.39～546.65MPa）升高。此外，发现 Si 通过瞬时液相烧结促进了 Mg_2SiO_4 相的合成过程，而 Fe_2O_3 通过活化烧结效应加速了烧结过程。因此，**镁橄榄石相**的存在有效地提高了 $MgO-Mg_2SiO_4$ 复合陶瓷的密度和强度，同时降低了热导率。该工作为菱镁尾矿提供了一种潜在的再利用策略，所制备的产品有望在建筑、冶金和化工等领域得到应用。

The recycling of solid waste is a win-win solution for humans and nature. For this purpose, magnesite tailings and silicon kerf waste were employed to prepare MgO-Mg2SiO4 composite ceramics by solid-state reaction synthesis in the present work. Then, effects of sintering temperature and raw material ratio on as-prepared ceramics were systematically studied. As-prepared ceramics showed improvement in their relative density (from 47.55%-68.12% to 90.96%-95.25%) and cold compressive strength (from 7.34-118.66MPa to 303.39-546.65MPa) with the increase in sintering temperature from 1300℃ to 1600℃. In addition, it was found that Si promoted synthesis process of Mg2SiO4 phase through transient liquid phase sintering and Fe2O3 accelerated sintering process through activation sintering. Consequently, the presence of **forsterite** phase effectively improved the density and strength of MgO-Mg$_2$SiO$_4$ composite ceramic, while reducing its thermal conductivity. This work provides a potential reutilization strategy for magnesite tailings, and as-prepared products are expected to be applied in fields of construction, metallurgy, and chemical industry[28].

不同烧结温度的海泡石多孔陶瓷试样的 XRD 图谱如图 4 所示。可以看出，700℃时的主晶相为**滑石**（Mg$_3$(OH)$_2$Si$_4$O$_{10}$）和**方解石**（CaCO$_3$），以及少量**石灰**（CaO）。此时矿物中的主要结晶海泡石相已转变为**滑石**相，结晶方解石相是矿物中的主要杂质。与700℃相比，800℃的烧结产品基本没有变化，但滑石的反射峰变尖，并出现董青石相（Mg$_2$Al$_4$Si$_5$O$_{18}$）。当烧结温度达到900℃时，**滑石**的衍射峰消失，主晶相转变为顽辉石（MgSiO$_3$）和硅酸镁钙（Ca$_3$Mg(SiO$_4$)$_2$），其中顽辉石由滑石转变而来，此外还有少量的董青石和透辉石相（CaMgSi$_2$O$_6$）。当烧结温度达到1000℃时，主相变为镁黄长石（Ca$_2$Mg(Si$_2$O$_7$)）和透辉石，以及少量的硅酸镁钙相。在1100℃的烧结温度下，硅酸镁钙相消失。而在1200℃时，试样中只有镁黄长石和透辉石相，其他相几乎可以忽略不计。随着这两个相的含量增加，其衍射峰强度也随之增大。

The XRD patterns of the sepiolite porous ceramic samples with different sintering temperatures were shown in Fig. 4. It could be seen that the main crystal phase at 700℃ was **talc**(Mg$_3$(OH)$_2$Si$_4$O$_{10}$) and **calcite**(CaCO$_3$), along with a small amount of **lime** (CaO). At this point the main crystalline sepiolite phase in the mineral had transformed into the **talc** phase and the crystalline calcite phase was the main impurity in the mineral. Compared to the 700℃ basically no difference in the sintering products of 800℃, but the reflection peaks of talc becomes sharper and the emergence of cordierite (Mg$_2$Al$_4$Si$_5$O$_{18}$). When the sintering temperature reached 900℃, the reflection peaks of **talc** disappeared, the main crystal phase changed to enstatite(MgSiO$_3$) and magnesium calcium silicate(Ca$_3$Mg(SiO$_4$)$_2$) where the enstatite was transformed from talc, also a small amount of cordierite and diopside(CaMgSi$_2$O$_6$). When the sintering temperature reached 1000℃, the main phase became akermanite(Ca$_2$Mg(Si$_2$O$_7$)) and diopside, with a small amount of magnesium calcium silicate present. In the case of sintering temperature of 1100℃, Magnesium calcium silicate was absent from the final products as compared with the temperature of 1000℃. At 1200℃, there were only the akermanite and diopside phases in the samples and the other phases were negligible. The corresponding intensity of the two-phase reflections was greatly increased, and the reflections became more intense[29].

铬渣是铬酸盐和金属铬生产的副产品，可溶性铬含量高，属于危险固体废物。在中国，每年排放约 0.8Mt 的铬渣，大部分被直接填埋处理。然而，铬渣中所含的六价铬是对人体的致癌物，很容易随着雨水浸出，从而污染土壤和地下水。因此，铬渣的处理和利用不仅是资源开发的迫切需要，也是环境保护的迫切需要。铬渣主要由 MgO、SiO₂、Al₂O₃ 组成，其成分与**滑石**相似，可用于替代天然材料制备泡沫陶瓷。此外，由于烧成过程中发生的化学包封，铬渣中的铬可以有效地凝固在陶瓷基体中。

Chromium slag is a by-product of chromate and metallic chromium production, which has a high content of dissolvable Cr and, thus, is a hazardous solid waste. In China, about 0.8Mt chromium slag is discharged annually and mostly disposed of directly in landfills. In particular, Cr(Ⅵ) contained in chromium slag is a carcinogen to human, which is easy leaching out with the rain and contaminates the soil and groundwater. Hence, the disposal and utilization of chromium slag are urgent not only for the resource development but also for the environment protection. Chromium slag is mainly comprised of MgO, SiO_2, Al_2O_3, which makes it similar to **talc** by composition and suggests its usage for the replacement of nature materials in the ceramic foam preparation. Moreover, Cr in chromium slag could be effectively solidified in the ceramic matrix, owing to the chemical encapsulation occurring in the firing process[30].

本文报道了泰国以**滑石**和菱镁矿为原料，采用机械活化和随后的煅烧制备**镁橄榄石**粉末的合成、表征、微观结构和性能。合成**镁橄榄石**粉末通过使用**滑石**和菱镁矿以 1∶5 摩尔比混合。在行星式氧化锆球磨机中，最大研磨时间为 24h。然后，将混合物在电炉中分别在 900℃、1000℃、1100℃、1200℃ 和 1300℃ 下煅烧 1h。通过 X 射线衍射（XRD）、X 射线荧光（XRF）、扫描电子显微镜（SEM）和物理性质对合成的粉末进行表征。合成**镁橄榄石**的物理性质结果表明，随着煅烧温度的升高，其密度增加。相反，孔隙率随着煅烧温度的升高而降低。因此，在 1300℃ 下煅烧的镁橄榄石提供了最佳结果，其真密度为 2.96g/cm³，真孔隙率为 15.41%。合成粉末的 XRD 结果表明**镁橄榄石**结晶是稳定的，机械活化 5h 后出现锐化。1000℃ 煅烧 1h 后出现**镁橄榄石**，随着煅烧温度的升高，**镁橄榄石**相含量增加。基于上述特性，由泰国**滑石**和菱镁矿生产的镁橄榄石表现出绝缘体的特性，可用作耐火装置。

This paper reports the synthesis, characterizations, microstructure and properties of **forsterite** powder produced in Thailand from **talc** and magnesite as raw materials by using mechanical activation with subsequent calcination. The synthesis **forsterite** powder was mixed by using **talc** and magnesite at 1∶5 mole ratio. The maximum milling time was

24h in a planetary zirconia ball mill. Afterward, the mixtures were calcined in an electric furnace for 1h at 900 ℃, 1000 ℃, 1100 ℃, 1200 ℃ and 1300℃ respectively. The synthesized powder was characterized by X-ray diffraction (XRD), X-ray fluorescence (XRF), scanning electron microscopy (SEM) and physical properties. Results of the physical properties of synthesized **forsterite** showed an increased in density as the calcining temperature increased. In contrast, porosity was decreased with an increase of the calcining temperature. Therefore, forsterite that was calcined at 1300℃ provided the best results which were 2.96g/cm^3 of true density and 15.41% of true porosity. Results of XRD of synthesized powder indicated that the forsterite crystallization was constant for which sharpen appeared after 5h of mechanical activation. Fraction of **forsterite** was appeared after being calcined at 1000℃ for 1h with an increasing of calcination temperature, the fraction of **forsterite** phase increased. Based on the mentioned characteristics, the **forsterite** produced from Thai **talc** and magnesite exhibited properties of an insulator and can potentially be used as refractory devices[31].

废**蛇纹石**、天然菱镁矿和苛性镁砂用于生产烧结镁橄榄石。在1650℃烧结的试样中**镁橄榄石**相含量（质量分数）达到93.6%。测试了添加剂、粒度和煅烧原料等各种参数对烧结行为的影响。在1650℃下烧制的试样在800℃下的热膨胀系数为9.54×10^{-6}K^{-1}，pH值为8.99。这些结果表明，**烧结镁橄榄石**具有与**天然橄榄石**相似的性质，并提出了解决阿布达什铬铁矿数百万吨废石所带来的环境问题的实际解决方案。

The waste **serpentine**, natural magnesite, and caustic magnesia were used for the production of the sintered forsterite. The **forsterite-phase** content of the samples sintered at 1650℃ reached to 93.6wt.%. The effects of various parameters such as additives, particle size, and calcined raw materials were tested on the sintering behavior. The thermal expansion of 9.54×10^{-6} K^{-1} at 800℃ and pH of 8.99 were obtained from the samples fired at 1650℃. These results reveal that **sintered forsterite** has properties similar to those of **natural olivine** and propose an actual solution to overcome the environmental problem resulting from several million tons of waste rocks in Abdasht chromite mines[32].

自现代炼钢技术发展以来，耐火**镁橄榄石**陶瓷发挥了重要作用。由于**镁橄榄石**耐火陶瓷的高熔点及其在高温下不与铁反应，它们主要用作冶金工业部门的熔炉和蓄热室的耐火衬里。它们还被用于水泥和石灰生产行业，作为回转窑的耐火衬里。在过去的几十年中，**镁橄榄石**陶瓷也被用于陶瓷金属接头的电工工程。**镁橄榄石**陶瓷具有较高的热膨胀系数，与用于接合的金属的系数相当。

Refractory **forsterite** ceramics have played a significant role since the development of modern steelmaking technology. Due to the high melting point of **forsterite** refractory ceramics and their non-reactivity with iron at high temperatures, they have been predominantly implemented as a refractory lining of furnaces and regenerators in the metallurgical industrial sector. They have also been utilized in the cement and lime production industries as refractory lining for rotary kilns. In the past decades, **forsterite** ceramics have also been utilized in electrotechnical engineering for ceramic-metal joints. **Forsterite** ceramics have relatively high coefficients of thermal expansion, which is comparable to the coefficient of metals used for joining[33].

植入材料的细菌感染是骨组织工程中最大的挑战之一。在这项研究中，结合3D 打印和聚合物衍生陶瓷（PDCs）策略，制备了具有抗菌活性的**多孔镁橄榄石支架**，有效避免了 $MgSiO_3$ 和 MgO 杂质的产生。镁橄榄石支架在氩气气氛中烧结，可在支架中产生游离碳，具有优异的光热效应，可抑制金黄色葡萄球菌和大肠杆菌的体外生长。此外，**镁橄榄石支架**具有均匀的大孔结构、高抗压强度（>30MPa）和低降解率。因此，结合 3D 打印和 PDCs 策略制造的镁橄榄石支架将是骨组织工程的有希望的候选者。

Bacterial infection of the implanting materials is one of the greatest challenges in bone tissue engineering. In this study, **porous forsterite scaffolds** with antibacterial activity have been fabricated by combining 3D printing and polymer-derived ceramics（PDCs）strategy, which effectively avoided the generation of $MgSiO_3$ and MgO impurities. **Forsterite scaffolds** sintered in an argon atmosphere can generate free carbon in the scaffolds, which exhibited excellent photothermal effect and could inhibit the growth of *Staphylococcus aureus*（*S. aureus*）and *Escherichia coli*（*E. coli*）in vitro. In addition, forsterite scaffolds have uniform macroporous structure, high compressive strength（>30MPa）and low degradation rate. Hence, forsterite scaffolds fabricated by combining 3D printing and PDCs strategy would be a promising candidate for bone tissue engineering[34].

在过去的几十年里，**镁橄榄石**耐火材料是由不同的矿物如**蛇纹石**、**橄榄石**等制备的。**镁橄榄石**耐火材料也是通过铁尾矿等废料制备的。制备陶瓷的另一种原料是"稻壳"（RH），因为它是一种可再生的二氧化硅来源，含有大量二氧化硅，而且供应充足。此外，从 RH 中提取二氧化硅并不困难。从 RH 中发现的二氧化硅用于制备硼硅酸盐、堇青石、碳硅石、铝硅酸盐和莫来石。许多其他研究也研究了稻壳作为优质无定形二氧化硅的极好来源的潜力。此外，这种无定形二氧化硅可用于制备太阳能级硅、碳化硅、镁-氧化铝-二氧化硅和锂-铝-二氧化硅。

In the last few decades **forsterite** refractories were prepared by different minerals such as **serpentine**, **olivine** etc. **Forsterite** refractories were also prepared through waste like iron ore tailing. Another raw material for preparation of ceramics is "rice husk" (RH) because this is a renewable source of silica that contains high amount of silica, and it is abundantly available. Also, extraction of silica from RH is not so difficult. Silica found from RH was used for preparation of borosilicate, cordierite, carbosil, alumino-silicate and mullite. The potential of rice husk as an excellent source of high-grade amorphous silica has also been investigated in many other studies. This amorphous silica can be utilized for preparation of solar grade silicon, silicon car-bide, magnesium-alumina-silica, and lithium-aluminum-silica[35].

以 Al_2O_3 和 MgO 为原料，以 SiC 为高温成孔剂，利用超塑性成功制备了具有高封闭孔隙率和低显气孔率的 Al_2O_3-$MgAl_2O_4$ 耐火骨料。超塑性变形能力随着**氧化镁**添加量的增加而提高。由 SiC 在高温下氧化引起的高内部气体压力促使晶界滑动以包围气体，然后形成封闭孔隙。闭孔率随着氧化镁和碳化硅添加量的增加而增加，但显气孔率保持在一个较低的水平。特别是在含有 10% 的 MgO 和 2% 的 SiC 的试样中，封闭孔隙和表观孔隙分别为 16.0% 和 1.8%。与不含 SiC 的试样相比，显气孔率高，闭合孔隙率低。这就是说，SiC 不仅提供气体形成封闭孔隙，而且还能促进材料的烧结性。随着氧化镁和 SiC 添加量的增加，孔隙大小逐渐增大。在含 15% 氧化镁的试样中，由于晶界滑动过大，部分闭孔尺寸超过 10μm，一些闭孔是相互连接的。

Al_2O_3-$MgAl_2O_4$ refractory aggregates with high closed porosity and low apparent porosity have been successfully fabricated by utilizing superplasticity with Al_2O_3 and MgO as raw materials and SiC as high temperature pore-forming agent. The superplastic deformation ability improved with increasing the addition of **MgO**. High internal gas pressure caused by the oxidation of SiC at high temperature drove the grain boundaries sliding to enclose the gases and then formed closed pores. The closed porosity increased with increasing the addition of MgO and SiC, but the apparent porosity maintained at a low level. Especially in the sample with 10% MgO and 2% SiC, the closed and apparent porosity were 16.0% and 1.8%, respectively. Compared to the sample without SiC, there was a high apparent porosity and low closed porosity. That is to say that SiC not only provides gases to form closed pores, but also can promote the sinterability of the materials. With the additions of MgO and SiC increased, the pore sizes increased gradually. In the sample with 15% MgO, due to the excessive grain boundary sliding, partial of the closed pore size was over 10μm and some closed pores were interconnected[36].

由于**镁橄榄石**具有熔点高、化学稳定性好、热传导率低以及抗熔融金属侵蚀性能好等优点，而被广泛应用于冶金、热工以及铸造行业。我国**镁橄榄石**原料资源丰富，利用其合成和制造性能优良、价格低廉的镁质原材料对于高效利用我国的矿产资源，扩大其在相关领域的应用具有重要意义。本文结合目前**镁橄榄石**研究及应用过程中存在的问题开展相关基础和应用研究。利用热力学对**镁橄榄石**合成中所涉及的物相体系进行分析，通过实验研究**镁橄榄石**的合成过程、反应机理及反应动力学，分析添加剂对于**镁橄榄石**生成和烧结过程的影响，考察以**镁橄榄石**为原料制备的中间包干式料的性能。获得的主要结论如下：（1）在 MgO-SiO_2-TiO_2 三元体系内，化合物的吉布斯标准生成自由能顺序为：$MgTiO_3 > MgSiO_3 > MgTi_2O_5 > Mg_2SiO_4 > Mg_2TiO_4$；在 MgO-SiO_2-Y_2O_3 三元体系内，$MgSiO_3 > Mg_2SiO_4 > Y_2SiO_5 > Y_2Si_2O_7$；在 MgO-SiO_2-Ce_2O_3 三元体系内，Ce_2O_3 与 SiO_2 反应形成 $Ce_2Si_2O_7$、Ce_2SiO_5 以及 $Ce_{0.36}Si_{0.23}O$；在 MgO-SiO_2-ZrO_2 三元体系内，MgO 和 ZrO_2 组分间形成固溶体；ZrO_2 与 $2MgO \cdot SiO_2$ 反应形成 ZrO_2-$2MgO \cdot SiO_2$ 二元系；在 MgO-SiO_2-Al_2O_3 三元体系内，$Al_6Si_2O_3 > MgSiO_3 > MgAl_2O_4 > Mg_2SiO_4$。（2）在镁橄榄石合成过程中，$MgO$ 组分是扩散相，其在合成初期向 SiO_2 组分扩散反应生成镁橄榄石和顽火辉石；在合成过程的中期，MgO 组分通过顽火辉石晶格及低熔点物相扩散至顽火辉石表面并与其反应生成镁橄榄石，在已经生成的镁橄榄石表面析出和沉积，使镁橄榄石颗粒间相互连接实现致密化；在合成过程的后期，MgO 全部从颗粒部位扩散迁移，在颗粒原位留有气孔，通过烧结使气孔缩小弥合。（3）镁橄榄石合成反应过程由 Mg^{2+} 扩散控制；MgO 与 SiO_2 间固相反应动力学可由杨德尔方程描述，其表观反应活化能为（220 ± 20）kJ/mol。（4）致密化镁橄榄石的制备条件为：加热温度高于 1550℃，保温时间大于 3h，素坯成型压力超过 150MPa。（5）添加 TiO_2、Y_2O_3 和 CeO_2 对于镁橄榄石的合成和烧结过程具有显著的促进作用，添加 ZrO_2 和 Al_2O_3 时的效果不明显；Ti^{4+} 向 MgO 晶格的固溶、含 Ti 低熔点化合物的生成以及 TiO_2 从含 Ti 化合物中析出有利于反应质点的扩散和迁移；含 Y 低熔点化合物的生成以及由于 Y_2O_3 添加而使 $MgSiO_3$ 相在高温下存在促进了反应质点的扩散和迁移；添加的 CeO_2 在高温下转化为 Ce_2O_3，含 Ce 低熔点化合物的生成以及由于 Ce_2O_3 和 SiO_2 间的反应而使 $MgSiO_3$ 相在高温下有利于反应质点的扩散和迁移；ZrO_2 以含 MgO 固溶体的形式存在于颗粒晶界和以复合化合物形式存在于基质中，存在于基质中的复合化合物由 $MgSiO_3 + ZrSiO_4 \rightarrow Mg_2SiO_5 + MgSiO_3 + ZrSiO_4 \rightarrow Mg_2SiO_5 + ZrSiO_4 + ZrO_2$ 在颗粒晶界析

出 ZrO_2。(6)随着镁橄榄石添加量的增加,镁橄榄石干式料的体积密度和强度降低,抗渣侵性能无明显变化,添加烧结镁橄榄石和生镁橄榄石试样的性能无明显差别;添加硼砂、氧化铁、三聚磷酸钠和九水硅酸钠均可以促进镁橄榄石干式料的烧结,有利于改善镁橄榄石干式料的强度和抗渣侵性能,但当添加量超过一定范围后,会引起镁橄榄石干式料的性能下降[37]。

Forsterite have been widely used in metallurgy, heat engineering and cast, because of its excellent properties such as high melting point, good chemical stability, low thermal conductivity and strength corrosion resistance to molten metal. **Forsterite** is abundant in natural resources in our country, so it is very significant to synthesizing the forsterite raw materials and then to manufacturing the forsterite products with excellent properties in low cost for utilizing efficiently the mineral resources in China and expanding the utilization in related fields. To resolve the present problems existed in the research and application of **forsterite**, the research on the related basic and applied have been carried out in this paper. The reaction systems related to the **forsterite** synthesis were analyzed by using thermodynamic, the synthesis process, reaction mechanism and reaction kinetics of **forsterite** were studied by experiments, and the effect of additives on the generating and sintering of **forsterite** and the properties of tundish dry materials added with **forsterite** were investigated. The obtained results are as follows. (1) The sequence for Gibbs standard free energy of compound formation in MgO-SiO_2-TiO_2 system from high to low is as bellows, $MgTiO_3 > MgSiO_3 > MgTi_2O_5 > Mg_2SiO_4 > Mg_2TiO_4$. In MgO-SiO_2-Y_2O_3 system is $MgSiO_3 > Mg_2SiO_4 > Y_2SiO_5 > Y_2Si_2O_7$. In MgO-SiO_2-Ce_2O_3 system, Ce_2O_3 reacts with SiO_2 to form $Ce_2Si_2O_7$, Ce_2SiO_5 and $Ce_{0.36}Si_{0.23}O$. In MgO-SiO_2-ZrO_2 system, MgO and ZrO_2 reacts to form solid solution, and ZrO_2 reacts with $2MgO \cdot SiO_2$ to form ZrO_2-$2MgO \cdot SiO_2$ system. In MgO-SiO_2-Al_2O_3 system, the sequence is $Al_6Si_2O_3 > MgSiO_3 > MgAl_2O_4 > Mg_2SiO_4$. (2) In the synthesis process of forsterite, MgO was diffusion phase and it diffused to SiO_2 and then reacted to form forsterite and enstatite in the early synthesis. In the middle period, MgO diffused through crystal lattice of enstatite and low melting point phase to the particle surface of enstatite and then reacted with it to form forsterite, and which separated out in the particle surface of the early formed forsterite and made the join of forsterite particles and densification of forsterite materials. In the late, all MgO diffused from the in-situ particle and left pores, the pores would be decreased by sintering process. (3) The synthesis process be controlled by Mg^{2+} diffusion, the reaction kinetic between MgO and SiO_2 could be described by Yang Deer equation and the apparent activation energy was (220 ± 20) kJ/mol. (4) The preparation condition of densification forsterite was as that heating temperature was high than 1550℃, holding time was long more than 3 hours and forming pressure was big more than 150MPa. (5) The synthesis and sintering process of forsterite could be efficiently improved by TiO_2, Y_2O_3 and CeO_2 additive, however the effect of ZrO_2 and Al_2O_3 additive was not remarkable. It is propitious to diffuse that solid solution of Ti^{4+} into MgO lattice, formation of Ti containing compound with low melting point and separation of TiO_2 from Ti containing compounds. The formation of Y containing compound with low melting point and the existence of $MgSiO_3$ phase caused by Y_2O_3 additive improved the diffusion of reactant. The added CeO_2 changed to Ce_2O_3 at high temperature, and which reacted with SiO_2 to form Ce containing compound with low melting point. The compound with low melting point and $MgSiO_3$ phase increased the diffusion of reactant. ZrO_2 existed in MgO containing solid solution in grain boundaries and in composite compounds in matrix, and the composite compounds existing in matrix changed as $MgSiO_3+ZrSiO_4 \rightarrow Mg_2SiO_5+MgSiO_3+ZrSiO_4 \rightarrow Mg_2SiO_5+ZrSiO_4+ZrO_2$ and then to separated out ZrO_2 in grain boundaries. (6) The bulk density and strength of tundish dry materials decreased with the increase of forsterite addition, however no change in the slag resistance, the properties of tundish dry materials added with sintered forsterite and raw forsterite showed a little change. The sintering and slag resistance of tundish dry materials could be improved by borax, ferric oxide, sodium tripolyphosphate and sodium silicate additive, however, it will cause the performance reduction of tundish dry materials if the additive amount is high much than an appropriate amount[37].

参考文献

[1] 马北越, 慕鑫, 高陟, 等. 低碳镁碳耐火材料抗热震性研究进展 [J]. 耐火材料, 2021, 55 (2): 174-177, 181.

[2] Liu Zhaoyang, Yu Jingkun, Yue Shuaijun, et al. Effect of carbon content on the oxidation resistance and kinetics of MgO-C refractory with the addition of Al powder [J]. Ceramics International, 2020, 46 (3): 3091-3098.

[3] Gardeh M, Kistanov A A, Nguyen H, et al. Exploring mechanisms of hydration and carbonation of MgO and Mg(OH)$_2$ in reactive magnesium oxide-based cements [J]. The Journal of Physical Chemistry C, 2022, 126 (14): 6196-6206.

[4] Kuenzel C, Zhang F, Ferrándiz-Mas V, et al. The mechanism of hydration of MgO-hydromagnesite blends [J]. Cement and Concrete Research, 2018, 103: 123-129.

[5] 马北越, 张誉忠, 高陟, 等. 致密与多孔氧化镁材料研究进展 [J]. 耐火材料, 2021, 55 (2): 169-173.

[6] 高陟, 马北越, 任鑫明, 等. 利用菱镁矿制备先进镁质材料的研究进展 [J]. 耐火材料, 2020, 54 (1): 88-92.

［7］ Liu Zhaoyang, Yu Jingkun, Wang Xiangnan, et al. Activated treatment of magnesite for preparing size-controlled $Mg_2B_2O_5$ whiskers and their microwave dielectric properties［J］. Ceramics International, 2021, 47（21）: 30471-30482.

［8］ Jin Endong, Yu Jingkun, Wen Tianpeng, et al. Effect of cerium oxide on preparation of high-density sintered magnesia from crystal magnesite［J］. Journal of Materials Research and Technology, 2020, 9（5）: 9824-9830.

［9］ Jin Endong, Yu Jingkun, Wen Tianpeng, et al. Effects of the molding method and blank size of green body on the sintering densification of magnesia［J］. Materials, 2019, 12（4）: 647.

［10］ Jin Endong, Yu Jingkun, Wen Tianpeng, et al. Fabrication of high-density magnesia using vacuum compaction molding［J］. Ceramics International, 2018, 44（6）: 6390-6394.

［11］ Guo Zongqi, Ding Qiang, Liu Lei, et al. Microstructural characteristics of refractory magnesia produced from macrocrystalline magnesite in China［J］. Ceramics International, 2021, 47（16）: 22701-22708.

［12］ Hou Qingdong, Luo Xudong, Xie Zhipeng, et al. Preparation and characterization of microporous magnesia based refractory［J］. International Journal of Applied Ceramic Technology, 2020, 17（6）: 2629-2637.

［13］ 马鹏程. 高活性氧化镁和高密度烧结镁砂的研究［D］. 沈阳: 东北大学, 2014.

［14］ 朱张磊, 印万忠, 李振, 等. 表面粗糙度对水滴在方解石表面黏附的影响［J］. 矿产保护与利用, 2022, 42（1）: 8-14.

［15］ Liu Tao, Ma Pengcheng, Yu Jingkun, et al. Preparation of MgO by thermal decomposition of $Mg(OH)_2$［J］. Journal of the Chinese Ceramic Society, 2010, 38（7）: 1337-1340.

［16］ 徐平坤. 回转窑用耐火材料的技术进步［J］. 工业炉, 2018, 40（3）: 1-6.

［17］ 陶林. 方解石表面亚稳态的电化学性质和热演变机制［D］. 鞍山: 辽宁科技大学, 2021.

［18］ Cui Yan, Qu Dianli, Luo Xudong, et al. Effect of La_2O_3 addition on the microstructural evolution and thermomechanical property of sintered low-grade magnesite［J］. Ceramics International, 2021, 47（3）: 3136-3141.

［19］ Tong Shanghao, Zhao Jizeng, Zhang Yucui, et al. Corrosion mechanism of Al-MgO-$MgAl_2O_4$ refractories in RH refining furnace during production of rail steel［J］. Ceramics International, 2020, 46（8）: 10089-10095.

［20］ Kerber F, Zienert T, Hubálková J, et al. Effect of MgO grade in MgO-C refractories on the non-metallic inclusion population in Al-treated steel［J］. Steel Research International, 2022, 93（6）: 2100482.

［21］ Sadik C, Moudden O, El Bouari A, et al. Review on the elaboration and characterization of ceramics refractories based on magnesite and dolomite［J］. Journal of Asian Ceramic Societies, 2018, 4（3）: 219-233.

---

［22］ Díaz-Tato L, López-Perales J F, Contreras J, et al. Hydration resistance and mechano-physical properties improvement of a magnesia-dolomite dense refractory by hercynite spinel ［J］. Materials Chemistry and Physics, 2022, 287: 126314.

［23］ Kumar P, Nath M, Roy U, et al. Improvement in thermomechanical properties of off-grade natural magnesite by addition of Y_2O_3 ［J］. International Journal of Applied Ceramic Technology, 2017, 14 (6): 1197-1205.

［24］ Liu Jing, Dong Yingchao, Dong Xinfa, et al. Feasible recycling of industrial waste coal fly ash for preparation of anorthite-cordierite based porous ceramic membrane supports with addition of dolomite ［J］. Journal of the European Ceramic Society, 2016, 36 (4): 1059-1071.

［25］ Booth F, Garrido L, Aglietti E, et al. $CaZrO_3$-MgO structural ceramics obtained by reaction sintering of dolomite-zirconia mixtures ［J］. Journal of the European Ceramic Society, 2016, 36 (10): 2611-2626.

［26］ 孟维. 基于中低品位原料的稳定性镁白云石材料制备与性能研究 ［D］. 郑州: 郑州大学, 2016.

［27］ 祁欣, 罗旭东, 李振, 等. 高硅菱镁矿的选矿提纯与应用研究进展 ［J］. 硅酸盐通报, 2021, 40 (2): 485-492.

［28］ Ren Xinming, Ma Beiyue, Fu Gaofeng, et al. Facile synthesis of MgO-Mg_2SiO_4 composite ceramics with high strength and low thermal conductivity ［J］. Ceramics International, 2021, 47 (14): 19959-19969.

［29］ Tian Li, Wang Lijuan, Wang Kailei, et al. The preparation and properties of porous sepiolite ceramics ［J］. Scientific Reports, 2019, 9: 7337.

［30］ Ge Xuexiang, Zhou Mingkai, Wang Huaide, et al. Preparation and characterization of ceramic foams from chromium slag and coal bottom ash ［J］. Ceramics International, 2018, 44 (10): 11888-11891.

［31］ Kullatham S, Thiansem S. Synthesis, characterization and properties of forsterite refractory produced from Thai talc and magnesite ［J］. Materials Science Forum, 2018, 940: 46-50.

［32］ Nemat S, Ramezani A, Emami S M. Recycling of waste serpentine for the production of forsterite refractories: The effects of various parameters on the sintering behavior ［J］. Journal of the Australian Ceramic Society, 2018, 55 (2): 425-431.

［33］ Nguyen M, Sokolář R. Corrosion resistance of novel fly ash-based forsterite-spinel refractory ceramics ［J］. Materials, 2022, 15 (4): 1363.

［34］ Zhu Tanglong, Zhu Min, Zhu Yufang. Fabrication of forsterite scaffolds with photothermal-induced antibacterial activity by 3D printing and polymer-derived ceramics strategy ［J］. Ceramics International, 2020, 46 (9): 13607-13614.

［35］ Hossain S, Mathur L, Singh P, et al. Preparation of forsterite refractory using highly abundant amorphous rice husk silica for thermal insulation ［J］. Journal of Asian Ceramic Societies,

2018，5（2）：82-87.

［36］Yuan Lei，Zhang Xiaodong，Zhu Qiang，et al. Preparation and characterisation of closed-pore Al$_2$O$_3$-MgAl$_2$O$_4$ refractory aggregate utilising superplasticity ［J］．Advances in Applied Ceramics，2017，117（3）：182-188.

［37］陈勇．镁橄榄石合成及应用研究［D］．沈阳：东北大学，2014.

5 尖晶石族耐火材料原料

5.1 镁铝质耐火材料原料

5.1.1 术语词

烧结尖晶石 sintered spinel
菱镁矿 magnesite, giobertite
镁砂 magnesia
含铝的（含铝土的）aluminous

铝矾土 bauxite
人造刚玉 boule
刚玉 corundum
电熔刚玉 electro-corundum

5.1.2 术语词组

活性尖晶石粉末 active spinel powder
铝镁尖晶石 alumina magnesia spinel
富铝尖晶石 alumina-rich spinel
刚玉–尖晶石复合浆料 alumina-spinel slurry
铝尖晶石 aluminous spinel
高铝矾土基尖晶石 bauxite-based spinel
电熔镁尖晶石 electrocast magnesia spinel, fused
　magnesia-alumina spinel
熔融尖晶石 fused spinel
高纯电熔尖晶石 high purity fused spinel
高铝尖晶石 high-alumina spinel
富镁尖晶石 magnesia enriched spinel
镁尖晶石合成砂 magnesia spinel clinker
镁铝尖晶石 magnesite-alumina spinel, magnesia-

　alumina spinel
尖晶石的预合成 preliminary synthesis of spinel
预合成尖晶石 pre-synthesized spinel
硅酸盐尖晶石结合剂 silicate spinel bond
尖晶石结合剂 spinel binder
尖晶石铁氧体 spinel ferrite
尖晶石粉料 spinel mass
尖晶石矿物 spinel mineral
尖晶石原料 spinel raw material
烧结尖晶石 sintered spinel
合成尖晶石 synthesized spinel
盐湖镁砂（卤水镁砂）brine magnesia
烧成镁砂 burned magnesite
轻烧镁砂（苛性镁砂）light burned magnesia,

caustic calcined magnesia

重烧镁砂（死烧镁砂）dead burned magnesia

致密烧结镁砂 dense sintered magnesia

冶金镁砂（补炉镁砂）fettling magnesia

细晶镁砂 fine-grained magnesia

细分散镁砂（高分散镁砂）finely dispersed magnesia

电熔镁砂 fused magnesia

电熔镁砂制品 fused magnesia product, electrocast magnesia product

高纯烧结镁砂 high-purity dead-burned magnesia

电熔镁砂高纯晶体 high-purity crystals of fused magnesia

高纯度高密度镁砂 high-purity and high-dense magnesite

高纯镁砂 high-purity magnesia

高钙镁砂 high-calcium magnesia

大结晶镁砂 large crystal magnesia

大结晶电熔镁砂 large crystallined fused magnesia

低硅高钙电熔镁砂 low-silica and high-calcia fused magnesia

低铁镁砂 low-iron magnesia

低温煅烧镁砂 low-temperature sintered magnesia

冶金镁砂 metallurgical magnesia

中档镁砂 medium grade sintered magnesia

非晶质菱镁矿 non-crystalline magnesite

菱镁矿原矿 raw magnesite

海水镁砂 sea-water magnesia

烧结镁砂 sintered magnesia

黑褐色的刚玉 adamantine spar

人造刚玉（人造金刚砂）artificial corundum

黑刚玉 black fused alumina

棕刚玉 brown adamantine spar

轻烧刚玉 caustic-burned corundum

铬刚玉 chrome corundum

刚玉熟料 corundum clinker

刚玉球 corundum granule

刚玉喷涂料 corundum gunning mixes

刚玉空心球 corundum hollow granule

刚玉微粉 corundum micro-powder

刚玉粉 corundum powder

致密刚玉 dense corundum

致密电熔刚玉 dense fused alumina

致密白刚玉 dense white fused alumina

电熔棕刚玉 electrically fused brown corundum

电熔铬刚玉 electrocast chrome-corundum

电熔锆刚玉 electrocast zirconia corundum

游离刚玉 free corundum

电熔刚玉 fused alumina, fused corundum, electromelting corundum

电熔致密刚玉 fused dense corundum

低碳高铝棕刚玉 low carbon and high alumina brown fused alumina

低碳亚白刚玉 low carbon vice-white corundum

粗晶刚玉 macrocrystalline corundum

微晶刚玉 microcrystalline alumina, microcrystal fused alumina

改性烧结刚玉 modified sintered corundum

普通电熔刚玉 normal electrocorundum

倾倒炉电熔低碳高铝刚玉 pouring furnace fused low-carbon high alumina corundum

赛隆结合刚玉 Sialon bonded corundum

烧结刚玉 sintered corundum

电熔亚白刚玉 sub-white electrically fused corundum

亚白刚玉 sub-white fused alumina, vice-white corundum, vice-white fused alumina

人造刚玉（合成刚玉）synthetic corundum

板状刚玉 tabular alumina, tabular corundum

5.1.3 术语例句

涂覆尖晶石色料及其生产工艺与使用。

Coated spinel color pigments process for their production and use.

锌铬铁矿棕色尖晶石色料及其制备方法和应用。

Brown spinel pigments based on zinc chromite method of their production and use.

合成**铝镁尖晶石**的两个主要方法是烧结法和电熔法。

Synthesis of **magnesium aluminium spinel** and fused two main methods are sintering method.

所有这些都影响**尖晶石熟料**和浇注料耐火材料的抗渣性。

All of them influence the slag resistances of **spinel clinker** and castable refractory.

加入**镁铝尖晶石**的镁质浇注料在强度方面要好于加入铬矿砂的。

The strength of castable containing **MA spinel** was better than that of the one containing chrome ore.

铝镁尖晶石广泛用于耐火材料、钢铁冶炼、水泥回转窑及玻璃工业窑炉上。

Magnesium aluminum spinel is widely used in refractory material, steel smelting, cement rotary kiln and glass kiln.

在所有可能的尖晶石矿族成员中，**铁铝尖晶石**和**锰铝尖晶石**被发现是最有效的。

From all possible members of the spinel groups, **hercynite** and **galaxite**, were found to be the most efficient.

研究表明，用白云石砖、**尖晶石砖**、镁锆钙砖代替镁铬砖，符合碱性砖无铬化的发展趋势。

These chrome free bricks include dolomite bricks, **magnesia spinel** bricks and magnesia calcium zirconate bricks.

尖晶石锂锰化合物成本低、性能好、无污染，是适合水溶液锂离子电池的电极材料。

Spinel lithium manganese complex is preferable because of low cost, good performance and environmental friendliness.

展望了**反尖晶石**离子掺杂在锂离子电池锰酸锂正极材料中的发展前景。

The development foreground of **inverse spinel** ions doping in positive electrode materials of lithium manganese oxide was viewed.

辉石岩中**尖晶石**在薄片下呈绿色，为铝含量高、铬含量低的铝尖晶石。

The spinels in pyroxenite xenoliths are aluminum **spinel** with higher Al content and lower Cr content and green in color in thin sections.

红宝石是对红色矿物**刚玉**的俗称，它是一种珍贵的宝石。

Ruby, the common name for the mineral **corundum** in its red form is a precious gemstone.

由铝和氧两种元素构成的**刚玉**，会呈现出黄色、灰色或者棕色。

Made up of the elements aluminum and oxygen, **corundum** also can be yellow, gray, or brown.

用于高炉、热风炉砌筑**刚玉**砖、莫来石砖、硅线石砖、低蠕变高铝砖。

Used for **corundum** bricks, mullite bricks, silicate bricks masonry in blast furnaces, hot air stoves.

铝是一种从**铝土矿**提炼出的金属。

Aluminium is a metal which is produced from **bauxite**.

铀矿、**铝矾土**、锰矿储量居世界第三位。

Uranium ore, **bauxite**, manganese ore reserves, ranks third in the world.

中国拥有丰富的铁矿石、**铝土矿**、锌、镍、煤炭和原油储量。

China is rich in iron ore, **bauxite**, zinc, nickel, coal and crude oil deposits.

中国境内矿产资源丰富，主要有煤、铁、**铝矾土**、石灰石、大理石、黏土等。

The territory rich in mineral resources, mainly coal, iron, **bauxite**, limestone, marble, clay and so on.

5.1.4 术语例段

以粉煤灰和铝灰为原料，通过原位铝热还原氮化制备了**镁铝尖晶石-刚玉-**

赛隆复合材料。采用 XRD 和 SEM-EDS 研究了复合材料的物相组成、含量和显微结构，并表征其性能。结果表明，当铝灰过量 50% 时，在 1550℃、3h 下成功制备得到了**镁铝尖晶石-刚玉-赛隆复合耐高温材料**，其中，尖晶石为**富铝尖晶石**，含量为 45%，呈八面体结构；**刚玉**含量为 25%，呈板片状；赛隆相含量为 26%，其 Z 值为 4，呈柱状形貌。此时其抗折强度最大，为 183MPa，显气孔率最小，为 5.3%，体积密度为 $2.6g/cm^3$，洛氏硬度 HRB 的值为 123[1]。

The **spinel-corundum**-Sialon composites were prepared from fly ash and aluminum dross by in-situ aluminothermic reduction-nitridation. The phase composition, content and microstructure of the composites were investigated by XRD and SEM with EDS. The properties of the composites were also characterized. The results showed that when aluminum dross exceeded 50%, the **spinel-corundum**-Sialon composites were prepared successfully at 1550℃ for 3h, in which the content of phases were 45% of **aluminum-rich spinel**, 25% of **corundum**, and 26% of Sialon with $Z = 4$. The patterns of phases were octahedral structure of **aluminum-rich spinel**, platy corundum, and columnar crystals of Sialon. The bending strength of the prepared spinel-corundum-Sialon composites reached the maximum of 183MPa, the apparent porosity was the minimum of 5.3%, the bulk density was $2.6g/cm^3$, and HRB hardness was 123[1].

以**板状氧化铝**、**活性氧化铝粉**、锆莫来石、锆刚玉、鳞片石墨为原料，纳米 ZrO_2、纳米 TiO_2 和纳米 Al_2O_3 粉末为添加剂，在石墨包埋条件下，在 1500℃ 条件下制备了纳米氧化物低碳 Al_2O_3-ZrO_2-C 耐火材料。研究了纳米 ZrO_2、纳米 TiO_2 和纳米 Al_2O_3 粉末作为单一添加剂对 Al_2O_3-ZrO_2-C 材料性能和微观结构的影响。结果表明，纳米氧化物在碳化硅纳米线（NWs）的形成和生长中起催化作用，导致碳化硅纳米线的长径比（L/D）增加。添加纳米氧化物的试样具有更好的烧结性、抗热震性和抗氧化性。

Low carbon Al_2O_3-ZrO_2-C refractories with nano-oxides are prepared at 1500℃ in graphite embedded condition using **tabular alumina**, **reactive alumina powder**, zirconia-mullite, zirconia-corundum and flake graphite as well as nano-ZrO_2, nano-TiO_2 and nano-Al_2O_3 powders as additives. The effects of nano-ZrO_2, nano-TiO_2 and nano-Al_2O_3 powders as single additive on the properties and microstructure of Al_2O_3-ZrO_2-C material are investigated. The results show that nano-oxides take the catalytic effect in SiC nanowires(NWs)formation and growth, leading to increasing in length-diameter ratio (L/D) of SiC NWs. Samples with nano-oxides addition possess improved sinterability, thermal shock resistance and oxidation resistance[2].

氧化铝-尖晶石耐火浇注料的工业应用对使用性能有着至关重要的要求。因此，研究了不同尺寸的脱硅氧化锆颗粒对浇注料微观结构、热力学性能和高温弹性模量的影响。氧化锆颗粒尺寸（d_{50}）从 1000μm 到 2.5μm 不等。观察到更细的（小于 88μm）氧化锆颗粒有利于提高常温抗折强度（CMOR）和高温抗折强度（HMOR），但不能有效提高抗热震性。细小的氧化锆颗粒可以均匀地分散在基体中并显著促进烧结过程。伴随着氧化锆的相变，在耐火浇注料中发现了高密度的基体裂纹和强的陶瓷结合（粗晶粒和基体之间），这是导致 CMOR 增加的原因。然而，结合特性也会产生高储存弹性能量，不利于抗热震性，而过多的预先存在的基体裂纹会在热震过程中引起更多的微损伤。此外，有人提出，当引入细小的 ZrO_2 颗粒时，第二相分散增强和高度陶瓷结合导致了优异的 HMOR。结合特性还可能导致高储存弹性能量，这不利于抗热震性，而过多的预先存在的基体裂纹会在热震过程中引起更多的微损伤。

The industrial application of **alumina-spinel** refractory castables has crucial requirements on the service performance. Thus, the effects of different sized desilicated zirconia particles on the castables microstructure, thermal-mechanical properties and high temperature elastic modulus have been investigated. The zirconia particle sizes (d_{50}) were varied from 1000μm to 2.5μm. It was observed that the finer (below 88μm) zirconia particles were beneficial to improve the cold modulus of rupture (CMOR) and the hot modulus of rupture (HMOR) but could not effectively enhance the thermal shock resistance. Fine zirconia particles can homogeneously disperse in the matrix and significantly promote the sintering process. Accompanied with the phase transformation of zirconia, both the high density of matrix cracks and the strong ceramic bonding (between the coarse grains and the matrix) were found in the refractory castables, which was responsible for an increase of CMOR. However, the binding characteristic could also give rise to the high stored elastic energy that was averse to the thermal shock resistance, and the excessive amount of preexisting matrix cracks could induce more microdamage during the thermal shock. Additionally, it was proposed that the second-phase dispersion reinforcement and the highly ceramics bonding resulted in the superior HMOR when introducing fine ZrO_2 particles[3].

研究了 Sm_2O_3 作为添加剂对预合成的**富镁尖晶石**致密化的影响。通过分析其膨胀特性、线收缩率、体积密度、显气孔率、相结构和微观结构来表征材料的烧结行为。此外，还测量了尖晶石的常温抗折强度（CMOR）和显微硬度。研究结果表明，氧化钐与尖晶石反应生成 $SmAlO_3$（一种稀土铝酸盐），提高了尖晶石的烧结性能和力学性能。菱镁矿相的形成降低了尖晶石的断裂强度。但这种 MgO

相镶嵌在尖晶石晶界中，具有钉扎作用，可以提高尖晶石的烧结性。

The effects of Sm_2O_3 as an additive on the densification of a pre-synthesized **magnesia-rich spinel** were investigated. The sintering behavior of the material was characterized by analyzing its dilatometric characteristics, linear shrinkage, bulk density, apparent porosity, phase structure, and microstructure. Moreover, the cold modulus of rupture(CMOR) and microhardness of the spinel were measured. The results show that samarium oxide reacts with the spinel to form $SmAlO_3$ (a rare-earth aluminate) to improve the sintering performance and mechanical properties of the spinel. The formation of magnesite phases decreases the rupture strength of the spinel. However, this MgO phase is inlaid in spinel grains boundary, has a pinning effect which can improve the sinterability of the spinel[4].

本文提出了提高水泥窑用耐火材料的机械强度和水泥熟料耐腐蚀性的新策略。为此，使用微孔氧化镁骨料制备了具有高力学性能和优异的水泥熟料抗性的先进轻质**方镁石铝酸镁尖晶石**耐火材料（LPSR）。然后将它们的微观结构和性能与含有烧结氧化镁骨料的传统致密方镁石－铝酸镁尖晶石耐火材料（DPSR）进行比较。微孔和致密氧化镁骨料的显气孔率分别为 43.1% 和 5%。扫描电镜观察结果表明，微孔聚集体与基体的界面较好，显著提高了 LPSR 的强度和抗热震性。此外，微孔氧化镁骨料从基质中吸收了一些渗透的炉渣，从而阻止了进一步的渗透。因此，在用微孔氧化镁骨料代替致密氧化镁骨料制造 LPSR 后，与 DPSR 相比，LPSR 获得了几乎相同的水泥熟料阻力。同时，体积密度从 2.87g/cm³ 降低到 2.61g/cm³，而抗压强度和抗弯强度分别从 48.7MPa 提高到 73.1MPa 和 6.0MPa 提高到 10.1MPa。同时，与 DPSR 相比，LPSR 获得了几乎相同的水泥熟料阻力。

A new strategy to improve the mechanical strength and the cement clinker corrosion resistance of refractories for cement kiln was proposed in this work. For this purpose, advanced lightweight **periclase magnesium aluminate spinel** refractories(LPSR) with high mechanical properties and excellent cement clinker resistance were fabricated using microporous magnesia aggregates. Their microstructures and properties were then compared to conventional dense periclase-magnesium aluminate spinel refractories (DPSR) containing sintered magnesia aggregates. The apparent porosities of the microporous and dense magnesia aggregates were 43.1% and 5%, respectively. Scanning electron microscopy observation results showed that the interface between the microporous aggregates and the matrix was better, which significantly improved the strength and thermal shock resistance of the LPSR. Additionally, the microporous magnesia aggregates absorbed some of the penetrating slag from the matrix, which prevented further infiltration. Thus, after substituting the microporous magnesia aggregates for dense magnesia aggregates to fabricate LPSR, the practically same cement clinker resistance was obtained for the LPSR compared to DPSR. Meanwhile, the bulk density was decreased from 2.87g/cm^3 to 2.61g/cm^3, while the compressive strength and flexural strength increased from 48.7MPa to 73.1MPa and 6.0MPa to 10.1MPa, respectively. Meanwhile, practically the same cement clinker resistance was obtained for the LPSR compared to DPSR[5].

铝酸镁尖晶石（MAS）是一种重要的耐火材料，具有高耐火度、高抗热震性、高温力学和热性能、高抗化学侵蚀性和优异的抗水化性等优点。铬铁矿基耐火材料用于水泥回转窑和钢包，因为铬铁矿基耐火材料会带来环境问题。MAS薄膜在燃气轮机热段部件的叶片和轮叶的热障涂层（TBCs）中具有潜在用途。稀土氧化物掺杂剂在 MAS 的形成和烧结中起重要作用。研究表明，可以在稀土掺杂剂的帮助下定制 MAS 的微观结构，从而显著改善其烧结行为和力学性能。本文尝试介绍稀土氧化物掺杂 MAS 的烧结方面以及机械和热机械行为。

Magnesium aluminate spinel(MAS) is an important refractory material known for its high refractoriness, high thermal shock resistance, high-temperature mechanical and thermal properties, high resistance to chemical attack, superior hydration resistance, etc. MAS has been employed to replace traditional chromite-based refractories in cement rotary kilns and steel ladles because chromite-based refractories pose environmental issues. Thin films of MAS have potential use in thermal barrier coatings(TBCs) for the blades and vanes in the hot section components of gas turbines. Rare-earth oxide dopants play a significant role in the formation and sintering of MAS. Research has shown that the microstructure of MAS can be tailored with the help of rare-earth dopants, leading to a significant improvement in their sintering behaviour and mechanical properties. This review paper attempts to present the sintering aspects, and the mechanical and thermo-mechanical behaviour of rare-earth oxide-doped MAS[6].

在这项工作中，研究了 1%、2% 和 3%（质量分数）的 CeO$_2$ 作为添加剂对**富铝尖晶石和富镁尖晶石**粉末在 1600℃、1650℃、1700℃ 和 1750℃ 温度下烧结行为的影响。根据膨胀测量、线性收缩、体积密度、相分析和微观结构研究了所制陶瓷的烧结行为。研究结果表明，CeO$_2$ 通过与从尖晶石中脱溶出 Al$_2$O$_3$ 反应形成散布在尖晶石晶粒之间的片状 CeAl$_{11}$O$_{18}$ 颗粒，从而阻碍了富铝尖晶石的烧结进程。与此同时，CeO$_2$ 的存在促进了富镁尖晶石的烧结过程，它以孤立的形式分布在尖晶石晶粒中。

In this work, the effects of 1wt.%, 2wt.%, and 3wt.% CeO_2 as an additive on the sintering behavior of **alumina-rich spinel** and **magnesia-rich spinel** powders subjected to sintering at temperatures of 1600℃, 1650℃, 1700℃, and 1750℃ were investigated. The sintering behavior of the ceramics was investigated according to dilatometry measurements, linear shrinkage, bulk density, phase analysis, and microstructure. It was demonstrated that CeO_2 hindered the sintering process in alumina-rich spinel by reacting with Al_2O_3 exsolved from the spinel to form platelet-shaped particles of $CeAl_{11}O_{18}$ interspersed between the spinel grains. Meanwhile, the presence of CeO_2 promotes the sintering process in magnesia-rich spinel by being distributed in an isolated form among the spinel grains[7].

通过一步固态反应烧结（SRS）工艺共掺杂 Sm_2O_3 和 La_2O_3 成功制备了棒状微晶增强 **$MgAl_2O_4$**（MA）基陶瓷，该陶瓷具有高强度和良好的抗热震性。研究了 Sm_2O_3 和 La_2O_3 的添加量对 MA 基陶瓷的相组成、显微组织、收缩率、显气孔率、体积密度、高温抗压强度和抗热震性的影响。结果表明，共掺杂添加剂后，陶瓷中可发现 MA、$SmAlO_3$、$La_{10}Al_4O_{21}$、$Sm_{4.67}(SiO_4)_3O$ 和 $La_{4.67}(SiO_4)_3O$ 相。新形成的稀土化合物可以阻止 MA 晶粒的生长，导致 MA 基陶瓷的微结构致密化。$Sm_{4.67}(SiO_4)_3O$ 和 $La_{4.67}(SiO_4)_3O$ 是由添加剂与原料中的杂质反应生成的，这种呈棒状的微晶能有效清除 MA 基陶瓷晶界中的杂质。因此，MA 基陶瓷的烧结性能、高温抗压强度和抗热震性显著提高。

A rod-like microcrystalline reinforced **$MgAl_2O_4$**（MA）based ceramic with high strength and good thermal shock resistance has been successfully prepared by co-doping Sm_2O_3 and La_2O_3 via a single-stage solid-state reaction sintering (SRS) process. The effects of addition amounts of Sm_2O_3 and La_2O_3 on the phase compositions, microstructures, shrinkage ratio, apparent porosity, bulk density, high temperature compressive strength and thermal shock resistance of the MA based ceramics have been investigated. The results showed that MA, $SmAlO_3$, $La_{10}Al_4O_{21}$, $Sm_{4.67}(SiO_4)_3O$ and $La_{4.67}(SiO_4)_3O$ phases could be found in the ceramics after co-doping additives. The new formed rare earth compounds could prevent the growth of MA grains leading to a densification microstructure of the MA based ceramics. The $Sm_{4.67}(SiO_4)_3O$ and $La_{4.67}(SiO_4)_3O$ were formed by the reaction between additives and impurities of raw materials, which presented as rod-like microcrystallines to effectively clean the impurities in the grain boundaries of the MA based ceramics. Accordingly, the sintering properties, high temperature compressive strength and thermal shock resistance of the MA based ceramics were improved markedly[8].

在 1923K 和 1873K 下，采用高速相机（1000 帧/s）座滴法，在**镁铝尖晶石**上使用非饱和熔渣和饱和熔渣，研究了熔渣和耐火材料之间的润湿、铺展和渗透现象。采用单晶尖晶石作为基底来确定与溶解反应等现象相关的内在值。采用富镁尖晶石、化学计量尖晶石和富铝尖晶石等具有 35% 显气孔率的工业尖晶石基体，研究耐火材料的化学成分和孔隙率对这些现象的影响。在化学计量尖晶石的情况下，与单晶尖晶石相比，表观接触角和液滴高度的值较低。当向尖晶石中添加 MgO 和 Al_2O_3 时，并且富镁尖晶石似乎具有更大的熔渣渗透性。另外，富铝尖晶石表现出更出色的抗渣性。采用球冠模型分析了渣与基体接触的表观体积变化。

The wetting, spreading and penetration phenomena between slags and refractories were investigated by using a dispensed drop technique with a high-speed camera(1000 frame/s) at 1923K and 1873K and using non-saturated slag and saturated slag on **$MgAl_2O_4$ spinel**. Single crystal spinel was adopted as a substrate to determine the intrinsic values associated with the phenomena including dissolution reaction. Industrial spinel substrates with 35% apparent porosity such as MgO-rich spinel, stoichiometric spinel and Al_2O_3-rich spinel were used to study the influence of chemical composition and porosity of refractories on those phenomena. In case of the stoichiometric spinel, the values of apparent contact angle and droplet height were found lower in comparison with single crystal spinel. When MgO and Al_2O_3 content were added to the spinel, the MgO-rich spinel appeared to have larger permeability of slag. In addition, the Al_2O_3-rich spinel showed larger resistance to slags. The change of apparent volume of slag in contact with substrate was analyzed using a spherical cap model[9].

微米级 $MgCO_3$ 被用作浇注料中的 MgO 前驱体，用于生成亚微级 MgO 并用于生产随后的**原位铝酸镁（$MgAl_2O_4$，MA）尖晶石**。研究了添加 0~2.0%（质量分数）的微量 $MgCO_3$ 对 1000℃和 1550℃烧制后浇注料的体积稳定性和热力学性能的影响。分别通过 X 射线衍射（XRD）和扫描电子显微镜（SEM）表征了热处理后不同微米尺寸 $MgCO_3$ 含量的浇注料基质的原位尖晶石形成及其对微观结构演变的影响。同时讨论了浇注料的体积稳定性和热力学性能对微尺寸 $MgCO_3$ 添加量的依赖性。

Micro-sized $MgCO_3$ was used in castables as the MgO-precursor for generating sub-micro sized MgO and for producing subsequent **in-situ magnesium aluminate (MgAl$_2$O$_4$, MA) spinel**. The influence of 0-2.0wt.% micro-sized $MgCO_3$ addition on the volumetric stability and thermo-mechanical properties of castables after firing at 1000℃ and 1550℃ was investigated. The in-situ spinel formation and its influence on the microstructure evolution in castable matrices with different micro-sized $MgCO_3$ contents after heat-treatment were characterized by X-ray diffraction(XRD) and scanning electron microscopy(SEM), respectively. The dependence of the volume stability and thermo-mechanical properties of castables on the micro-sized $MgCO_3$ addition was discussed with respect to the in-situ spinel formation[10].

镁铝尖晶石（化学式为 $MgAl_2O_4$ 或 $MgO \cdot Al_2O_3$，简写为 MA）具有优异的高温力学性能，优异的耐剥离性和耐腐蚀性，是 Al_2O_3-MgO 体系中最为典型的耐高温陶瓷材料。而六铝酸钙（化学式为 $CaAl_{12}O_{19}$ 或 $CaO \cdot 6Al_2O_3$，简写为 CA_6）晶粒的沿基面优先生长的特性，使得其生长为片状或针状形貌，可以增强材料的韧性。二铝酸钙（$CaAl_4O_7$ 或 $CaO \cdot 2Al_2O_3$，简写为 CA_2）具有较低的热膨胀系数，与其他高熔点、高膨胀系数材料复合时能很好地抵抗热冲击导致的损毁。因此，MA-CA_6 复合材料因其综合两者的性能，在高温行业中作为新型的耐高温陶瓷材料受到大家广泛的关注。本论文采用高温固相烧结法制备了 MA 陶瓷材料、MA-CA_2-CA_6 陶瓷复合材料和 MA-CA_6 陶瓷复合材料，研究了矿化剂对上述陶瓷材料性能的影响规律，并探讨了矿化剂对陶瓷材料性能的强化机理，得到了以下的研究成果：(1) 研究发现随着烧结温度的升高，MA 陶瓷材料的体积密度和抗折强度逐渐增大。经 1600℃保温 2h 烧结后，MA 材料的烧结性能较差，其体积密度为 $3.17g/cm^3$，抗折强度值为 133.31MPa；随着矿化剂 Fe_2O_3 添加量的增加，MA 陶瓷材料的体积密度逐渐增大，抗折强度则出现先增大后减小的变化，当添加量（质量分数）为 3% 时，抗折强度达到最大值 209.3MPa。(2) 研究表明 MA-CA_6 陶瓷材料的性能和相组成与 $CaCO_3$ 原料的粒度、α-Al_2O_3 的纯度、α-Al_2O_3 原料的粒度、合成的温度和保温时间有关。使用粒径小的 $CaCO_3$ 和纯度大的 α-Al_2O_3 为原料，经 1600℃烧结，保温 2h，合成的 MA-CA_6 陶瓷材料有最大的抗折强度。$CaCO_3$ 的粒度对 MA-CA_6 陶瓷材料中 CA_6 相的形成和晶粒的生长发育起着重要的作用。α-Al_2O_3 中杂质 Si 在高温下将形成瞬态液相，使得 CA_6 晶粒形貌由薄片状向等轴状演变。(3) 研究了矿化剂 ZnO 和 $Mg(BO_2)_2$ 对 MA-CA_6 陶瓷材料性能的影响规律和强化机理。发现矿化剂 ZnO 和 $Mg(BO_2)_2$ 所形成的 $(Mg_{1-x}Zn_x)Al_2O_4$ 固溶体和含硼的液相使得 MA 的晶粒变小和 MA 的含量增多。这些致密相包覆着微晶 MA 颗粒形成分散区域的致密体，导致了 CA_6 晶粒向等轴状晶粒转变，从而促进了 MA-CA_6 陶瓷材料的致密化和提高了其抗折强度。(4) 通过用分析纯的 Al_2O_3 代替 α-Al_2O_3，完全使用分析纯原料合成了 MA-CA_2-CA_6 陶瓷材料。研究了矿化剂 SnO_2 和 H_3BO_3 对材料的物理和力学性能，微观结构和相组成的影响。结果表明，添加矿化剂 SnO_2 和 H_3BO_3 后陶瓷材料中分别出现了固溶体和含硼的瞬态液相，使得 CA_2 相向 CA_6 相转变，从而加速了 MA 和 CA_6 的形成，提高了陶瓷材料的烧结活性。多余的 Ca 所形成的致密相，使得 MA 和 CA_6 晶粒之间黏结紧密，提高了陶瓷材料的力学性能[11]。

Magnesium aluminum spinel ($MgAl_2O_4$, $MgO \cdot Al_2O_3$ or MA) has superior high temperature mechanical properties, excellent peeling resistance and corrosion resistance. It is the most typical high temperature ceramic in the Al_2O_3-MgO system. The preferential growth of calcium hexaaluminate ($CaAl_{12}O_{19}$, $CaO \cdot 6Al_2O_3$ or CA_6) crystal grains along the basal plane makes it grow into platelet or needle morphology, which can greatly enhance the toughness of the material. Calcium dialuminate ($CaAl_4O_7$ or $CaO \cdot 2Al_2O_3$, CA_2) has a low coefficient of thermal expansion. When CA_2 is compounded with other materials with high melting point and high coefficient of expansion, it can well resist damage caused by thermal shock. Therefore, MA-CA_6 composites have received extensive attention as a new type of high temperature ceramic material in the high-temperature industry because of its comprehensive properties of CA_6 and MA. In this paper, MA ceramic, MA-CA_2-CA_6 ceramic composites and MA-CA_6 ceramic composites were prepared by high temperature solid-phase sintering, and the influence of mineralizers on the properties of these ceramic materials were studied. The strengthening mechanism of mineralizers on the performance of ceramics were discussed, and the following research results were obtained: (1) The results showed that the bulk density and flexural strength of MA ceramic materials increased gradually with the increase of sintering temperature. After sintering at 1600℃ for 2h, the sintering performance of MA ceramic was poor, with a bulk density of 3. 17g/cm^3 and a flexural strength value of 133. 31MPa. With the increase of the mineralizer Fe_2O_3 , the bulk density of MA ceramic materials increased gradually, and the flexural strength first increased and then decreased. When the addition amount was 3wt.%, the flexural strength reached the maximum 209. 3MPa. (2) The performance and phase composition of MA-CA_6 ceramic are related to the particle size of $CaCO_3$ and α-Al_2O_3 raw materials, the purity of α-Al_2O_3 , the temperature of synthesis and the holding time. Using small particle size $CaCO_3$ and high purity α-Al_2O_3 as raw materials, after sintering at 1600℃ and holding for 2h, the synthesized MA-CA_6 ceramic has great flexural strength. The particle size of $CaCO_3$ plays an important role in the formation of CA_6 phase and the growth and development of crystal grains in MA-CA_6 ceramic materials. At high temperature, the impurity Si in α-Al_2O_3 will form a transient liquid phase, which makes the morphology of CA_6 grains evolve from platelet to equiaxed. (3) The effect of mineralizers ZnO and

Mg(BO$_2$)$_2$ on the properties of MA-CA$_6$ composites and the strengthening mechanism were investigated. It is found that (Mg$_{1-x}$Zn$_x$) Al$_2$O$_4$ solid solution and boron-containing liquid phase formed by the mineralizers ZnO and Mg(BO$_2$)$_2$ make the grain size of MA smaller and the content of MA increased. These dense phases are coated with microcrystalline MA particles to form regional dispersed dense bodies, which leads to the transformation of CA$_6$ grains into equiaxed grains, thus promoting the densification of MA-CA$_6$ ceramic materials and improving its flexural strength. (4) By using analytically pure Al$_2$O$_3$ instead of α-Al$_2$O$_3$, MA-CA$_2$-CA$_6$ ceramic composites were synthesized from analytically pure raw materials. The effects of mineralizers SnO$_2$ and H$_3$BO$_3$ on the physical and mechanical properties, microstructure and phase composition of the composites were studied. The results show that solid solution and boron-containing transient liquid phase appear in the ceramic material after adding mineralizers SnO$_2$ and H$_3$BO$_3$ respectively, it makes the CA$_2$ phase change to CA$_6$ phase and accelerates the formation of MA and CA$_6$, thus improving the sintering activity of the ceramic material. The dense phase formed by the excess Ca makes the bond between MA and CA$_6$ grains tight, which improves the mechanical properties of ceramic materials[11].

5.2 镁铬质耐火材料原料

5.2.1 术语词

水镁铬石 barbertonite
铬铁矿 chromite
镁铬尖晶石 picrochromite

镁铬铁矿 magnesiochromite
菱镁矿 magnesite, giobertite
镁砂 magnesia

5.2.2 术语词组

铬矿 chrome ore
氧化铬 chromic oxide
铬砂 chromite sand
低铁铬矿 chromite with low iron content
精选铬矿 enriched chromite

电熔镁铬砂 fused magnesite-chrome clinker
高品位铬矿 high-grade chromite ore
电熔高纯铬 high-purity fused chrome
疏松铬矿 loose chromite ore
低品位铬矿 low-grade chromite ore

低硅铬精矿 low-silicon chrome concentrate

粗晶铬矿 macrocrystalline chromite

镁铬合成砂 magnesite-chrome clinker

镁铬尖晶石 magnesite-chrome spinel, magnesia-chromite spinel

中级铬矿 medium-grade chromite

中等颗粒铬矿 medium-grained chromite

方镁石铬质合成砂 periclase-chrome clinker

粉状铬矿 powdery chromite

烧结氧化铬 sintered chromia

南非铬矿 south African chrome ore

盐湖镁砂（卤水镁砂） brine magnesia

烧成镁砂 burned magnesite

轻烧镁砂（苛性镁砂） light burned magnesia, caustic calcined magnesia

重烧镁砂（死烧镁砂） dead burned magnesia

致密烧结镁砂 dense sintered magnesia

冶金镁砂（补炉镁砂） fettling magnesia

细晶镁砂 fine-grained magnesia

细分散镁砂（高分散镁砂） finely dispersed magnesia

电熔镁砂 fused magnesia

电熔镁砂制品 fused magnesia product, electrocast magnesia product

高纯烧结镁砂 high-purity dead-burned magnesia

电熔镁砂高纯晶体 high-purity crystals of fused magnesia

高纯度高密度镁砂 high-purity high-dense magnesite

高纯镁砂 high-purity magnesia

高钙镁砂 high-calcium magnesia

大结晶镁砂 large crystal magnesia

大结晶电熔镁砂 large crystallined fused magnesia

低硅高钙电熔镁砂 low-silica and high-calcia fused magnesia

低铁镁砂 low-iron magnesia

低温煅烧镁砂 low-temperature sintered magnesia

冶金镁砂 metallurgical magnesia

中档镁砂 medium grade sintered magnesia

非晶质菱镁矿 non-crystalline magnesite

菱镁矿原矿 raw magnesite

海水镁砂 sea-water magnesia

烧结镁砂 sintered magnesia

5.2.3 术语例句

目前为止，不锈钢行业是对**铬矿**需求量最大。

The stainless-steel industry is by far the largest user of **chromite ore**.

冶炼中**烧结铬矿**的搭配比例以 60%～65% 为佳。

The optimal burden ratio of **sintered chromite** is 60%-65%.

本文综合分析了不锈钢冶炼和**铬矿**还原的理论和工艺。

The theory and technology of stainless-steel smelting and **chrome ores** reduction were discussed in this thesis.

原砖由电熔镁铬颗粒和**铬矿颗粒**组成，其显微结构是不均匀的。

The original brick consists of fused magnesia-chromite grains and raw **chromite grains** and its microstructure is non-uniformity.

结果表明转炉内用**铬矿**熔融还原直接合金化是可行的，且必须外加还原剂。

The results show that smelting reduction and direct alloying by **chromium ore** in the converter are feasible, and externally applied reducing agent is indispensable.

通过生产实践，对用石灰和**铬矿**洗炉进行了分析，并介绍了洗炉后转炼硅铬合金的方法。

According to the practice, the paper analyzes washing lime and **chromite**, and introduces the means of tunning Cr-Si alloy after washing furnace.

结合含碳**铬矿粉**在微波场中的比热容、介电性质等热及物理性质的变化，通过计算模拟得出微波耗散功率与温度的关系，进而拟合出含碳铬矿粉的升温曲线。

Combining the changes in thermal and physical properties such as specific heat capacity and dielectric properties of carbon-containing **chromite fines** in the microwave field, the relationship between microwave power dissipation and temperature is calculated and simulated, and the temperature rise curve of carbon-containing chromite fines is then fitted.

铬铁矿是生产铬铁合金的重要原料。

Chromite is the main raw material for smelting ferrochromium.

选用碱性树脂**铬矿砂**生产出合格的叶轮。

Alkaline resin **chromite sand** is used to produce qualified impeller casting.

当前**无铬**碱性砖还在不断开发和更新之中。

There still exhibits development and innovation of **chromite-free** basic bricks.

铬砂制作的砂模可让铸件有较快的冷却速率，但也产生较多的微孔隙。

The cooling rate is faster in **chromite sand** mold, but also causes more porosity in castings.

希腊生产耐火级**铬铁矿石**已有百余年历史。

Greece has been producing refractory grade **chromite ore** for more than a hundred years.

5.2.4　术语例段

镁铬耐火材料是以方镁石和镁、铁尖晶石（Mg,Fe）（Cr,Al,Fe）$_2$O$_4$ 为主晶相的碱性耐火材料，具有耐火度高、高温强度大、热震稳定性优良以及抗熔渣侵蚀性和经济性等优点，在钢铁、有色、水泥等高温工业装备领域有着广泛的应用，是诸多高温装备炉衬关键部位的主导材料。然而镁铬耐火材料在氧化性气氛或与碱性氧化物如 K$_2$O、Na$_2$O、CaO 等共存时，在一定温度下其 Cr^{3+} 会部分转化为 Cr^{6+}，由此带来六价铬污染问题，与高温工业绿色、环保、高效的发展理念相违背。通过法规对镁铬耐火材料产业链进行规范引导、采取系列措施抑制镁铬耐火材料六价铬化合物的形成均无法从根本上消除 Cr^{6+} 带来的隐患，而最根本的办法是开展耐火材料无铬化研究与应用[12]。

Magnesia-chromium refractory is an alkaline refractory material with periclase and spinel (Mg, Fe) (Cr, Al, Fe)$_2$O$_4$ as the main crystalline phases. It has high refractoriness, high strength under high temperature, excellent thermal shock resistance, fine slag corrosion resistance and good economic efficiency, and has been widely used in iron and steel metallurgy, non-ferrous smelting, cement and other industries. Magnesium chromium refractories are the dominant materials in many key parts of high-temperature furnace. However, Cr^{3+} of Cr$_2$O$_3$ will be partly converted to Cr^{6+} at a certain temperature in oxidizing atmosphere or with alkaline oxides such as K$_2$O, Na$_2$O, CaO, etc., resulting in hexavalent chromium pollution, which is not line with green, environmentally friendly and efficient for high temperature industry. It is impossible to eliminate the hidden danger caused by Cr^{6+} in magnesium chromium refractories by regulating the industrial chain and taking a series of measures to restrain the formation of hexavalent chromium compounds. However, the most consistent and permanent way is to carry out the research and application of chromium-free refractories[12].

镁铬耐火材料适合在循环温度或大气条件下工作，因为**铬矿石**的一些氧化物在加热时容易释放氧（被还原），而在冷却或改变气氛时会吸收氧（被氧化）。多年来，它们一直是 RH 脱气机的理想内衬材料，尽管在使用后存在 Cr^{6+} 处理问题。镁铬耐火材料的品种取决于其原材料和燃烧温度，这可以从它们的微观结构中得到很好的证明。但印度铬精矿由于烧结密度降低，不能直接用于镁铬砖的生产。硅酸盐结合镁铬砖是用低纯度镁砂在相对较低的温度下烧制而成，导致在铬矿颗粒周围形成硅酸盐结合的液膜。直接结合和再结合的电熔镁铬耐火材料以高纯镁砂或电熔镁铬颗粒为主要特征，形成晶间次生铬铁矿尖晶石。二次尖晶石的数量随着燃烧温度的升高而增加，从而导致高温抗折强度和整体热性能增加。在弱还原气氛下烧成镁铬砖很重要。800~1650℃燃烧气氛中氧含量最好控制在0.5%以下，以提高燃烧强度，避免镁铬砖内裂和黏结松动，而气氛中 O_2 含量为3%~6%是最经济的燃烧条件。

Magnesia-chrome refractories are appropriate to work under cyclic temperatures or atmospheric conditions because some oxides of **chrome ore** readily release oxygen(are reduced)upon heating and pick up oxygen(are oxidized)upon cooling or upon changing the atmosphere. They have been ideal lining materials for RH degassers for many years, despite challenging of Cr^{6+} disposal issue after using. The varieties of magnesia-chrome refractories depend on their raw materials and burning temperatures, which could be well demonstrated by their microstructures. But Indian chrome concentrate cannot be directly used in the production of magnesia-chrome bricks because of reducing sintering density. Silicate bonded magnesia-chrome bricks are produced with low purity magnesia after burning at relatively lower temperatures, resulting in liquid film forming as silicate bond around chrome ore particles. Direct bonded and rebonded fused-grains magnesia-chrome refractories are made of high purity magnesia or fused magnesia-chrome grains, forming euhedral and intergranular secondary chromite spinels as the main feature. The amount of secondary spinel increases with the rising burning temperature, leading to increasing hot modulus of rupture and overall hot properties as well. It is important to burn magnesia-chrome bricks under weakly reducing atmosphere. The oxygen content of burning atmosphere from 800℃ to 1650℃ would better controlled below 0.5% to increase the burnt strength and to avoid inner cracks and loose bonding of magnesia-chrome bricks, while the atmosphere contains 3%-6% O_2 under most economical firing condition[13].

氧气底吹冶炼技术（SKS）是一种利用低品位和复杂多金属铜矿石的有前途的方法。为了揭示炉衬的磨损机制，本文通过化学和微观结构分析对使用过的耐火材料（**电熔再结合镁铬酸盐耐火材料**）进行了事后研究。研究结果表明，在近渣侧的炉衬处产生了较厚的腐蚀层（约4.0cm），测定为连续分布的镁橄榄石相，因此此处的化学相互作用特别强。这主要是由于高注入压力和熔渣中存在 $Cu-Cu_xO$ 熔体，极大地加速了熔渣的渗透和方镁石的溶解。在这种情况下，发现具有高含量晶间尖晶石的熔凝晶粒更难被熔渣腐蚀。直接结合的显著损失和耐火材料的软化被认为是 SKS 炉衬的主要退化机制。炉渣中的 Fe_xO 与铬铁矿尖晶石的相互作用很明显，但只发生在靠近热面的区域，因此对腐蚀行为的影响有限。

Oxygen bottom-blown smelting technology(SKS) is a promising way to utilize the low-grade and complex multimetallic copper ores. To reveal the wear mechanisms of the furnace linings, a post-mortem study of the used refractories(**fused-grain rebonded magnesia-chromite refractory**) was carried out by chemical and microstructural analysis. The results showed that the chemical interaction at the bath-lining interface was particularly strong since thick corrosion layers(around 4.0cm) were generated, where a continuously distributed phase of forsterite was identified. This was mainly owing to the high injection pressure and the presence of $Cu-Cu_xO$ melts in the slag, which greatly accelerated the slag infiltration and periclase dissolution. In this case, fused grains with a high content of intergranular spinel were found more difficult to be corroded by the slag. The significant loss of direct bonds and the softening of refractory are considered as the main degradation mechanisms for the SKS furnace linings. The interaction between Fe_xO in slag and chromite spinel was evident, but it only occurred in the areas near the hot face, showing limited effect on the corrosion behavior[14].

本文报道了添加纳米级 $MgCr_2O_4$ 和 $FeCr_2O_4$ 对**直接结合镁铬耐火材料**的微观结构和力学性能的影响。通过 XRD、BET 和 TEM 对通过柠檬酸盐–硝酸盐路线合成并在几种不同温度下煅烧的纳米结构添加剂进行了表征。将 0.5% 和 1%（质量分数）的这些纳米结构氧化物添加到镁铬耐火材料中，并在梭式窑中于 1650℃下煅烧。通过 SEM/EDX 分析它们的微观结构，并根据各自的 DIN 标准确定它们的物理和力学性能永久线性变化（PLC）、体积密度、显气孔率、冷压碎强度（CCS）和高温抗折强度（HMOR）。在镁铬耐火材料中添加纳米结构的氧化物促进了次生尖晶石的形成，从而影响了物理和力学性能。由于在 $FeCr_2O_4$ 纳米粉末中存在磁铁矿杂质时会形成液相，$FeCr_2O_4$ 的添加增加了次生尖晶石的尺寸。在耐火材料的基础配方中添加纳米级 $MgCr_2O_4$ 和 $FeCr_2O_4$ 使 CCS 分别从 67.4MPa 提高到 82.8MPa 和 81.0MPa，而纳米级 $MgCr_2O_4$ 将 HMOR 值从 5.48MPa 提高到 5.91MPa 和纳米级 $FeCr_2O_4$ 的 HMOR 从 5.48MPa 增加到 5.72MPa。XRD 结果表明，与加入 $FeCr_2O_4$ 相比，这一增幅较小的原因是在磁铁矿存在下形成了液相。

The effect on the microstructure and mechanical properties of **direct-bonded magnesia-chrome refractories** of additions of nanostructured $MgCr_2O_4$ and $FeCr_2O_4$ is reported. The nanostructured additives, synthesized by the citrate-nitrate route and calcined at several different temperatures, were characterized by XRD, BET and TEM. Additions of 0.5wt.% and 1wt.% of these nanostructured oxides were made to magnesia-chrome refractories and calcined at 1650℃ in a shuttle kiln. Their microstructures were analyzed by SEM/EDX and their physical and mechanical properties permanent linear change(PLC), bulk density, apparent porosity, cold crushing strength(CCS) and hot modulus of rupture (HMOR) were determined according to the respective DIN standards. The addition of the nanostructured oxides to the magnesia-chrome refractories facilitated the formation of secondary spinels, influencing the physical and mechanical properties. $FeCr_2O_4$ additions increased the size of the secondary spinel due to liquid phase formation in the presence of magnetite impurities in the $FeCr_2O_4$ nano-powder. The addition of nano-sized $MgCr_2O_4$ and $FeCr_2O_4$ to the base formulation of the refractory increased the CCS from 67.4MPa to 82.8MPa and 81.0MPa respectively, while nano-sized $MgCr_2O_4$ increased the HMOR value from 5.48MPa to 5.91MPa and nano-sized $FeCr_2O_4$ increased the HMOR from 5.48MPa to 5.72MPa. This smaller increase than that obtained with $FeCr_2O_4$ additions is attributed to liquid phase formation in the presence of magnetite, as observed by XRD[15].

收集来自二级铜冶炼厂的使用过的**直接结合镁铬耐火砖**，以表征在应用过程中发生的降解。基于这种事后分析，使用直接黏结耐火材料和电熔晶粒镁铬铁矿耐火材料设计了实验室规模的实验。首先，研究了 Cu-Cu$_x$O 和 Cu-Cu$_x$O-PbO 混合物的渗透行为。其次，确定了温度和 Cu-Cu$_x$O 混合物的预先渗透对含 ZnO 铁橄榄石渣渗透的影响。事后研究和实验室规模测试相结合，可以更全面地了解降解过程中耐火材料微观结构的演变。同样，它允许评估某些参数，如温度和耐火材料类型，以及如何影响渗透和降解行为。因此，可以得出关于减少铜冶炼炉衬里耐火材料磨损的措施的结论。

Used **direct-bonded magnesia-chromite refractory bricks** from a secondary Cu smelter were collected to characterize the degradation occurring during application. Based on this post-mortem analysis, lab scale experiments were designed using the direct-bonded refractory type and a fused grain magnesia-chromite refractory. Firstly, the infiltration behavior of a Cu-Cu$_x$O and a Cu-Cu$_x$O-PbO mixture was investigated. Secondly, the influence of temperature and the prior infiltration of a Cu-Cu$_x$O mixture on the infiltration of the ZnO containing fayalite slag was determined. The combination of the post-mortem study and the lab scale tests allowed a more comprehensive understanding of the evolution of the refractory microstructure during degradation. Likewise, it allowed to evaluate how certain parameters, such as temperature and refractory type, affect the infiltration and degradation behavior. As a consequence, conclusions can be drawn about measures to minimize refractory wear in copper smelter linings[16].

多年来，**镁铬 (MgO-Cr) 耐火材料**一直是铜和铅有色金属冶金内衬的关键材料。它们的成分主要是铬铁矿，具有高含量的氧化铬（Cr_2O_3 质量分数高达

50%的）和氧化镁熟料作为 MgO 的来源。到目前为止，已经开发了多种类型的 MgO-Cr 产品，包括熔铸、共烧、化学结合、硅酸盐和直接结合。自 20 世纪 50 年代以来生产的直接结合产品已被证明具有最佳性能，即 MgO 晶体与铬铁矿晶粒直接结合，因此晶粒间没有任何额外的相。在这种类型的耐火材料中，Cr_2O_3 的典型含量达到 18% ~ 35%。随着 MgO-Cr 耐火材料中 Cr_2O_3 含量的增加，耐腐蚀性和抗热震性越大。此外，研究发现将纳米 $MgCr_2O_4$ 和 $FeCr_2O_4$ 掺入此类产品的基体中可提高其致密性和力学性能。添加微米 $\alpha\text{-}Al_2O_3$ 也显示出很好的效果。

For many years, **magnesia-chromite (MgO-Cr) refractories** have been critical materials used for linings in the non-ferrous metallurgy of Cu and Pb. Their composition is based on chromite ores with a high concentration of chromium oxide (up to 50wt.% Cr_2O_3) and magnesia clinker as a source of MgO. A few types of MgO-Cr products have been developed so far, including fusion cast, co-burned, chemically bonded, silicate, and direct-bonded. The best performance has been proven for the direct-bonded products, being manufactured since the 1950s, in which MgO crystals are bonded directly with chromite grains, thus, without any additional phases in between. In this type of refractory typical concentration of Cr_2O_3 reaches 18% -35%. The increased Cr_2O_3 content in MgO-Cr refractories the greater both corrosion and thermal shock resistance. The incorporation of nano-$MgCr_2O_4$ and $FeCr_2O_4$ into the matrix of such products was found to enhance their densification and mechanical properties. The addition of micro α-Al_2O_3 also showed promising results[17].

为了研究 ZnO 在**镁铬耐火材料**降解中的作用，采用旋转镁铬耐火材料指状腐蚀试验，研究了在 1200℃ 的还原气氛中与含 ZnO 铁橄榄石渣 （$FeO\text{-}SiO_2\text{-}ZnO\text{-}Al_2O_3$） 的化学腐蚀行为。研究结果表明，与不含 ZnO 的铁橄榄石渣一样，方镁石主要受到腐蚀。ZnO 和 FeO 均扩散到原始方镁石和铬铁矿晶粒中，从而分别形成 （Zn, Fe, Mg）O 固溶体和 （Zn, Fe, Mg）（Cr, Al, Fe）$_2O_4$ 尖晶石。同时，（Zn, Fe, Mg）$_2SiO_4$ 橄榄石在其形成过程中也掺入了 ZnO 和 FeO。这些相的形成都没有导致耐火材料中产生新的裂纹。考虑到渗入的炉渣在距热面 4mm 处几乎完全耗尽了 ZnO，因此可以认为炉渣的渗透并不是由初始炉渣中存在 ZnO 引起的。

In order to study the role of ZnO in the degradation of **magnesia-chromite refractories**, the chemical corrosion behaviour in a ZnO-containing fayalite slag (FeO-SiO_2-ZnO-Al_2O_3) is investigated using a rotating magnesia-chromite refractory finger corrosion test under a reducing atmosphere at 1200℃. The results show that, likewise to ZnO-free fayalite slag, periclase is predominantly corroded. Both ZnO and FeO diffused into the original periclase and chromite grains thereby forming, respectively, $(Zn,Fe,Mg)O$ solid solution and $(Zn,Fe,Mg)(Cr,Al,Fe)_2O_4$ spinel. Concurrently, ZnO and FeO were also incorporated into the $(Zn,Fe,Mg)_2SiO_4$ olivine during its formation. None of these phase formations resulted in new crack generation in the refractory. Considering the infiltrated slag is almost completely depleted in ZnO at 4mm from the hot face, the severe slag penetration up to the center of the refractory sample is not caused by the ZnO presence in the initial slag[18].

再结合镁铬耐火材料用于真空氧脱碳钢包和不锈钢的二次精炼。由于所施加的严格的化学、热和机械条件,它们遭受剧烈磨损。尽管磨损耐火材料的用后调查对于研究降解机制是必不可少的,但由于原位冷却过程中发生的结晶现象,其应用经常受到评估困难的阻碍。为了研究高温下的实际微观结构,工业磨损和原始耐火试样在圆柱形单模微波炉中重新加热和淬火。管状基座的使用允许试样的可重复混合加热高达1800℃。用SEM和EPMA-EDS分析淬火试样。同时,进行了元素线扫描、X射线映射和定量图像分析。这有助于解释次生铬铁矿键合相的"高温失活"机制:(1)溶解在方镁石相中,(2)溶解在液态渣中,其中(2)占主导地位。最后,得出关于尖晶石键合相贡献者的耐火度的结论:MgO·Cr_2O_3(镁铬矿)>MgO·Al_2O_3(尖晶石)≫ MgO·Fe_2O_3(镁铁矿)。此外,混合微波加热被证明是对耐火试样进行常规熔炉实验的有趣替代方案。

Rebonded magnesia-chromite refractories are used in vacuum oxygen decarburization ladles for the secondary refining of stainless steel. They suffer from acute wear due to the stringent chemical, thermal and mechanical conditions imposed. Although post-mortem investigation of worn refractories is indispensable to study the degradation mechanisms, its application is often hampered by evaluation difficulties due to crystallization phenomena occurring during in situ cooling. To study the actual microstructures at elevated temperatures, industrially worn and virgin refractory samples were reheated and quenched in a cylindrical single-mode microwave furnace. Usage of a tubular susceptor allowed reproducible hybrid heating of the samples up to 1800℃. The quenched samples were analyzed with SEM and EPMA-EDS. Concurrently, elemental line scans, X-ray mappings and quantitative image analyses were performed. This allowed the description of the "high-temperature inactivation" mechanisms of the secondary chromite bonding phase: dissolution(1) in the periclase phase and (2) in the liquid slag, with (2) being predominant. Conclusions are drawn with respect to the refractoriness of the contributors to the spinel bonding phase: MgO·Cr_2O_3(magnesiochromite)>MgO·Al_2O_3(spinel) ≫ MgO·Fe_2O_3(magnesioferrite). Hybrid microwave heating is shown to be an interesting alternative for conventional furnace experiments on refractory samples[19].

通过在真空感应炉中旋转手指试验研究了富含氧化铝（质量分数为 15% ~ 20%）不锈钢渣对**镁铬耐火材料**的腐蚀行为。讨论了工艺温度、腐蚀时间，特别是熔渣中高 Al_2O_3 含量对耐火材料磨损的影响。两种不同的机制导致初级铬铁矿降解：由于低氧势导致 FeO_x 和 Cr_2O_3 分解，以及由于高 Al_2O_3 而被渗透熔渣溶解渣含量。分解后，在原生铬铁矿内部均匀地产生小的金属颗粒和孔隙。在耐火材料/熔渣界面处，形成了相对连续的固体（Mg，Mn）（Al，Cr）$_2O_4$ 尖晶石层。它的密度和稳定性随着温度的升高和湍流的增加而降低。尖晶石的形成是通过从富含尖晶石形成化合物的炉渣中非均匀原位沉淀而产生的。渣中较高的 Al_2O_3 含量促进尖晶石层的形成，这可能会限制渣的渗透。最后，表明本实验程序是模拟工业过程中耐火材料磨损的极好工具，可降低与工厂试验相关的风险。

The corrosion behaviour of **magnesia-chromite refractory** by an alumina-rich (15wt.%-20wt.%) stainless steelmaking slag is investigated by rotating finger tests in a vacuum induction furnace. The influence on the refractory wear, of the process temperature, corrosion time and, in particular, the high Al_2O_3 content in the slag, is discussed. Two distinct mechanisms cause primary chromite degradation: FeO_x and Cr_2O_3 decomposition because of low oxygen potentials and dissolution by infiltrated slag due to the high Al_2O_3 slag content. Upon decomposition, small metallic particles and pores are homogeneously generated inside the primary chromite. At the refractory/slag interface, a relatively continuous solid(Mg,Mn)(Al,Cr)$_2O_4$ spinel layer is formed. Its density and stability decrease with higher temperatures and more turbulent conditions. The spinel formation arises through heterogeneous in situ precipitation from a slag rich in spinel forming compounds. Higher Al_2O_3 levels in the slag promote the spinel layer formation, which may limit slag infiltration. Finally, it is shown that the present experimental procedure is an excellent tool to simulate refractory wear in industrial processes, diminishing the risks associated with plant trials[20].

几种商业耐火砖的微观结构已经使用扫描和透射电子显微镜进行了表征。"直接键" 显示在可达到的分辨率水平（约 0.3nm）是直接的，并且大量此类直接接触发生在**镁铬砖**微结构中的非硅酸盐相之间。然而，由于 TEM 试样的不均匀离子变薄，检查镁尖晶石砖中直接键合的尝试没有成功。

The microstructures of several commercial refractory bricks have been characterized using scanning and transmission electron microscopy. The "direct bond" is shown to be direct at the level of resolution attainable(about 0.3nm), and a significant number of such direct contacts occur between non-silicate phases in the microstructures of **magnesia-chromite bricks**. However, attempts to examine the direct bond in magnesia-spinel bricks were unsuccessful due to nonuniform ion thinning of TEM specimens[21].

在火法炼铜工艺中，转炉吹炼的工作环境对所用耐火材料有苛刻的要求：炉渣以及金属铜熔体的侵蚀，较大的温度波动，流体的冲刷，以及 SO_2 气氛的影响等均对炼铜转炉用耐火材料有较高的要求。目前，我国依然采用**镁铬质耐火材料**作为炼铜转炉炉衬的主要材料，而研究炼铜转炉用镁铬质耐火材料的损毁机理不论是对炼铜转炉用镁铬质耐火材料的改进抑或是最终达到炼铜转炉用耐火材料的无铬化都是有积极意义的。关于炼铜转炉用镁铬质耐火材料的损毁机理，前人已经进行了较为详细的研究，然而目前对镁铬质耐火材料研究集中在对残砖的分析上，没有侵蚀机理详细研究，尤其是 SO_2 气氛对镁铬质耐火材料的影响研究不足，仅仅是在残砖分析中发现了针状的 Mg_2SO_4，并没有详细的研究。本课题的目的就是在对铜转炉用后镁铬砖进行分析的基础上，通过在铜转炉炉渣中加入不同量的 S 单质，在氧化气氛进行侵蚀实验来研究 SO_2 气氛对镁铬质耐火材料抗转炉铜渣侵蚀性能的影响，并且在镁铬砖中添加镁铝尖晶石与 ZrO_2 微粉以改进镁铬砖抗加 S 转炉铜渣侵蚀的性能。通过研究发现，由于砖体结构成分的不同，不同工艺的镁铬质耐火材料在不同温度下抗加 S 铜渣的侵蚀能力不同；添加镁铝尖晶石微粉可以很好地改善镁铬砖的物理性能和抗加 S 铜渣的侵蚀性；同时添加镁铝尖晶石和 ZrO_2 微粉反而会增大镁铬质耐火材料的气孔率，然而 ZrO_2 自身抗铁硅渣侵蚀方面表现优良[22]。

The requirements of refractory are very high in copper smelting process, because of the harsh working environment of refractories, such as the erosion of copper slag and metal melt, large temperature fluctuations, the scouring of fluid and the influence of SO_2. Up to now, the refractories used in the copper smelting converter in our country are mainly **magnesite-chrome refractory**. Thus, study on the damage mechanism of magnesite-chrome refractory is beneficial to the improvement of magnesite-chrome refractory and still achieving the object of adopting chrome-free refractory in the copper convertor. There are many former studies on the damage mechanism of magnesite-chrome refractory used in copper smelting convertor while little study on erosion mechanism of magnesite-chrome refractory. Especially the research on the effect of SO_2 is insufficient. The needle-like Mg_2SO_4 crystal is found in the residual brick in former studies. In this work, the effect of SO_2 on the copper convertor slag-resistance of magnesite-chrome refractory is studied by adding sulphur into copper converter slag in the erosion experiment in oxidizing atmosphere. In addition, magnesia-alumina spinel and ZrO_2

powder were added into magnesite-chrome brick to improve copper converter slag-resistance of magnesite-chrome brick. The results indicate that the ability of resisting copper converter slag containing sulphur varies with kinds of magnesite-chrome brick because of different brick structure and chemical compositions. Adding magnesia-alumina spinel powder can improve physical properties and copper converter slag-resistance of magnesite-chrome brick. While adding both magnesia-alumina spinel and ZrO_2 powder leads to the increase in porosity of magnesite-chrome brick. The resistance of FeO-SiO$_2$ slag of the bricks containing ZrO_2 is excellent[22].

参考文献

[1] 黄军同, 黄朝晖, 吴小贤, 等. 利用铝灰和粉煤灰铝热还原氮化制备镁铝尖晶石-刚玉-Sialon 复相材料 [J]. 稀有金属材料与工程, 2009, 38 (S2): 1255-1258.

[2] Wang Zihao, Su Kai, Gao Jinxing, et al. Preparation, microstructure and properties of Al$_2$O$_3$-ZrO$_2$-C slide plate material in presence of nanoscale oxides [J]. Ceramics International, 2022, 48 (7): 10126-10135.

[3] Liu Yanshan, Han Bingqiang, Zhang Ting, et al. Effect of zirconia particle size on the properties of alumina-spinel castables [J]. Ceramics International, 2016, 42 (15): 16961-16968.

[4] Quan Zhenghuang, Wang Zhoufu, Wang Xitang, et al. Effects of Sm$_2$O$_3$ addition on sintering behavior of pre-synthesized magnesia-rich magnesium aluminate spinel [J]. Journal of Rare Earths, 2021, 39 (11): 1450-1454.

[5] Peng Wangding, Chen Zhe, Yan Wen, et al. Advanced lightweight periclase-magnesium aluminate spinel refractories with high mechanical properties and high corrosion resistance [J]. Construction and Building Materials, 2021, 291: 123388.

[6] Baruah B, Sarkar R. Rare-earth oxide-doped magnesium aluminate spinel-an overview [J]. Interceram, 2020, 69: 40-45.

[7] Quan Zhenghuang, Wang Zhoufu, Wang Xitang, et al. Effect of CeO$_2$ addition on the sintering behavior of pre-synthesized magnesium aluminate spinel ceramic powders [J]. Ceramics International, 2019, 45 (1): 488-493.

[8] Yuan Lei, Ma Beiyue, Zhu Qiang, et al. Preparation and properties of MgAl$_2$O$_4$ based ceramics reinforced with rod-like microcrystallines by co-doping Sm$_2$O$_3$ and La$_2$O$_3$ [J]. Ceramics International, 2017, 43 (18): 16258-16263.

[9] Yoon T, Lee K, Lee B, et al. Wetting, spreading and penetration phenomena of slags on MgAl$_2$O$_4$ spinel refractories [J]. ISIJ International, 2017, 57 (8): 1327-1333.

[10] Hu Shuhe, Ye Guotian, Chen Liugang, et al. Effect of micro-sized MgCO$_3$ addition on properties of MgAl$_2$O$_4$ spinel containing castables [J]. Ceramics International, 2017, 43 (13): 9891-9895.

[11] 王晓军. MgAl$_2$O$_4$ 及 MgAl$_2$O$_4$-CaAl$_{12}$O$_{19}$ 陶瓷材料的固相合成和性能调控研究 [D]. 太原: 太原科技大学, 2021.

[12] 钱凡, 段雪珂, 杨文刚, 等. 镁铬耐火材料及高温装备绿色化应用研究进展 [J]. 材料

导报，2019，33（12）：3882-3891.

[13] Guo Zongqi, Ma Ying, Li Yong. Sintering complexity of magnesia-chrome refractories [J]. China's Refractories, 2022, 31（1）：16-23.

[14] Xu Lei, Chen Min, Wang Nan, et al. Chemical wear mechanism of magnesia-chromite refractory for an oxygen bottom-blown copper-smelting furnace：a post-mortem analysis [J]. Ceramics International, 2021, 47（2）：2908-2915.

[15] Lotfian N, Nourbakhsh A, Nezamoddin Mirsattari S, et al. A comparison of the effect of nanostructured $MgCr_2O_4$ and $FeCr_2O_4$ additions on the microstructure and mechanical properties of direct-bonded magnesia-chrome refractories [J]. Ceramics International, 2020, 46（1）：747-754.

[16] Chen Liugang, Li Shuangliang, Jones P T, et al. Identification of magnesia-chromite refractory degradation mechanisms of secondary copper smelter linings [J]. Journal of the European Ceramic Society, 2016, 36（8）：2119-2132.

[17] Ludwig M, Śnieżek E, Jastrzębska I, et al. Corrosion of magnesia-chromite refractory by PbO-rich copper slags [J]. Corrosion Science, 2022, 195：109949.

[18] Chen Liugang, Guo Muxing, Shi Huayue, et al. The influence of ZnO in fayalite slag on the degradation of magnesia-chromite refractories during secondary Cu smelting [J]. Journal of the European Ceramic Society, 2015, 35（9）：2641-2650.

[19] Jones P T, Vleugels J, Volders I, et al. A study of slag-infiltrated magnesia-chromite refractories using hybrid microwave heating [J]. Journal of the European Ceramic Society, 2002, 22（6）：903-916.

[20] Guo M, Jones P T, Parada S, et al. Degradation mechanisms of magnesia-chromite refractories by high-alumina stainless steel slags under vacuum conditions [J]. Journal of the European Ceramic Society, 2006, 26（16）：3831-3843.

[21] Gotod K, Lee W. The " Direct Bond" in magnesia chromite and magnesia spinel refractories [J]. Journal of the American Ceramic Society, 1995, 78（7）：1753-1760.

[22] 张原. 炼铜转炉用镁铬质耐火材料侵蚀机理的研究 [D]. 郑州：郑州大学，2014.

6 碳复合耐火材料原料

6.1 碳−氧化物系耐火材料原料

6.1.1 术语词

石墨 graphite

石墨的 graphitic

石墨化 graphitization

高温石墨 pyrographite

碳 carbon

木炭 charcoal

焦炭 coke

镁砂 magnesia

刚玉 corundum

方解石（冰洲石）calcite，calcspar

6.1.2 术语词组

土状石墨 amorphous graphite

人造石墨 artificial graphite，synthetic graphite

隐晶石墨 cryptocrystalline graphite

膨胀石墨 expanded graphite

细分散石墨 finely dispersed graphite

鳞片石墨 flake graphite

石墨碎屑 graphite fragment

石墨粉 graphite powder

高纯石墨 high-purity graphite

低灰分石墨 low-ash graphite

天然鳞片石墨 natural flake graphite

天然石墨 natural graphite

天然石墨粉 natural graphite powder

造粒石墨 pelleting graphite

显晶石墨 phanerocrystalline graphite

粉状石墨 powdered graphite

球状石墨 spheroidal graphite

鳞片石墨微粉 ultrafine flake graphite powder

超纯石墨 ultra-pure graphite

精制石墨 washed graphite

不定型碳 agraphitic carbon，amorphous carbon

表观残碳量 apparent residual carbon

炭黑 carbon black

盐湖镁砂（卤水镁砂）brine magnesia

烧成镁砂 burned magnesia

轻烧镁砂（苛性镁砂）light burned magnesia，

caustic calcined magnesia

重烧镁砂（死烧镁砂）dead burned magnesia

致密烧结镁砂 dense sintered magnesia

冶金镁砂（补炉镁砂）fettling magnesia

细晶镁砂 fine-grained magnesia

细分散镁砂（高分散镁砂）finely dispersed magnesia

电熔镁砂 fused magnesia

高纯烧结镁砂 high-purity dead-burned magnesia

电熔镁砂高纯晶体 high-purity crystals of fused magnesia

高纯度高密度镁砂 high-purity high-dense magnesite

高纯镁砂 high-purity magnesia

高钙镁砂 high-calcium magnesia

大结晶镁砂 large crystal magnesia

大结晶电熔镁砂 large crystalline fused magnesia

低硅高钙电熔镁砂 low-silica and high-calcia fused magnesia

低铁镁砂 low-iron magnesia

低温煅烧镁砂 low-temperature sintered magnesia

冶金镁砂 metallurgical magnesia

中档镁砂 medium grade sintered magnesia

海水镁砂 sea-water magnesia

烧结镁砂 sintered magnesia

人造刚玉（人造金刚砂）artificial corundum

黑刚玉 black fused alumina

棕刚玉 brown adamantine spar

轻烧刚玉 caustic-burned corundum

铬刚玉 chrome corundum

刚玉熟料 corundum clinker

刚玉球 corundum granule

刚玉微粉 corundum micro-powder

刚玉粉 corundum powder

致密刚玉 dense corundum

致密电熔刚玉 dense fused alumina

致密白刚玉 dense white fused alumina

电熔棕刚玉 electrically fused brown corundum

电熔铬刚玉 electrocast chrome-corundum

电熔锆刚玉 electrocast zirconia corundum

游离刚玉 free corundum

电熔刚玉 fused alumina, fused corundum, electromelting corundum

电熔致密刚玉 fused dense corundum

低碳高铝棕刚玉 low carbon and high alumina brown fused alumina

低碳亚白刚玉 low carbon vice-white corundum

粗晶刚玉 macrocrystalline corundum

微晶刚玉 microcrystalline alumina, microcrystal fused alumina

改性烧结刚玉 modified sintered corundum

普通电熔刚玉 normal electrocorundum

赛隆结合刚玉 Sialon bonded corundum

烧结刚玉 sintered corundum

电熔亚白刚玉 sub-white electrically fused corundum

亚白刚玉 sub-white fused alumina, vice-white corundum, vice-white fused alumina

人造刚玉（合成刚玉）synthetic corundum

板状刚玉 tabular alumina, tabular corundum

稳定氧化锆 stabilized zirconia

部分稳定氧化锆 partially-stabilized zirconia

电熔氧化锆 fused zirconia

高纯化学氧化锆 high-purity chemical zirconia

6.1.3 术语例句

石墨是一种高效的电导体。

Graphite is a highly efficient conductor of electricity.

活塞表面涂有**石墨**以减少摩擦。

The pistons are **graphite**-coated to reduce friction.

这种**石墨**可以在自然状态下用来书写。

This **graphite** could be used in its natural state for writing.

研究者使用普通胶布从**石墨**中提取石墨烯。

The researchers extracted graphene from **graphite** using ordinary adhesive tape.

碳纳米管是由比人的头发细千倍的**石墨**板卷曲成的管状结构。

Carbon nanotubes are rolled-up sheets of **graphite** thousands of times thinner than a human hair.

炭黑的氧含量、表面酸性含氧基团决定了**炭黑**的表面化学性质。

The oxide content and surface acidic groups determine the surface chemical property of **carbon black**.

通过接枝高聚物对**炭黑**表面改性，可显著提高炭黑与基质的相容性。

Grafting polymer onto surface can markedly improve the compatibility of **carbon black** with substrates.

炭黑是一种微观结构、粒子形态和表面性能都极为特殊的炭素材料。

Carbon black is special kind of carbonaceous material on microstructure, particle configuration and surface behavior.

将导电**炭黑**混入热塑性聚氨酯中，经熔融纺丝制得导电纤维。

Carbon black was blended into thermoplastic polyurethane, the electrically conductive fiber was obtained by melting spinning.

氧化锆增韧莫来石（ZTM）是优良的高温结构陶瓷材料。

Zirconia Toughened Mullite (ZTM) ceramics is one of finest Performance high temperature structure ceramics.

本文对氧化镁部分稳定**氧化锆**超细粉末的制备工艺和性能进行了实验研究。

The synthesis and characteristics of MgO Partially stabilized **zirconia** ultrafine powder are investigated in this work.

氧化锆是一种重要的催化剂及载体。

Zirconia is an important catalyst and support.

对**氧化锆**涂层的显微结构进行了探查。
The microstructure of the as sprayed **zirconia** coatings was analyzed.

用氧化硅掺杂硫酸化氧化锆可以增强硫酸化**氧化锆**的酸性。
Doping sulfated zirconia with silica increases the acidity of the sulfated **zirconia**.

6.1.4   术语例段

镁碳耐火材料因其中含**鳞片石墨**等碳组分而具有优异的抗热震性和抗高温熔渣侵蚀能力,被广泛用作钢铁冶炼用转炉、电炉、钢包等炉衬材料。为满足洁净钢冶炼技术、钢铁冶金节能减排、资源高效利用的需求,**镁碳耐火材料**的低碳化具有重要的现实意义。然而,单纯地降低镁碳耐火材料中**鳞片石墨**的含量,势必会造成材料高温使用性能下降,尤其是抗热震性下降。为此,近年来国内外研究者主要从碳源的选择、结合剂的改性、抗氧化剂的优化和原位形成陶瓷相的控制等方面开展了大量卓有成效的工作。主要从纳米结构基质的构筑、结合剂碳结构的优化及外加剂的引入与陶瓷相的调控出发,综述了国内外**低碳化镁碳耐火材料**的研究进展,提出了高强韧复合化/轻量化镁质骨料的制备、新型抗氧化剂的研发及不同抗氧化技术的结合使用、复合纳米碳/陶瓷相协同强韧化的调控、含纳米碳复合粉体的制备以及**鳞片石墨**替代材料的选择等发展方向,有望为低碳化镁碳耐火材料的研究和工业应用提供参考[1]。

MgO-C refractories are widely used as the lining materials in steelmaking industries such as basic oxygen furnace, electric arc furnaces and steel ladles owing to their excellent thermal shock resistance and corrosion resistance originated from the presence of **flake graphite** and other carbon sources. In order to meet the requirements of clean steel smelting, low carbon society and effective usage of resources, it is greatly important to develop low-carbon **MgO-C refractories**. Nevertheless, the high temperature performance (especially thermal shock resistance) will be decreased significantly when the content of **flake graphite** in MgO-C refractories is reduced. Therefore, much effort based on the selection of carbon sources, modification of binders, optimization of antioxidants and control of in-situ formed ceramic phases was performed to improve the properties of the refractories in recent years. In this review, the recent progress on **low-carbon MgO-C refractories** is summarized, and the future investigation work including the fabrication of composite/lightweight magnesia-based aggregates with high strength and toughness, the development of new antioxidants, the combined utilization of different antioxidation techniques, the control of synergistic strengthening and toughening of composite nano-carbon/ceramic phase, the preparation of composite powders containing nano-carbon and the selection of **flake graphite** substitute materials is also proposed, with the aim to provide the guidance for the development and application of low-carbon MgO-C refractories[1].

　　在过去的几十年里，钢铁制造技术发生了翻天覆地的变化，以满足钢铁用户对高纯度和高质量钢铁的需求。炼钢工艺的进步也需要高质量的耐火材料来承受恶劣的操作条件并生产出所需的钢材质量。镁碳耐火材料是钢铁制造和加工必不可少的材料，自问世以来经历了多次改进，仍然是耐火材料开发的主要挑战领域之一。就在几年前，人们还认为碳的使用量更高对耐火材料的性能和寿命有益，但这一概念已被证明是错误的，**低碳 MgO-C 耐火材料**正成为关注的焦点。许多研究人员已经通过多种方式尝试了在不影响其性能的情况下降低 **MgO-C 耐火材料**中的碳含量，其中纳米碳的使用广泛流行。本综述讨论了在 **MgO-C 耐火材料**中使用**纳米碳**的前景以及正在进行的各种研究工作以及低碳镁碳耐火材料的开发。

Steel manufacturing technology has changed drastically in last few decades to meet the demand of high purity and quality of steel from the steel users. The advancements in steel making process have also demanded high quality refractories to withstand severe operating conditions and to produce desired steel quality. Magnesia carbon refractories, being essential for making and processing of steel, have gone through many modifications since its inception and still are one of the major challenging areas for refractories development. Only a few years ago higher use of carbon was assumed to be beneficial for the refractory's performances and life, but the concept has been proven wrong and **low-carbon MgO-C refractories** are becoming the point of attention. Reduction in carbon content in **MgO-C refractories** without much affecting their properties have been tried by many ways by many researchers, amongst them use of nano carbon is widely popular. This review discusses the prospects for using **nano carbons** in **MgO-C refractory** and various research works that are going on and also on the development of low carbon containing magnesia carbon refractories[2].

　　提钒转炉受炉渣腐蚀严重。**MgO-C 耐火材料**由于氧化性高、CaO 含量低、熔渣晶相分散分布等原因，尚未采用溅渣法对 MgO-C 耐火材料进行维护。本研究提出了添加氧化镁和减少氧化铁来进行组分改性。在减少腐蚀和确保合理的熔化温度的前提下，结晶行为得到了预期的优化。研究结果表明，随着 MgO 含量的增加，渣中的 [FeO_4]—四面体从 0 增加到 19.1%，MgO 对 FeO 向 Fe_2O_3 的转变起到了促进作用。铁镁尖晶石（$MgO \cdot Fe_2O_3$）和固溶体（MgO-FeOss）产生了高熔点和 $FeO \cdot V_2O_3$ 析出减弱。熔融温度随着 MgO 含量的增加而升高，随着 T_{Fe} 含量的降低而降低。氧化镁的添加降低了渣的聚合度，T_{Fe} 的降低减少了晶相的析出，导致结晶活化能降低。MgO 含量（质量分数）为 12%、T_{Fe}（质量分数）为 16% 的钒渣满足熔渣飞溅对熔融温度和结晶倾向的要求。显微组织由分散分布转变为块状晶体结合带状固溶体，大大提高了溅渣层的耐蚀性。

Vanadium extracting BOF suffers serious corrosion from slag. Maintenance on **MgO-C refractory** based on slag splashing has not been applied because of high oxidizability, low CaO content and dispersed distribution crystalline phase of slag. The present study proposed MgO addition and iron oxides reduction for component modification. Crystallization behaviors were expectantly optimized on the premise of reducing corrosion and ensuring reasonable melting temperature. The results showed [FeO_4]-tetrahedral increased from 0 to 19. 1% in the slag structure with the increase of MgO content, and MgO played a role in motivating FeO change into Fe_2O_3. Pleonaste ($MgO \cdot Fe_2O_3$) and solid solution(MgO-FeOss) with high melting temperature generated and $FeO \cdot V_2O_3$ precipitation weakened. The melting temperature increased with the increase of MgO content and decreased with the decrease of T_{Fe} content. MgO addition reduced the polymerization degree of slag and T_{Fe} decrease reduced the precipitation of crystalline phases, which led to decreasing of crystallization activation energy. Vanadium slag with MgO = 12wt.% and T_{Fe} = 16% satisfied the demands on melting temperature and crystallization tendency for slag splashing. Microstructure changed from dispersed distribution to blocky crystals combined with banded solid solution which greatly promoted the corrosion resistance of slag splashing layer[3].

采用座滴法研究了在低氧分压的氩气氛下纯氧化铝（Al_2O_3）和**碳键合氧化铝（Al_2O_3-C）** 在 1625℃ 下与铁的相互作用（加热显微镜内进行 1h）。从纯氧化铝和铁水的实验中没有观察到晶须的形成。而在 Al_2O_3-C/Fe 体系的情况下，观察到了几种现象，包括 Al_2O_3-C 的脱碳、Al_2O_3-C 在铁水中的溶解、铁滴上 Al_2O_3 层的形成和 Al_2O_3 晶须的形成。为了阐明铁对 Al_2O_3 晶须形成的可能影响，将不存在铁的空白 Al_2O_3-C 试样在半密封铂坩埚中加热并于 1550℃ 保持 2h，检测到了 Al_2O_3 晶须的形成。最后，实验研究结果与热力学模拟相一致。

Pure alumina (Al_2O_3) and **carbon-bonded alumina (Al_2O_3-C)** are used to investigate their interaction with iron (Armco) at 1625℃, applying the sessile drop method. The investigations are conducted inside a heating microscope for 1h under argon atmosphere with low oxygen partial pressure. No whiskers formation is observed from the experiments with pure Al_2O_3 and molten iron. In the case of Al_2O_3-C/Fe system, several phenomena including the decarburization of Al_2O_3-C, Al_2O_3-C dissolution in molten iron, Al_2O_3 layer formation on the iron droplet, and Al_2O_3 whiskers formation are observed. To clarify the possible influence of iron on Al_2O_3 whiskers formation, blank Al_2O_3-C samples without the presence of iron are heated and held in a half-sealed platinum crucible at 1550℃ for 2h. Thereby, the Al_2O_3 whiskers formation is detected. Finally, the experimental investigations are supported with thermodynamic simulations[4].

Al_2O_3-C 耐火材料广泛用于钢的连铸工艺。在这项研究中，通过添加 Al_4SiC_4 粉末，获得了新型原位片状 Al_4O_4C 和多壁碳纳米管（MWCNTs）增强的 Al_2O_3-C 耐火材料。研究了 Al_4SiC_4 对 Al_2O_3-C 耐火材料显微组织和力学性能的影响，并讨论了 Al_4O_4C 和 MWCNTs 的生长机理。板状 Al_4O_4C 和 MWCNTs 均匀分布并与刚玉很好地结合，而不是仅仅在表面形成互锁结构，因此结合改善的界面结合力、载荷传递能力和裂纹扩展阻力。

The **Al_2O_3-C refractories** were used widely in the continuous casting process of the steel. In this research, novel in-situ plate-like Al_4O_4C and multi-walled carbon nanotubes(MWCNTs) reinforced Al_2O_3-C refractories were achieved with the addition of Al_4SiC_4 powder. The effect of Al_4SiC_4 on microstructure and mechanical performance of Al_2O_3-C refractories were studied, and the growth mechanism of Al_4O_4C and MWCNTs was discussed. The plate-like Al_4O_4C and MWCNTs homogenously distributed and bonded well with corundum rather than merely on the surface to form an interlocked structure, thus the mechanical properties of Al_2O_3-C specimens were enhanced by combining the advantages of improved interfacial bond, load transferring capacity and crack propagation resistance[5].

碳结合氧化铝（Al_2O_3-C）耐火材料具有高强度、低渣腐蚀和高抗热震性等优点，广泛用于功能部件，如滑动浇口、整体塞、长水口和浸入式水口。Al_2O_3-C 耐火材料的性能直接影响钢铁生产的连续性和稳定性。在高温下，Al_2O_3-C 耐火材料的碳含量较高，导致钢水中的碳吸收量较高，同时随着 Al_2O_3-C 耐火材料与钢水接触时间的延长而增强。因此，考虑到当前全球"低碳经济"环境和对洁净钢的严格要求，开发低碳/超低碳 Al_2O_3-C 耐火材料至关重要。但是，如果直接降低碳含量，可能会降低 Al_2O_3-C 耐火材料的断裂韧性、抗热震性和抗渣腐蚀性能。以往的研究表明,纳米碳能够降低耐火材料的总碳含量,同时保持,甚至一定程度上提高耐火材料的综合性能。

Carbon-bonded alumina (Al_2O_3-C) refractories are widely used in functional components, such as slide gates, monobloc stoppers, long nozzles, and submerged entry nozzles, because of their high strength, low slag corrosion, and high thermal shock resistance. The performance of Al_2O_3-C refractories can directly affect steel production continuity and stability. At high temperatures, higher carbon content of Al_2O_3-C refractories results in higher carbon pick-up in molten steel, which is intensified with the contact time between the Al_2O_3-C refractories and the molten steel. Therefore, considering the current global "low carbon economy" environment and the strict requirement for clean steel, developing low-/ultra-low-carbon Al_2O_3-C refractories is critical. However, the fracture toughness, thermal shock resistance, and slag corrosion resistance of Al_2O_3-C refractories may be degraded if the carbon content is reduced directly. Previous studies have shown the capability of nanocarbons to decrease the total carbon content while maintaining even improving the comprehensive properties of refractories[6].

浸入式水口是钢铁连铸工序中关键的功能耐火材料,其中以渣线部位的工作环境最为恶劣。目前,最适合的渣线材料是 **ZrO_2-C 材料**。为了提高浸入式水口的性能,本文以氧化锆与鳞片石墨为主要原料,添加增强材料氧化锆纤维及金属硅粉等,以酚醛树脂为结合剂制备 ZrO_2-C 复合材料。比较了 1000℃、1200℃ 和 1500℃ 三种热处理温度对 ZrO_2-C 材料的性能及显微结构的影响。结果表明,在热处理温度高于 1200℃ 时,ZrO_2-C 材料中的硅粉与石墨发生反应生成碳化硅,大量晶须状碳化硅与 ZrO_2 纤维交错在一起形成网络结构,提高了材料的力学性能和抗热震性[7]。

Submersed entry nozzle is critical functional refractories in continuous casting and the slag line parts has the worst working environment. At present, the most suitable slag line material is **ZrO_2-C material**. In order to improve the performance of submersed entry nozzle, ZrO_2-C material was prepared by using zirconia and flake graphite as the main raw materials, zirconia fiber and metal silica powder as reinforcement materials, and phenolic resin as binder. The effects of three heat treatment temperatures, 1000℃, 1200℃ and 1500℃, were compared on the properties and microstructure of ZrO_2-C material. The results show that when the heat treatment temperature is higher than 1200℃, the silicon powder reacts with carbon to form silicon carbide, and a large number of whisker silicon carbide interleaves with ZrO_2 fiber to form a network structure, which improves the mechanical properties and thermal shock resistance of the material[7].

Al₂O₃-MgO-C（AMC）耐火材料的主要优点是通过掺入石墨和氧化铝和氧化镁之间的固体反应形成尖晶石来实现的。关于氧化物-C 耐火材料的其他成员（例如 MgO-C 耐火材料）和其他性能（例如抗渣腐蚀性或 PLC），有关此类耐火材料的力学行为的信息很少。在这项工作中，通过在室温和 1000℃（氮气气氛）下的压缩应力-应变曲线研究了用于炼钢钢包的商用 AMC 砖的力学行为。在机械测试之前，通过多种技术对 AMC 材料进行了全面表征：XRD、DTA/TGA、SEM/EDS、骨料尺寸分布分析以及密度、孔隙率和热膨胀测量。与断裂的主要特征一起确定了断裂强度和应变、屈服应力和杨氏模量等力学参数。为了研究在高温停留期间发生的转变，在 1000℃下测试的试样采用与原样砖表征相同的技术进行分析（热膨胀分析除外）。发现 AMC 耐火材料在力学性能及其对测试温度的依赖性方面表现出差异。最后，从两种耐火材料的成分和微观结构及其热转变的差异解释了这些结果。

The advantages of **Al₂O₃-MgO-C(AMC) refractories** are achieved mainly by the incorporation of graphite and the formation of spinel by solid reaction between alumina and magnesia. Regarding other members of oxide-C refractories(such as MgO-C bricks) and other properties(such as the slag corrosion resistance or the PLC), the information about the mechanical behavior of this type of refractories is scarce. In this work, the mechanical behavior of commercial AMC brick used in steelmaking ladles was studied by stress-strain curves in compression at RT and 1000℃(nitrogen atmosphere). Before the mechanical testing, a comprehensive characterization of AMC materials was performed by several techniques:XRD,DTA/TGA,SEM/EDS, aggregate size distribution analysis and densities, porosities and thermal expansion measurements. Mechanical parameters such as fracture strength and strain, yield stress and Young modulus were determined together with the main characteristics of the fracture. In order to study the transformations occurred during the stay at high temperature, the specimens tested at 1000℃ were analyzed by the same techniques used for the as-received bricks characterization(with the exception of the thermal expansion analysis). The AMC refractories displayed differences in the mechanical behavior and its dependence on the testing temperature. These results were explained considering the differences in the composition and microstructure of both refractories and in their thermal transformations[8].

Al₂O₃-MgO-C 耐火材料是 Al₂O₃-C 复合体系中一类重要的多相材料。这些耐火材料主要应用在二次炼钢容器中，特别是在钢包中。这些多相耐火材料的复合结构构成了两个主要元素：（1）骨料；（2）基质相。骨料由粗和中等尺寸的氧化铝颗粒组成，而基体相由更细粒度的氧化铝、氧化镁、金属抗氧化剂和片状石墨粉末组合而成。相关文献中充分记录了组成相在与氧化铝骨料相邻的基体区域中的形成和生长行为对于控制这些耐火复合材料的微观结构评估和随后的结构性能具有决定性作用。有研究表明，Al_2O_3-MgO-C 耐火材料的结构性能在 1400 ~ 1600℃的使用温度范围内严重退化。这种结构性能的下降表明了耐火材料的微观结构损伤。这种弱化的微观结构导致裂纹从基体区域快速扩展到整体，从而导致强度低得多的耐火材料失效。

Al_2O_3-MgO-C refractories are the one important class of multiphase materials in Al_2O_3-C composite system. These refractories find applications in secondary steel making vessels, particularly in the steel ladles. The composite structure of these multiphase refractories constitutes two primary elements. They are：（1）aggregate phase; and （2）matrix phase. The aggregate phase is composed of coarse and medium size alumina particles. Whereas, the matrix phase is an assemblage of finer size alumina, magnesia, metal antioxidants and flake graphite powder. It is well-documented in the ceramic literature that the formation and growth of constituent phases throughout the matrix region adjoining alumina aggregates can have a decisive role in controlling the microstructure evaluation and consequent structural properties of these refractory composites. A series of studies showed that the structural properties of Al_2O_3-MgO-C refractories were severely degraded in the service temperature range of 1400-1600℃. This structural property degradation seriously indicates the microstructure damage in the refractory. Such a weakened microstructure is responsible for quick propagation of cracks from the matrix region through the bulk, thereby leading to refractory failure at much lower strength[9].

本文以电熔镁砂、电熔镁铝尖晶石、金属 Al 粉、Zn 粉和鳞片石墨为主要原料，以热固性酚醛树脂为结合剂，制备了 Al/Zn 复合低碳 **MgO-Al_2O_3-C 材料**，研究了金属 Al/Zn 对低碳 MgO-Al_2O_3-C 材料高温力学性能、抗氧化性影响以及它们与物相组成和显微结构的关系。研究了在埋碳加热过程中 Al/Zn 复合低碳 MgO-Al_2O_3-C 材料热态抗折强度及结构的变化。结果表明：加入 Al（质量分数为 2% ~6%）和 Al/Zn（质量分数为 4% ~6%）可以提高该材料的热态抗折强度，在高温阶段尤其显著；在 1400℃时，热态抗折强度从 2MPa（未加金属的试样）提高到 16 ~31MPa（Al/Zn 加入量为 4% ~6%）。Al 复合低碳 MgO-Al_2O_3-C 材料的热态抗折强度的变化可划分为三段：200 ~800℃，强度降低；800 ~1200℃，

强度提高；1200～1400℃，强度快速提高。Al 的最佳加入量（质量分数）为
4%。Al/Zn 复合低碳 MgO-Al$_2$O$_3$-C 材料的热态抗折强度也可划分为三段：200～
600℃，强度降低；600～1000℃，强度提高；1000～1400℃，强度快速提高。
Al/Zn 复合的最佳加入量（质量分数）为 Al 4%、Zn 1%。加入 Al 后，试样在
900℃和1000℃时原位反应生成了粒状 Al$_4$C$_3$ 和纤维状 AlN。到1200℃时 AlN 发
育长大呈针状，分布在方镁石骨架中，使材料的结合方式由原来的碳结合为主
（200～800℃）开始转变为非氧化物结合。而 Al/Zn 复合加入时，在 600℃出现
Al-Zn 熔融液相，液相中的 Al 分别与 C、N$_2$ 反应，生成 Al$_4$C$_3$ 和 AlN，Zn 在反应
中起到了催化作用；其反应温度比单加 Al 时的分别降低了 200℃和100℃，且使
材料开始转变为非氧化物结合的温度降低，从单加 Al 的1200℃降低至1000℃。
反应生成了大量纤维状 AlN 晶体，Zn 加入量（质量分数在 0.5%～2% 范围内）
越多，AlN 晶体就越细。它们交叉连锁形成网络，分布在方镁石骨架内，起到了
增强增韧的作用，提高了试样的热态抗折强度。在低碳 MgO-Al$_2$O$_3$-C 材料中加入
Al 和 Al/Zn 后，试样经受热震试验后（$\Delta T = 1100℃$），风冷原位生成了 AlN 纤
维，起到了增韧作用，提高了试样的抗热震性。风冷一次后，残余强度保持率从
60%（未加金属的试样）提高到62%～79%（Al/Zn 2%～6%）；风冷三次后，残
余强度保持率从48%（未加金属的试样）提高到66%～95%（Al/Zn 2%～6%）。
研究 Al/Zn 复合低碳 MgO-Al$_2$O$_3$-C 材料在 1500℃的抗氧化性和显微结构，结果
表明：加入 Al 和 Al/Zn 可以提高该材料在空气中的抗氧化性。未加入 Al 粉时，
试样全部被氧化。加入 Al 粉后，氧化层厚度随 Al 加入量的增加（质量分数
2%～6%）而减薄。在加 4% Al 的基础上再加入 Zn（质量分数 0.5%～2%）
后，试样的抗氧化性又有显著改善。加入 0.5% Zn 时氧化层厚度为 0.8mm，加
入 1%～2%Zn 时，氧化层厚度为 0.3～0.4mm。试样 MCA0（未加金属）、MCA4
（Al 4%）和 MCA4Z1（Al/Zn 5%）在 1500℃下分别经3h、6h 和 9h 氧化后，其
常温物理性能和高温力学性能发生了较大的变化。氧化 3h 后，三种试样的体积
密度、常温抗折强度和高温抗折强度衰减最大；氧化 3～9h 后，衰减变缓。显气
孔率的增加也符合这个规律。氧化 9h 后，试样的常温抗折强度和高温抗折强度
均符合以下次序：试样 MCA4Z1>试样 MCA4>试样 MCA0。试样 MCA4 和 MCA4Z1
在氧化最初的 3h 内，残余强度保持率增加了约 15%，继续延长氧化时间到 9h，
残余强度保持率开始降低，但始终高于未氧化试样。加入 Al 和 Al/Zn 能显著改
善低碳 MgO-Al$_2$O$_3$-C 材料在 1500℃抗氧化性的原因是：加入 Al 粉后，试样的孔
隙内生成了 MA 尖晶石，堵塞气孔，阻止了氧气的进入。而复合加入 Al/Zn 时，
不仅孔隙内有原位生成尖晶石堵塞气孔，更能阻止氧气进入的是试样的氧化层和
原砖层交界处生成的 MgO 致密层；该致密层产生的原因是：Zn 具有较高活性，
高温下它与砖内产生的还原性气体共同还原 MgO 生成 Mg(g)，Mg(g) 向试样外

部扩散，遇到氧气后生成 MgO，沉积在试样的孔隙中。根据实验室研究结果，开发了 Al/Zn 复合低碳 MgO-Al$_2$O$_3$-C 钢包渣线砖，其热态抗折强度、抗热震性和抗氧化性等明显优于高碳镁碳渣线砖；该砖在某钢厂 210t 精炼比为 15% 的钢包渣线进行了试用，使用寿命为 45 炉次，计算其平均蚀损率为 3.1mm/炉，比钢厂现用的高碳镁碳砖低 0.3mm/炉，初见成效[10]。

Al/Zn bearing low carbon MgO-Al$_2$O$_3$-C specimens have been prepared by using fused magnesia, fused spinel, Al powder, Zn powder and flake graphite as starting materials and phenolic resin as binder. The effect of Al/Zn addition on thermomechanical properties and oxidation resistance of low carbon **MgO-Al$_2$O$_3$-C materials** have been studied. Modulus of rupture vs temperature (MOR-T curves) and microstructure of Al/Zn bearing low-carbon MgO-Al$_2$O$_3$-C specimens under carbon embedded condition have been studied. The results show that addition of Al (2wt.% -6wt.%) or Al/Zn (4wt.%-6wt.%) to low carbon MgO-Al$_2$O$_3$-C specimens leads to noticeable increase of HMOR, especially at high temperatures, e. g. , at 1400℃, it increases from 2MPa (metal free specimens) to 16-31MPa (specimens with 4wt.% -6wt.% Al/Zn addition). Changes of MOR-T curves of Al bearing low-carbon MgO-Al$_2$O$_3$-C specimens can be divided into three stages: (1)200-800℃, HMOR decreases; (2)800-1200℃, HMOR increases; (3)1200-1400℃, HMOR increases dramatically. The optimum addition of Al is 4wt.%. Changes of MOR T curves of Al/Zn bearing low-carbon MgO-Al$_2$O$_3$-C specimens can be also divided into three stages: (1)200-600℃, HMOR decreases; (2) 600-1000℃, HMOR increases; (3) 1000-1400℃, HMOR increases dramatically. The optimum addition of Al/Zn is 4wt.% of Al and 1wt.% of Zn. Granular Al$_4$C$_3$ is in situ formed at 900℃ and fibrous AlN at 1000℃ by adding Al into low carbon MgO-Al$_2$O$_3$-C specimens. Prismatic AlN crystals fill in periclase skeleton structure after heating at 1200℃, indicating the mode of bonding transforms from carbon bonding (200-800℃) to monoxide bonding. When Al/Zn is added, the Al Zn low temperature eutectic mixture appears at 600℃, in which Al reacts with C and N$_2$ forming Al$_4$C$_3$ and AlN, and Zn acts as catalyst. Compared with single adding Al, adding Al/Zn decreases the reaction temperature between Al and C about 200℃, decreases the reaction temperature between Al and N$_2$ about 100℃, and decreases the temperature transforming to monoxide bonding from 1200℃ to 1000℃. A great deal of in situ formed fibrous AlN is generated by adding Zn. The more Zn (0.5wt.% -2wt.%) added, the thinner AlN fiber product. Fibrous AlN crystals interlock with each other and fill in periclase skeleton structure, leading to the increase of HMOR. Al/Zn bearing low carbon MgO-Al$_2$O$_3$-C specimens have been tested for thermal shock resistance at temperature difference of

1100℃. In situ formed AlN fibers intersperse in periclase skeleton structure, toughening the material and increasing the residual strength ratio from 60% (metal free specimen) to 62%-79% (specimens with 2wt.%-6wt.% Al/Zn addition) after one air cooling cycle and from 48% (metal free specimen) to 66%-95% (specimens with 2wt.%-6wt.% Al/Zn addition) after three air cooling cycles. Oxidation resistance and microstructure of Al/Zn bearing low carbon $MgO-Al_2O_3-C$ specimens at 1500℃ have been studied. The results show that addition of Al or Al/Zn to low carbon $MgO-Al_2O_3-C$ specimens is beneficial to reducing the thickness of oxidized layer. The specimen without additives has been oxidized completely, the more Al added (2wt.%-6wt.%), the thinner the generated oxidized layer. Oxidation resistance improves significantly with addition of 4wt.% Al and 0.5wt.%-2wt.% Zn, and the thickness of oxidized layer reduces to 0.8mm (0.5wt.% Zn), while 0.3-0.4mm is related to 1wt.%-2wt.% Zn addition. Physical properties at room temperature and thermal mechanical properties change dramatically when specimens MCA0 (metal free specimen), MCA4 (specimens with 4wt.% Al addition) and MCA4Z1 (specimens with 5wt.% Al/Zn addition) after oxidation at 1500℃ for 3h, 6h and 9h respectively. Bulk density, modulus of rupture and hot modulus of rupture of specimens decrease significantly after oxidation for the first three hours and then decrease slightly after oxidation for six more hours. The increase of apparent porosity is similar to these described above. The order of MOR after oxidation (at room temperature and 1400℃) for 9h is MCA4Z1>MCA4>MCA0. The residual strength ratio of specimens MCA4 and MCA4Z1 increases about 15% after oxidation for the first three hours and begins to decrease after six more hours oxidation, the residual strength ratio is still higher than that of the specimen unoxidized. Adding Al or Al/Zn improves oxidation resistance of $MgO-Al_2O_3-C$ specimens obviously. For adding Al, MA spinel forms and fills the pores, thus retarding O_2 infiltration. For adding Al/Zn, the formed MA spinel retards O_2 infiltration. What's more, Zn with high activity, together with the reducing gas generated in the brick, reduce MgO to Mg(g) at high temperatures, Mg(g) diffuses outsides and reacts with O_2 forming MgO, MgO deposits in the pores, the MgO dense layer forms, which retards the O_2 infiltration further. According to experimental results, Al/Zn bearing low carbon $MgO-Al_2O_3-C$ ladle slag line brick has been produced. Compared with conventional MgO C brick, Al/Zn bearing low carbon brick possesses higher hot strength, better thermal shock resistance and oxidation resistance. It has been successfully used in the slag line of a 210t ladle with 15% refining ratio, the service life reaches 45 heats and the average wear rate is 3.1mm/heat, 0.3mm/heat higher than the conventional MgO-C brick[10].

以 SiC 为抗氧化剂添加至 **MgO-CaO-C 耐火材料**中，研究了 SiC 对提高 MgO-CaO-C 耐火材料抗氧化性能的作用及抗氧化机理。通过 TG-DSC、XRD、显气孔率测定以及近似计算氧化层面积等进行分析和鉴定。结果表明：加入 SiC 后 MgO-CaO-C 耐火材料的抗氧化性能得到显著提高，结构更加致密，并确定 SiC 的最佳加入量为 4%。SiC 大约从 1210℃ 开始与氧气反应，反应后生成的 SiO$_2$ 继续和 MgO 反应生成镁橄榄石，填充了气孔形成致密氧化层阻止了碳的进一步氧化[11]。

The SiC added as an antioxidant in the **MgO-CaO-C refractories** to investigated on the antioxidant properties and antioxidant mechanism of MgO-CaO-C refractory. Apparent porosity, TG-DSC, XRD and approximate calculation of the oxide layer area showed that antioxidant property and density of MgO-CaO-C refractories has been significantly improved. The reasonable proportion of SiC is 4% in weight was determined. Moreover, the oxidation temperature of SiC was 1210℃, and the generation of forsterite prevents further oxidation of carbon by filling stoma[11].

随着洁净钢品质的提高和需求量的增加，对炉外精炼钢包用耐火材料要求越来越高。碱性耐火材料特别是镁钙材料对钢水有一定净化作用，但目前钢包使用的 **MgO-CaO-C 材料**碳含量一般较高（C 质量分数≥8%）会使钢水增碳。因此研究金属复合超低碳 MgO-CaO-C 材料（C 质量分数≤3%）具有重要的意义。本工作以烧结镁钙砂、电熔镁砂、Al 粉、Zn 粉、石墨及炭黑为原料制备了超低碳 MgO-CaO-C 材料，研究了金属 Al、Al/Zn 对超低碳 MgO-CaO-C 材料的高温抗折强度、热震稳定性以及抗氧化性能的影响，并探讨了材料结构与性能的关系。加入 2%~6%（质量分数）Al 能够显著提高超低碳 MgO-CaO-C 材料 1400℃ 高温抗折强度和 1100℃ 风冷 3 次后的残余强度，表现在高温抗折强度提高了 2~4.6 倍，残余强度提高了 0.5~1.7 倍；残余强度保持率保持在较好的水平（≥66%）；加入 Al 后超低碳 MgO-CaO-C 材料的抗氧化性能有明显改善，未加 Al 试样 1000℃、1500℃均全部氧化脱碳，随 Al 加入量的增多试样的氧化层厚度逐渐变薄[12]。

With the improving quality and the demand quantity of clean steel, the requirements of refractories for ladle refining are more and more highly. As we know, the steel liquid can be purified by basic refractories especially **MgO-CaO-C materials**. But MgO-CaO-C materials used at present contain too much carbon (C≥8wt.%), which has bad effect on steel purifying. So, it is significantly important to study metal containing ultra-low carbon MgO-CaO-C materials (C≤3wt.%). In this work, ultra-low carbon MgO-CaO-C materials were prepared using sintered magnesia-calcium clinker, fused magnesia and Al powder. The effect of Al and Al/Zn addition on ultra-low carbon MgO-CaO-C materials was inspected. The physical properties, hot modulus of rupture (HMOR), thermal shock resistance (TSR) and oxidation resistance of synthesized MgO-CaO-C materials were researched, also the relation of structure and property was investigated. The HMOR (1400℃) and the residual strength ($\Delta T = 1100℃$, air-cooling three times) of ultra-low carbon MgO-CaO-C materials added with 2wt.%-6wt.% can be significantly improved by adding Al powder. The HMOR of specimens improved by 2 to 4.6 times and the residual strength improved by 0.5 to 1.7 times after Al addition. The residual strength ratio maintains at a high level ($\geq 66\%$). Also, oxidation resistance of specimens containing Al powder is improved remarkably. Compared with complete oxidation of carbon in specimens without Al powders at 1000℃ and 1500℃, the oxidation resistance of specimens with Al content is quite good and with the Al content increasing, the thickness of oxidization layer becomes thinner[12].

以板状氧化铝、活性氧化铝粉、锆莫来石、锆刚玉、鳞片石墨为原料，纳米 ZrO_2、纳米 TiO_2 和纳米 Al_2O_3 粉末为添加剂，在石墨包埋条件下，在1500℃条件下制备了纳米氧化物**低碳 Al_2O_3-ZrO_2-C 耐火材料**。详细研究了纳米 ZrO_2、纳米 TiO_2 和纳米 Al_2O_3 粉末作为单一添加剂对 **Al_2O_3-ZrO_2-C 耐火材料**性能和微观结构的影响。研究结果表明，纳米氧化物在碳化硅纳米线（NWs）的形成和生长中起催化作用，导致碳化硅纳米线的长径比（L/D）增加。添加纳米氧化物的试样具有更好的烧结性、抗热震性和抗氧化性。

Low carbon Al_2O_3-ZrO_2-C refractories with nano-oxides are prepared at 1500℃ in graphite embedded condition using tabular alumina, reactive alumina powder, zirconia-mullite, zirconia-corundum and flake graphite as well as nano-ZrO_2, nano-TiO_2 and nano-Al_2O_3 powders as additives. The effects of nano-ZrO_2, nano-TiO_2 and nano-Al_2O_3 powders as single additive on the properties and microstructure of **Al_2O_3-ZrO_2-C refractories** are investigated. The results show that nano-oxides take the catalytic effect in SiC nanowires (NWs) formation and growth, leading to increasing in length-diameter ratio (L/D) of SiC NWs. Samples with nano-oxides addition possess improved sinterability, thermal shock resistance and oxidation resistance[13].

以锆石、硼酸和活性炭为起始原料，在氩气气氛下，采用微波辅助碳/硼热还原法合成了 ZrB_2-SiC_w（w-晶须）复合粉末。通过调整加热温度、硼酸和活性炭的用量，得到了性能最佳的 ZrB_2-SiC_w 复合粉体。此外，研究了 ZrB_2-SiC_w 复合粉末添加对 **Al_2O_3-ZrO_2-C 滑板材料**对 O_2 和 Ca 处理钢的耐腐蚀性的影响。研究结果表明，ZrB_2-SiC_w 的相对含量随温度升高而增加。过量的硼酸有利于 ZrB_2-

SiC_w 的形成，硼酸的最佳用量过量至 30% ～ 45%（质量分数）。再添加 6%（质量分数）的 ZrB_2-SiC_w 复合粉末时可有效提高 Al_2O_3-ZrO_2-C 滑板材料的耐腐蚀性和热力学性能，这可归因于 ZrB_2-SiC_w 复合粉末优异的抗氧化和耐腐蚀性能。此外，ZrB_2-SiC_w 可通过减少裂纹的产生而提高滑板材料的抗热震性，最终防止腐蚀介质渗入滑板，从而进一步提高板的耐腐蚀性[14]。

ZrB_2-SiC_w (w-whisker) composite powders were synthesized by a microwave-assisted carbo/borothermal reduction method under argon atmosphere using zircon, boric acid, and activated carbon as starting materials. The optimized ZrB_2-SiC_w composite powder was obtained by adjusting heating temperature and amounts of boric acid and activated carbon. Further, the effect of the optimized ZrB_2-SiC_w composite powder addition on corrosion resistance of the **Al_2O_3-ZrO_2-C slide plate material** against O_2 and Ca-treated steel was investigated. The results showed that the relative content of ZrB_2-SiC_w increased with increasing temperature. The excess of boric acid favored the formation of the ZrB_2-SiC_w, and the optimum amount of boric acid was overdosed to 30wt.% - 45wt.%. The corrosion resistance and thermo-mechanical properties of Al_2O_3-ZrO_2-C slide plate material could be improved effectively with 6wt.% ZrB_2-SiC_w composite powder addition, which was attributed to the excellent oxidation and corrosion resistance of ZrB_2-SiC_w composite powder. In addition, ZrB_2-SiC_w could confer excellent thermal shock resistance and reduce the generation of cracks, ultimately preventing corrosive media from infiltrating into slide plates, further increasing corrosion resistance of the plates[14].

硅、微硅粉及其组合分别被用作含碳耐火材料的添加剂。研究了这些添加剂对 **Al_2O_3-ZrO_2-C 耐火材料**的微观结构和力学性能的影响。研究结果表明，在这类耐火材料中，硅是形成 SiC 晶须的原因；微硅粉主要促进莫来石的形成；而它们的组合（硅+微硅）会带来高分压的 SiO(g)，并导致更细的 SiC 晶须和针状莫来石共存。不同的微观结构无疑会导致 Al_2O_3-ZrO_2-C 耐火材料的力学性能不同。硅的引入主要导致机械强度的提高，而微硅粉的添加主要促进韧性的提高。当硅和微硅粉组合用作添加剂时，Al_2O_3-ZrO_2-C 耐火材料可以同时获得优异的强度和韧性，这主要是由于硅添加剂形成的 SiC 晶须和针状莫来石形成的协同效应。

Silicon, microsilica and their combination were used as additive in carbon containing refractories, respectively. The effects of such additive on microstructures and mechanical properties of **Al_2O_3-ZrO_2-C refractories** were investigated. The results show that in refractories of this kind, silicon is responsible for the formation of SiC whiskers; microsilica mainly dominates the formation of mullite; while their combination (silicon plus microsilica) brings high partial pressure of SiO(g) and leads to the co-existence of finer SiC whiskers and needle-like mullite. The various microstructures unquestionably cause differences in mechanical properties of Al_2O_3-ZrO_2-C refractories. The use of silicon mainly results in an enhanced mechanical strength, while the addition of microsilica primarily triggers the improvement in toughness. When the combination of silicon and microsilica is used as additive, excellent strength and toughness can be obtained simultaneously in Al_2O_3-ZrO_2-C refractories, which is mainly attributed to the synergistic effects of SiC whiskers formed from silicon additive and needle-like mullite formed from microsilica additive[15].

碳结合镁砂和镁铝砖是钢包的侧壁和底部的最先进的内衬材料。工业试验测试表明，新一代 **MgO-MgAl$_2$O$_4$-C 砖**，其中以新的铝酸钙镁原料的形式添加了预反应尖晶石，由于耐腐蚀性增强，使用寿命更长。在这项工作中，进行了实验室级别的腐蚀测试，以模拟保护渣层的形成并研究形成的动力学。研究结果发现，形成了高度依赖于熔体中存在的铁的量，这会导致褐铁矿相的沉淀。此外，该研究还扩展到通过有针对性地调整砖的成分来强制形成不同成分的保护渣层，这些保护渣层在更高的温度和更宽的温度范围内是可靠的。坩埚试验表明，通过有针对性地调整砖的成分，可以有效控制保护渣层的成分，尤其是来自砖或熔渣的 MgO、Al_2O_3、Fe_xO_y 和 SiO_2 在熔渣/耐火材料界面富集。

Carbon-bonded magnesia and magnesia-alumina bricks are the state-of-the-art lining materials of the sidewalls and the bottom of steel ladles. Industrial trial tests revealed that a new generation of **MgO-MgAl$_2$O$_4$-C bricks**, where pre-reacted spinel is added in form of a new calcium magnesium aluminate (CMA) raw material, exhibit a longer service life caused by an enhanced corrosion resistance due to the formation of a protective slag layer. In terms of this work, laboratory corrosion tests have been performed in order to mimic the protective slag layer formation and to study the kinetics of the formation. It has been found that the formation highly depends on the amount of iron present in the melt, which leads to the precipitation of the brownmillerite-phase. Furthermore, the study was extended to force the formation of protective slag layers of different composition which are reliable at higher temperatures and wider temperature ranges by targeted adaption of the brick composition. Cup tests showed that there is a potential to manipulate the composition of a protective slag layer by targeted adaption of the brick composition. Especially MgO, Al_2O_3, Fe_xO_y, and SiO_2 from the brick or slag enrich at the slag/refractory-interface[16].

为实现特钢的优质稳定生产，低碳 MgO-C 耐火材料的性能需要进一步优化。为此，通过引入 Al_2O_3 作为增强剂和 La_2O_3 作为改性剂，设计并成功制备了具有增强的抗热震性和抗渣性的低碳 **MgO-Al$_2$O$_3$-La$_2$O$_3$-C 耐火材料**。研究结果表明，添加添加剂的耐火试样比未添加添加剂的耐火试样表现出更好的综合性能。添加 10%（质量分数）的 Al_2O_3 和 La_2O_3 时，1400℃焦化的耐火材料试样的抗氧化性、抗热震性和抗渣性分别提高了 13.57%、17.75% 和 43.09%。分析发现，这主要归因于 $MgAl_2O_4$、Mg_2SiO_4 和 $2CaO \cdot 4La_2O_3 \cdot 6SiO_2$ 的形成以及随之而来的体积膨胀效应和晶间相增强效应。因此，本研究提出了一种低成本且可执行的低碳 MgO-C 耐火材料补强策略，有望在炼钢中得到应用。

In order to achieve high-quality and stable production of special steel, the performance of low-carbon MgO-C refractories needs to be further optimized. For this purpose, low-carbon **MgO-Al$_2$O$_3$-La$_2$O$_3$-C refractories** with enhanced thermal shock resistance and slag resistance were designed and successfully prepared by introducing Al_2O_3 as a reinforcer and La_2O_3 as a modifier. The results showed that the refractory samples with additives show better overall performance than those without additives. When 10wt.% of Al_2O_3 and La_2O_3 were added, the oxidation resistance, thermal shock resistance and slag resistance of the refractory samples coked at 1400℃ are increased by 13.57%, 17.75% and 43.09%, respectively. The analysis found that this can be mainly attributed to the formation of $MgAl_2O_4$, Mg_2SiO_4, and $2CaO \cdot 4La_2O_3 \cdot 6SiO_2$ and the consequent volume expansion effect and intergranular phase enhancement effect. Therefore, a low-cost and enforceable reinforcement strategy for low-carbon MgO-C refractories is proposed, which is expected to be applied in steelmaking[17].

近年来，随着超低碳钢和洁净钢冶炼需求的不断增加，对镁碳砖提出了更高的要求。为不向钢水中增碳，以及综合考虑钢水洁净度、冶炼热损耗和使用寿命，降低碳含量已成为镁碳砖发展的重要方向。然而，碳含量降低会引发一系列的问题，如易氧化、抗侵蚀能力下降、使用寿命降低等。Al_4SiC_4 材料具有优异的抗氧化、抗侵蚀性能，常温下物理化学性质稳定；将其引入低碳镁碳砖，有望弥补镁碳砖因石墨含量减少而导致的相关性能下降等问题，从而得到较为出色的使用性能。为此，本论文首先以金属 Al 粉、金属 Si 粉和炭黑为原料，采用固相反应烧结法合成出了纯净的 Al_4SiC_4 粉体，并对其合成机制进行了热力学计算和理论分析；而后对其在空气和 MgO-C 体系下的高温抗氧化性和稳定性等进行了研究；当反应温度达到 1500℃时，能够得到较为纯净的 Al_4SiC_4 粉体。Al_4SiC_4 粉体在空气条件下的氧化开始于 850℃，当温度在 1200℃以下时，Al_4SiC_4 的氧化主要是 Al_4SiC_4 表面的 Al 元素先行被氧化，导致 Al_4SiC_4 表面的 Al 元素含量减少，内部 Al 元素向外迁移，引起了 Al₄SiC₄ 结构的坍塌、劣变，而 Si 元素在此演变过程中较 Al 元素稳定，未被氧化而维持 SiC 结构；当温度高于 1200℃时，随着表面结构中 Al 元素的氧化和结构坍塌，Si 元素也明显被氧化，SiO₂ 的生成量不断提高，导致增重加剧，同时氧化产物进一步反应生成莫来石。热力学计算表明，在 MgO-C 体系中，随着温度的升高，体系内的 CO(g) 的分压不断升高，O₂(g) 的分压不断降低；在该氧分压下，Al₄SiC₄ 将被氧化，发生一系列反应。当反应温度低于 1400℃时，Al₄SiC₄ 氧化的产物趋向于形成 **Al₂O₃-Al₆Si₂O₁₃-C 体系**；而当温度继续升高时，Al₄SiC₄ 氧化的产物则趋向于 **Al₂O₃-SiC-C 体系**。通过对添加 Al₄SiC₄ 的镁碳体系在 1400～1600℃的分析，发现试样中的 MgAl₂O₄ 的数量和晶粒的尺寸也有所增加，在 1600℃的氧化温度下出现了 SiC 的衍射峰，这与热力学计算的结果是相符合的。试样中 Al₄SiC₄ 的生成进一步强化了 MgO-C 体系的力学性能，降低气孔率，提升抗氧化和抗侵蚀性能。鉴于 Al₄SiC₄ 的合成成本以及镁铝尖晶石的生成对低碳镁碳砖性能的积极作用，在对低碳镁碳砖中引入不同含量 Al₄SiC₄ 的研究之前，对低碳镁碳砖的微结构进行了优化实验，即通过尖晶石物相控制镁碳砖的微结构和抗熔渣渗透性。为此，分别以 α-Al₂O₃ 微粉、板状刚玉和电熔镁铝尖晶石等为添加剂进行了系列实验，最终优化得到 α-Al₂O₃ 微粉的添加量（质量分数）为 4% 时，低碳镁碳砖表现出最优的使用性能。在此基础上进行了 Al₄SiC₄ 不同添加量对低碳镁碳砖性能的影响。当 Al₄SiC₄ 的添加量（质量分数）为 8% 时，低碳镁碳砖的抗氧化性和抗熔渣侵蚀性能都得到了提升。将添加 Al₄SiC₄ 为 8% 的低碳镁碳砖进行工业化制备，并在国内某钢厂 210t 钢包精炼炉进行工业应用试验。尽管该试验的低碳镁碳砖的碳含量（质量分数）仅为 3%，却达到了碳含量为 12% 的传统镁碳砖的使用效果，抗熔渣侵蚀性和抗渗透性表现都很好，使用寿命达到了 50 次。添加 Al₄SiC₄ 的低碳镁碳砖在转炉、精炼钢包等具有非常大的应用潜力和商业价值[18]。

In recent years, with the increasing demand for ultra-low carbon steel and clean steel smelting, higher requirements have been placed on MgO-C bricks. In order not to add carbon to molten steel, as well as considering the cleanliness of molten steel, smelting heat loss and service life, reducing carbon content has become an important direction for the development of MgO-C refractories. However, the reduction of carbon content will cause a series of problems, such as easy oxidation, reduced corrosion resistance, and reduced service life. Al₄SiC₄ has excellent oxidation resistance and corrosion resistance, and its physical and chemical properties are stable at room temperature. The introduction of it into low-carbon MgO-C refractory materials is expected to make up for the related performance degradation of MgO-C bricks caused by the reduction of graphite content, and obtain excellent performance. The thesis first uses Al, Si and carbon black powders as raw materials to synthesize pure Al_4SiC_4 powder by solid-phase sintering, and conducts thermodynamic calculations and process analysis of its synthesis mechanism; the high-temperature oxidation resistance and stability under air and MgO-C systems were studied. When the reaction temperature reaches 1500℃, relatively pure Al_4SiC_4 powder can be obtained. The oxidation of Al_4SiC_4 powder under air conditions starts at 850℃; when the temperature is below 1200℃, the oxidation of Al_4SiC_4 is mainly due to the first oxidation of the Al on the surface of Al_4SiC_4, which leads to the reduction of the Al content on the Al_4SiC_4 surface and the immigration of internal Al to outside. The external migration causes the collapse and deterioration of the Al_4SiC_4 structure; during this evolution, the Si is more stable than the Al and maintains the SiC structure without being oxidized; when the temperature is higher than 1200℃, Al and Si element was obviously oxidized, and the content of SiO_2 continued to increase, resulting in weight increasing, and at the same time the oxidation product further formed mullite. Thermodynamic calculations show that in the MgO-C system, as the temperature rises, the partial pressure of CO(g) in the system continues to increase, and the partial pressure of O_2(g) continues to decrease; under this partial pressure, Al_4SiC_4 will is oxidized, a series of reactions occur. When the reaction temperature is lower than 1400℃, the oxidation product of Al_4SiC_4 tends to form **Al_2O_3-$Al_6Si_2O_{13}$-C system**; and when the temperature continues to rise, the oxidation product of Al_4SiC_4 tends to **Al_2O_3-SiC-C system**. Through the analysis of magnesia carbon bricks with Al_4SiC_4 added at 1400-1600℃, it is found that the amount of $MgAl_2O_4$ in the sample and the size of its crystal grains have increased, and the diffraction peak of SiC appears at the oxidation temperature of 1600℃, which is accordance with thermodynamics. The formation of $MgAl_2O_4$ in the sample further strengthens the mechanical properties of the MgO-C system, reduces porosity, and improves oxidation and corrosion resistance. In view of the synthesis cost of Al_4SiC_4 and the positive effect of the formation of $MgAl_2O_4$ on the performance of low-carbon MgO-C bricks, the microstructure of low-carbon MgO-C bricks should be studied before the introduction of different contents of Al_4SiC_4. The optimization experiment is to control the microstructure and slag penetration resistance of magnesia-carbon bricks through the spinel phase. For this reason, α-Al_2O_3 powder, tabular corundum and fused $MgAl_2O_4$ are used as additives to finally optimize. When the addition amount of α-Al$_2$O$_3$ micropowder is 4wt.%, the low-carbon magnesia-carbon brick shows the best performance. And on this basis, Al$_4$SiC$_4$ was introduced as an additive into low-carbon magnesium-carbon refractories. When the addition amount of Al$_4$SiC$_4$ is 8wt.%, the oxidation resistance and slag erosion resistance of low-carbon magnesia-carbon bricks have been improved. The low-carbon MgO-C bricks with Al$_4$SiC$_4$ added were industrially prepared and applied to a 210-ton ladle refining furnace in a domestic steel plant for industrial application tests. Although the carbon content of the low-carbon magnesia-carbon brick in this test is only 3wt.%, it achieves the similar use effect of the traditional magnesia-carbon brick with a carbon content of 12wt.%. The resistance to slag erosion and penetration is very good, and the service life is up to 50 times. Low-carbon magnesia-carbon bricks with Al$_4$SiC$_4$ have great application potential and commercial value in converters, refining ladle, etc[18].

6.2 碳-氧化物-非氧化物系耐火材料原料

6.2.1 术语词

石墨 graphite
碳 carbon

镁砂 magnesia
刚玉 corundum

6.2.2 术语词组

黑色碳化硅 black carborundum, black silicon carbide
结晶碳化硅 crystalline silicon carbide
立方碳化硅 cubic silicon carbide
绿碳化硅 green silicon carbide
反应烧结碳化硅 reaction sintered silicon carbide
二次碳化硅 secondary silicon carbide
自结合碳化硅 self-bonded silicon carbide
碳化硅粉 silicon carbide powder

鳞片石墨 flake graphite
轻烧镁砂（苛性镁砂）light burned magnesia, caustic calcined magnesia
重烧镁砂（死烧镁砂）dead burned magnesia
电熔镁砂 fused magnesia
烧结刚玉 sintered corundum
电熔刚玉 fused alumina, fused corundum, electromelting corundum

6.2.3　术语例句

综述了**碳化硅**反射镜材料常用制备方法的特点。
The fabrication process and application of **silicon carbide** mirror are presented.

采用燃烧合成技术制备多孔**碳化硅**基复合材料。
Porous **silicon carbide** based composite was prepared by combustion synthesis technique.

研究对无压烧结**碳化硅**陶瓷的实际生产提供参考依据，具有重要的应用前景。
The researches were important for offering the reference for the production of **silicon carbide** ceramic.

在**碳化硅**质耐火材料中，黏土结合碳化硅是生产工艺最简单、价格最低的制品。
Refractory in the **silicon carbide**, clayey SiC has the most simple production process and the lowest price.

细晶、裂纹偏转和晶粒桥联是**碳化硅**陶瓷的主要增韧机制。
The fine grains, crack deflexion and grain-bridging are main toughening mechanisms of **silicon carbide** ceramic.

该薄膜大致上可为**氮化硅**。
The film may be substantially **silicon nitride**.

氮化硅是优良的陶瓷材料，应用广泛。
Silicon nitride is a kind of excellent ceramic material.

氮化硅陶瓷是一种较好的滚动轴承材料。
Silicon nitride ceramic is a kind of fairly good material for rolling bearings.

烧结助剂是影响**氮化硅**陶瓷的显微结构和性能的关键因素之一。
Sintering aids were one of key factors affecting microstructure and properties of

silicon nitride ceramics.

浸渍强化处理是提高反应烧结**氮化硅**（RBSN）力学性能的有效措施。

Infiltration strengthening is an effective means to improve the mechanical properties of Reaction Bonded **Silicon Nitride**(RBSN).

6.2.4　术语例段

Al_2O_3-SiC-C(ASC)浇注料采用棕刚玉、SiC、超细 α-Al_2O_3、微硅粉、硅粉、铝酸钙水泥和改性煤焦油沥青(CP)作为起始材料。研究了 CP 含量和烧成气氛对 ASC 浇注料显微组织和性能的影响。结果表明，随着 CP 含量的增加，ASC 浇注料的冷强度、热强度、抗热震性和抗渣性略有提高。以含 2%（质量分数）CP 的浇注料为例，在氧化气氛和还原气氛中烧制后的试样的冷强度和热强度均高于在弱氧化气氛下烧制的试样，在还原气氛中烧制后的试样气氛具有较高的强度、良好的抗热震性和抗渣性等综合性能。不同 CP 添加量的 ASC 浇注料的性能差异与其相组成、微观结构(包括孔径分布)和原位形成的 SiC 晶须量密切相关。

Al_2O_3-SiC-C(ASC) castables were performed using brown alumina, SiC, ultrafine α-Al_2O_3, microsilica, Si powder, calcium aluminate cement and modified coal tar pitch (CP) as starting materials. Effect of CP content and firing atmosphere on the microstructure and properties of ASC castables were investigated. The results show that cold strength, hot strength, thermal shock resistance and slag resistance of ASC castables were slightly improved with increasing of CP content. As an example of castables containing 2wt.% CP, it can be found that the cold strength and hot strength of the sample after firing in oxidizing atmosphere and reducing atmosphere were higher than that of in weak oxidizing atmosphere, and the sample after firing in reducing atmosphere possess overall properties including higher strength, good thermal shock resistance and slag resistance. The differences in properties of ASC castables with different CP addition were closely related to their phase compositions, microstructure including pore size distribution, and the amount of in-situ formed SiC whiskers[19].

在腐蚀界面上形成致密层以抑制腐蚀总是需要的，但它受许多环境条件的控制。通过工业试验，对钢包炉金属浴区 **MgO/Al_2O_3-SiC-C** 耐火材料的腐蚀显微组织进行了研究。在含 6%（质量分数）粗/细 SiC 添加剂耐火材料的腐蚀界面上形成了由 MgO 层和液相层组成的隔离层。形成的抵抗钢/熔渣侵蚀的隔离层导致耐腐蚀性比具有 3%（质量分数）细 SiC 添加剂的耐火材料提高约 30%。更重要的是，液相隔离层阻断了钢水与耐火材料之间的直接传质，同时减少了耐火材料的外源污染。SiC 添加剂通过控制 Mg(g) 在耐火材料表面的生成/迁移来影响隔离层的形成过程。

Formation of a dense layer on corroded interface to suppress corrosion is always desired, but it is controlled by numerous environmental conditions. In this work, corroded microstructures of **MgO/Al$_2$O$_3$-SiC-C** refractories in metal bath area of ladle furnace were investigated after industrial trails. A liquid-phase isolation layer in which MgO islands and liquid phases was established on the corroded interface of refractories with 6wt.% coarse/fine SiC-additive. The formed isolation layer against steel/slag attacks led to an approximate 30% improvement in corrosion resistance than that of refractory with 3wt.% fine SiC-additive. More importantly, the liquid-phase isolation layer blocked the direct mass transfer between molten steel and refractories while it decreased exogenous pollution from refractories. SiC-additive affected the formation process of isolation layer by controlling the generation/migration of Mg(g) on refractory' surface. A further formation mechanism of liquid-phase isolation layer was discussed in detail and role of SiC was elucidated[20].

本文研究了 B$_2$O$_3$ 添加量（质量分数分别为 0.4%、0.8%、1.2% 和 1.6%）对 **Al$_2$O$_3$-SiC-SiO$_2$-C**（ASSC）耐火材料性能和微观结构的影响。研究结果表明，B$_2$O$_3$ 对促进低温下莫来石形成具有重要作用。莫来石晶体中的 Si 被 B 取代导致在 1450℃烧制后形成短而强的 B—O 键，导致形成孔体积较小的掺硼莫来石，导致整体膨胀减小 ASSC 耐火材料。添加 B$_2$O$_3$ 向 ASSC 耐火材料加速了莫来石沿平行 c 轴方向的生长，导致莫来石形貌改善（针状）。原位掺硼莫来石的存在提高了这类耐火材料的力学性能。在不同的热处理温度下，试样的力学性能显著提高。此外，B$_2$O$_3$ 的加入有利于提高试样的体积密度，因为富硼液相的形成可以加速烧结和致密化。B$_2$O$_3$ 的最佳用量试样中添加量（质量分数）小于 1.2%，超过该值，会形成过量的液相，导致高温性能下降。

This study investigates the effect of B$_2$O$_3$ addition(0.4wt.%, 0.8wt.%, 1.2wt.%, and 1.6wt.%) on the properties and microstructure of **Al$_2$O$_3$-SiC-SiO$_2$-C**(ASSC) refractories. The results indicate that B$_2$O$_3$ plays an important role in accelerating mullite formation at low temperature. The substitution with Si by B in mullite crystal resulted in the formation of short and strong B—O bonds after firing at 1450℃, causing the formation of boron-doped mullite with a smaller cell volume and leading to a decrease in the overall expansion of ASSC refractories. The addition of B$_2$O$_3$ to the ASSC refractories accelerated the growth of mullite along the direction parallel to the c axis, resulting in the improvement of mullite morphology (needle-like). The presence of in situ boron-doped mullite allowed the performance improvement of this class of refractory materials. The mechanical properties of the specimens improved significantly at various thermal-treatment temperatures. Moreover, the addition of B_2O_3 was beneficial for increasing the bulk density of the specimens, owing to the formation of a boron-rich liquid phase, which could accelerate sintering and densification. The optimal amount of B_2O_3 added in the specimens was< 1.2wt.% , as above this value, an excessive liquid phase was formed, which degraded the high temperature performance[21].

由于 SiO_2-MgO 基渣熔炼过程中对耐火材料的严重破坏，限制了渣制备高附加值产品的发展。在目前的研究中，选择 Al_2O_3-SiC-C 耐火材料在空气和还原气氛中对 SiO_2-MgO 基渣进行静态电阻测试。结果表明，在这两种气氛下都有良好的耐腐蚀性，但耐渗透性较差。渗透性差可归因于铝酸镁尖晶石形成过程中体积膨胀引起棕刚玉骨料开裂。此外，提出了骨料裂纹扩展模型来说明 Al_2O_3-SiC-C 耐火材料的腐蚀机理。通过对裂纹径向长度和法向长度的计算表明，骨料的裂纹扩展主要沿法向方向进行。

The development of high value-added products prepared by slag has been restricted by the severe damage to refractories occurring during SiO_2-MgO-based slag melting. In the current study, the **Al_2O_3-SiC-C** refractories were selected to conduct static resistance tests with SiO_2-MgO-based slag in air and reducing atmospheres. The results show favorable corrosion resistance but poor permeability resistance in both atmospheres. The poor permeability can be attributed to the cracking of brown fused alumina aggregates caused by volume expansion during the formation of magnesium aluminate spinel. In addition, a model of crack growth for the aggregate is proposed to illustrate the corrosion mechanism of the Al_2O_3-SiC-C refractories. It is indicated that the crack growth of the aggregate proceeds mainly along the normal direction based on the calculation of the radial and normal lengths of the crack[22].

以电熔镁砂、鳞片石墨和 Si 粉负载的 $Fe(NO_3)_3 \cdot 9H_2O$ 为原料，酚醛树脂为黏结剂，在 N_2 流动下制备 **Si_3N_4-MgO-C** 耐火材料。研究了不同氮化温度对未催化耐火材料中 Si_3N_4 形貌和生成的影响。β-Si_3N_4 形貌的影响对不同催化剂含量的耐火材料的性能进行了研究。还研究了耐火材料的抗氧化性。结果表明，氮化温度的升高有利于耐火材料中 β-Si_3N_4 的形成。然而，非常高的渗氮温度会在耐火材料中产生大量气孔并降低其性能。含1%（质量分数）催化剂的耐火材料表现出良好的力学性能和抗氧化性。然而，过量催化剂的加入抑制了 β-Si_3N_4 的形成，降低了耐火材料的性能。

Si₃N₄-MgO-C refractories were prepared using fused magnesia, flake graphite, and Si powder-supported $Fe(NO_3)_3 \cdot 9H_2O$ as the raw materials and phenolic resin as the binder under the flow of N_2. The effect of different nitriding temperatures on the morphology and production of Si_3N_4 in the uncatalyzed refractories was investigated. The effect of the morphology of β-Si_3N_4 on the properties of the refractories with different catalyst contents was investigated. The oxidation resistance of the refractories was also investigated. The results showed that the increase in the nitriding temperature was beneficial for the formation of β-Si_3N_4 in the refractories. However, very high nitriding temperatures generated a large number of pores in the refractories and deteriorated their properties. The refractory with 1wt.% catalyst exhibited good mechanical properties and oxidation resistance. However, the addition of excessive catalyst inhibited the formation of β-Si_3N_4 and deteriorated the performance of the refractories[23].

使用铝（Al）、硅（Si）和碳（C）的粉末混合物在1800℃氩气中合成高纯度 Al_4SiC_4 粉末。在1400~1600℃下研究了它们在 **MgO-C-Al₄SiC₄** 系统中的氧化行为和机理。采用 XRD、SEM 和能谱仪（EDS）分析微观结构和相演变。研究结果表明，氧化产物的组成与原子扩散速度密切相关，在 Al_4SiC_4 表面生成了复合氧化物层。此外，还通过热力学计算研究了不同 CO 分压对 Al_4SiC_4 晶体氧化的影响。这项工作证明了 Al_4SiC_4 在改进 MgO-C 耐火材料方面的巨大潜力。

Al_4SiC_4 powder with high purity was synthesized using the powder mixture of aluminum(Al), silicon(Si), and carbon(C) at 1800℃ in argon. Their oxidation behavior and mechanism in a **MgO-C-Al₄SiC₄** system was investigated at 1400-1600℃. XRD,SEM, and energy dispersive spectrometry(EDS) were adopted to analyze the microstructure and phase evolution. The results showed that the composition of oxidation products was closely related to the atom diffusion velocity and the compound oxide layer was generated on Al_4SiC_4 surface. In addition, the effect of different CO partial pressure on the oxidation of Al_4SiC_4 crystals was also studied by thermodynamic calculation. This work proves the great potential of Al_4SiC_4 in improving the MgO-C materials[24].

本文研究了 B_4C 的添加量（质量分数分别为 0.5%、1%、1.5%、2%）对 **Al₂O₃-SiC-C**（ASC）基铁钩浇注料性能和微观结构的影响。研究结果表明，当 B_4C 的添加量（质量分数）超过 1% 时，试样在 1450℃烧制后的体积密度降低。当处理温度超过 1100℃时，加入 B_4C 后试样的常温抗折强度显著提高。其中

B_4C 含量（质量分数）为 1.5% 的试样抗渣性能和抗热震性能最好。添加 B_4C 的 ASC 浇注料在 1450℃ 的空调器下焙烧后，在未氧化区形成 SiC 纤维。获得的纤维扎根于试样的内微孔壁中，直径为 50~200nm，长度为 10~50μm。实验表明 SiC 在 1000℃ 左右开始成核，然后在有利的条件下 SiC 纤维的生长在较高温度下发生。可以观察到垂直于纤维生长方向的阶梯孪晶结构。最后，通过热力学计算解释了生长机制。

The effect of B_4C addition（0.5wt.%，1wt.%，1.5wt.%，2wt.%）on properties and microstructure of **Al_2O_3-SiC-C**（ASC）based trough castables was studied in this paper. The results showed that the bulk density of samples after firing at 1450℃ decreased when the addition of B_4C was over 1wt.%. The cold modulus of rupture of samples improved significantly with the addition of B_4C when the treating temperature was over 1100℃. Among them, sample with 1.5wt.% of B_4C had the best anti-slag performance and thermal shock resistance. The SiC fibers were formed in the non-oxidized zone of ASC castables with addition of B_4C after firing at 1450℃ in air condition. The obtained fibers rooted in the walls of inner micropores of samples with the diameter of 50-200nm, and the lengths of 10-50μm. The experiment revealed the nucleation of SiC started at about 1000℃, and then the growth of SiC fibers happened at higher temperature in a favorable condition. A stepped twin structure perpendicular to the growth direction of fibers could be observed. Finally, the growth mechanism was interpreted by thermodynamic calculations[25].

研究了不同 Fe_2O_3 含量（质量分数分别为 0、0.5%、1.0%）的 **Al_2O_3-SiC-C** 耐火材料在 1400℃ 热处理后的显微组织演变和性能。此外，还讨论了菱形莫来石片晶的原位生长机理。实验结果表明，在没有添加剂的情况下，只有莫来石和 SiC 晶须形成。然而，除了 SiC 晶须外，晶须形和菱形莫来石相都可以生成，这与 Fe_2O_3 的添加具有令人满意的相关性。有趣的是，Fe_2O_3 的引入将莫来石晶须的形态从弯曲的结构（无添加剂）改变为直的细长形状（有添加剂）。此外，这些二维莫来石薄片得到了更好的发展，导致尺寸更大，产量更高，Fe_2O_3（质量分数）含量为 1.0%。XRD、SEM、TEM 和 HRTEM 分析表明，随着 Fe_2O_3 的加入，耐火材料试样中的莫来石相优先沿莫来石纳米晶的（130）和（$\overline{4}20$）方向晶面生长，菱形结构就生成了。随着发育良好的莫来石片晶的形成，常温断裂模量（CMOR）和常温耐压强度（CCS）分别从 6.5MPa 增加到 10.3MPa 和从 38.7MPa 增加到 55.7MPa。同时，经过三个热冲击循环后，含有 1.0%（质量分数）Fe_2O_3 的试样的 CCS 仅下降了 2.6MPa，CCS（CCSst）的残余强度比达到最大值 95.3%。研究结果表明，原位合成的片状莫来石相显著提高了 Al_2O_3-SiC-C

耐火材料的抗热震性。

The microstructure evolution and properties of **Al₂O₃-SiC-C** refractories with different Fe₂O₃ content (0, 0.5wt.%, 1.0wt.%) heat-treated at 1400℃ were investigated. Also, the in-situ growth mechanism of diamond-shaped mullite platelets was discussed. The experimental results indicate that there were only mullite and SiC whiskers formed in the absence of the additive. However, both whisker-shaped and diamond-shaped mullite phases could be generated aside from SiC whiskers, which were satisfactorily correlated with Fe₂O₃ addition. Interestingly, the introduction of Fe₂O₃ changed the morphology of mullite whiskers from the curved structure(without additive) to the straight and elongated shape(with additive). Furthermore, these two-dimensional mullite platelets were better developed resulting in the larger size and higher yield with 1.0wt.% Fe₂O₃. XRD, SEM, TEM and HRTEM analyses reveal that with the addition of Fe₂O₃, mullite phases in refractory samples grew preferentially toward the crystal facets along (130) and (4$\overline{2}$0) directions of mullite nanocrystals, thus, a diamond-shaped structure was generated. As the well-developed mullite platelets were formed, the cold modulus of rupture (CMOR) and the cold crushing strength (CCS) increased from 6.5MPa to 10.3MPa and from 38.7MPa to 55.7MPa, respectively. Meanwhile, after three thermal shock cycles, the CCS of samples containing 1.0wt.% Fe₂O₃ only decreased by 2.6MPa, and the residual strength ratio of CCS (CCSst) reached the maximum value of 95.3%. These results suggest that the in situ synthesized platelet-shaped mullite phase observably improved the thermal shock resistance of Al₂O₃-SiC-C refractories[26].

Al₂O₃-SiC-SiO₂-C 耐火材料广泛用作高炉炉缸衬里，但在 CO₂ 存在下抗氧化性低。碳素复合砖是一种新开发的 Al₂O₃-SiC-SiO₂-C 耐火材料，由于提高了抗侵蚀性和高导热性，在高炉炉缸中可以更好地发挥作用。本研究研究了碳复合砖、碳砖和刚玉砖在 CO₂ 气氛中的氧化行为和动力学。结果表明，碳复合砖的抗氧化性能优于碳砖，但比刚玉砖差。SiC 作为抗氧化剂在氧化过程中发挥重要作用。碳复合砖在界面反应控制阶段的氧化活化能为 141.72kJ/mol，在气体扩散控制阶段的氧化活化能为 161.24kJ/mol。与碳砖和刚玉砖相比，碳复合砖表现出更好的导热性和抗氧化性，这应该可以提高高炉炉床衬里的性能。

Al₂O₃-SiC-SiO₂-C refractories are widely used as blast furnace hearth lining but have low oxidation resistance in the presence of CO₂. Carbon composite brick is a newly developed Al₂O₃-SiC-SiO₂-C refractory that may function better in the blast furnace hearth due to improved erosion-resistance and high thermal conductivity. In this study, the oxidation behavior and kinetics of carbon composite brick, carbon brick, and corundum brick were investigated in CO_2 atmosphere. The results show the oxidation resistance of carbon composite brick was better than that of carbon brick but worse than that of corundum brick. SiC functions as an antioxidant to play an important role in the oxidation process. For carbon composite brick, the oxidation activation energy was 141.72kJ/mol at the step controlled by the interface reaction and 161.24kJ/mol at the step controlled by gas diffusion. Carbon composite brick exhibits improved thermal conductivity and oxidation resistance relative to carbon brick and corundum brick that should allow improved performance in blast furnace hearth lining[27].

以 Ti_3AlC_2 代替部分石墨制备了不同碳含量的 **Al_2O_3-Ti_3AlC_2-C** 耐火材料。研究了 Ti_3AlC_2 对 Al_2O_3-Ti_3AlC_2-C 耐火材料显微组织和热力学性能的影响。通过将碳含量（质量分数）从 10% 降低到 4%，有效地防止了 Al_2O_3-Ti_3AlC_2-C 耐火材料热力学性能的急剧下降。我们特别关注了 Ti_3AlC_2 的微观结构演变，开发了 Al_2TiO_5、Sialon 和 Al-Si-C-O 晶须的形成来解释我们工作中热冲击试验后强度的有限下降。本文解释了 Ti_3AlC_2 试样中新键合相的形成机制。

Al_2O_3-Ti_3AlC_2-C refractories with different carbon content were prepared by using Ti_3AlC_2 as substitute for partially graphite. The influences of Ti_3AlC_2 on microstructure and thermal mechanical properties of Al_2O_3-Ti_3AlC_2-C refractories were investigated. The drastic deterioration in thermal mechanical properties of Al_2O_3-Ti_3AlC_2-C refractories was effectively prevented by reducing carbon content from 10wt.% to 4wt.%. Particular attention was paid to the microstructure evolution of Ti_3AlC_2, the formations of Al_2TiO_5, Sialon and Al-Si-C-O whiskers were developed to explain the limited fall in strength after thermal shock test in our works. The formation mechanism of the new bonded phases in the specimen with Ti_3AlC_2 was explained in this paper[28].

Al_2O_3-SiC-C 耐火材料因其优异的性能被广泛应用于炼钢过程中，为了提高其抗渣腐蚀性能，常采用耐蚀材料。β-Sialon 和 Ti(C,N) 被指出是新型添加剂，具有提高耐火材料耐腐蚀性的合适性能。本研究采用 Ti(C,N)/β-Sialon 粉末、棕刚玉、SiC、微 α-Al_2O_3 粉末、硅灰球沥青、铝酸钙水泥、铝和硅为起始原料制备未烧制的 **Al_2O_3-SiC/β-Sialon/Ti(C,N)-C** 耐火材料。研究了 Ti(C,N)/β-Sialon/Ti(C,N) 粉体和 SiC 对耐渣渗透和耐火材料耐蚀性的影响，以及讨论了 Al_2O_3-SiC/β-Sialon/Ti(C,N)-C 耐火材料的抗渣机理。为了更好地了解渣阻效应和反应机理，进行了分形理论、扫描电子显微镜（SEM）和能量色散光谱仪（EDS）。研究结果表明，所有用 Ti(C,N)/β-Sialon 粉末制成的试样都具有优异的抗渣腐蚀性能。试样的抗渣腐蚀性是基于它们的低显气孔率和高体积密度。β-Sialon 和 SiC 粉末有助于提高复合耐火材料的抗渣腐蚀性能。

Al_2O_3-SiC-C refractories are widely used in the steelmaking process due to their outstanding properties, and in order to improve their slag corrosion resistance, the erosion resistant materials have often been used. β-Sialon and Ti(C,N) are point out as novel additives that present suitable properties to increase the corrosion resistance of refractories. In this study, unfired **Al_2O_3-SiC/β-Sialon/Ti(C,N)-C** refractories were prepared using Ti(C,N)/β-Sialon powders, brown alumina, SiC, micro α-Al_2O_3 powder, silica fume, ball pitch, calcium aluminate cement, aluminum and silicon as the starting materials. The effect of the Ti(C,N)/β-Sialon powders and SiC on the slag penetration and corrosion resistance of the refractories was investigated, and the slag resistance mechanisms of the Al_2O_3-SiC/β-Sialon/Ti(C,N)-C refractories were also discussed. Fractal theory, scanning electron microscopy (SEM) and energy dispersive spectroscope (EDS) were carried out in order to better understand the slag resistance effects and reaction mechanisms. The results showed that all samples made with Ti(C,N)/β-Sialon powders had excellent slag corrosion resistance. The slag corrosion resistance of the samples was based on their low apparent porosity and high volume density. The β-Sialon and SiC powders contributed to the improved slag corrosion resistance performance of the composite refractories[29].

含碳耐火材料因其优异的性能被广泛用于炼钢过程中，为了提高其抗氧化性，经常使用所谓的抗氧剂。Al_4SiC_4 被指出是一种新型添加剂，它具有合适的特性，如 Al，但没有其缺点。因此，在这项工作中，研究了在设计用于衬砌高炉槽的 **Al_2O_3-SiC-SiO_2-C** 浇注料中添加 Al_4SiC_4 的效果。为了更好地了解抗氧化作用和反应机理，进行了显气孔率、氧化、热重、X 射线衍射、热弹性模量测试和热力学计算。此外，将收集的结果与含有其他常用抗氧化剂（Si、B_4C 和硼硅酸钠玻璃）的组合物的结果进行比较。这种新型添加剂的性能被证明是有限的，因为大多数使用的碳源在 Al_4SiC_4 作用之前反应。结果表明，强烈的碳氧化以及冷却步骤期间各相之间的热膨胀失配加剧了耐火材料的劣化。

Carbon-containing refractories are widely used in the steelmaking process due to their outstanding properties and, in order to improve their oxidation resistance, the so-called antioxidants have often been used. Al_4SiC_4 is pointed out as a novel additive that presents suitable properties such as Al, but without its drawbacks. Therefore, the effect of Al_4SiC_4 addition to **Al_2O_3-SiC-SiO_2-C** castables designed for lining blast furnace troughs was investigated in this work. Apparent porosity, oxidation, thermogravimetric, X-ray diffraction, hot elastic modulus tests and thermodynamic calculations were carried out in order to better understand the antioxidant effects and reaction mechanisms. Additionally, the collected results were compared with those from the compositions containing other commonly used antioxidants (Si, B_4C and sodium borosilicate glass). The performance of the novel additive proved to be limited as most of the carbon source used reacted earlier than the Al_4SiC_4 action. As a consequence, intense carbon oxidation, along with the thermal expansion mismatch among the phases during the cooling step, intensified the deterioration of the evaluated refractory material[30].

文献表明，$MgAl_2O_4$ 可以加速 **Al_2O_3-$MgAl_2O_4$-SiC-C** 耐火浇注料中的 SiC 氧化。因此，在这项工作中，使用 FactSage® 软件进行了热力学计算，以探索、寻找和了解 $MgAl_2O_4$ 在 SiC 氧化中的作用。根据热力学预测，在1500℃和还原气氛下，没有证据表明尖晶石可能直接影响 SiC 氧化。Al_2O_3-SiC-C(AL) 浇注料组合物中 SiC 含量的增加主要与莫来石和碳之间的反应有关。另外，Al_2O_3-$MgAl_2O_4$-SiC-C(SP) 组合物中 SiC 的生成是液态 SiO_2 和耐火材料中的碳发生反应的结果。因此，SP 浇注料中较低的 SiC 含量是由耐火材料的相变造成的。也有人提出，在1500℃下热处理15次的试样未达到平衡条件，这解释了实验结果和热力学结果之间的差异。

The literature suggests that $MgAl_2O_4$ can accelerate SiC oxidation in **Al_2O_3-$MgAl_2O_4$-SiC-C** refractory castables. Thus, in this work thermodynamic calculations have been carried out using FactSage® software in order to explore, search for and understand the role of $MgAl_2O_4$ on the SiC oxidation. According to the thermodynamic predictions, at 1500℃ and under a reducing atmosphere, there is no evidence that spinel might directly affect SiC oxidation. The increase of SiC content in an Al_2O_3-SiC-C(AL) castable composition was mainly related to the reaction between mullite and carbon. Moreover, the SiC generation in the Al_2O_3-$MgAl_2O_4$-SiC-C(SP) composition was a result of the reaction involving liquid SiO_2 and carbon from the refractory. Therefore, the lower SiC content in the SP castable resulted from the refractory's phase transformations. It was also suggested that the samples thermally treated 15 times at 1500℃ did not reach the equilibrium condition, which explains the differences between experimental and thermodynamic results[31].

我国高炉炼铁工艺仍将在一段时间内占主导地位，高炉的长寿、平稳运行对于整个炼铁工序具有重要意义。虽然高炉的初期设计、设备配置、耐火材料选用等在一定程度上决定了高炉的寿命，但是，自高炉建成并投入运行后，对高炉进行的日常维护对于高炉长寿的作用也不可忽视。具体到耐火材料而言，主要涉及出铁沟浇注料和出铁口用炮泥的质量、应用和维修技术；而对于高炉本体部分耐火材料，则主要通过热态在线维修技术来延长耐火材料的服役时间，进而提高高炉的整体寿命。虽然硅溶胶结合的出铁沟用 **Al_2O_3-SiC-C 浇注料**正在开发，但是仍未得到系统理论研究与大规模工业应用，Al_2O_3-SiC-C 超低水泥结合浇注料仍是高炉出铁沟工作衬的首选材料。随着现代冶金技术的不断进步，高炉日益大型化、高效化，这意味着出铁频率和总量的增加，这对于承担输送铁水任务的"出铁沟"通道来说，无疑其将面临冲刷加剧、侵蚀更为严重、温度波动频繁等更为苛刻的工况条件。尽管高炉热态在线维修技术经过 20 余年的发展，无论是在施工装备，或是维修用材料方面都有了长足的进步，并基本能够满足基本的需要。然而，与美国、英国及日本等发达国家相比，其高性能在线热态维修用耐火材料的快干、速凝和速强方面仍存在很大的差距。本论文从结构和组分两方面入手，对传统 Al_2O_3-SiC-C 浇注料的性能进行了优化；研制了硅溶胶结合 Al_2O_3-SiC-C 泵送料及快干、速凝速强 Al_2O_3-SiC-C 喷注料，并进行了现场扩大试验，得出了以下结论：（1）基于正交试验设计与分析方法，重点研究了粒度分布（q）、聚丙烯酸钠（PAAS）、三聚磷酸钠（STPP）及萘系磺酸盐甲醛缩合物（FDN）对 Al_2O_3-SiC-C 浇注料流动性的作用效果，极差分析数据表明 PAAS 对该体系的流动性作用效果最明显，STPP 次之，FDN 最小。另外，上述因素对其常温抗折强度、常温耐压强度及显气孔率和体积密度也有影响，其中粒度分布（q）的影响最大，并借助对添加不同分散剂的 Al_2O_3-SiC-C 浇注料浆体黏度、zeta 电位及其养护过程中释放的表观热量测量等手段，研究了各分散剂对 Al_2O_3-SiC-C 浇注料性能产生影响的机理，认为分散剂对上述性能的影响主要是通过对水泥水化过程及产物的作用实现的。基于正交试验的最优试验方案为 q 值取 0.23，PAAS 的加入量（质量分数）为 0.05%，STPP 的加入量（质量分数）为 0.025%，FDN 的加入量（质量分数）为 0.1%。（2）固定正交试验最优方案中的分散剂种类和加入量不变，改变 Al_2O_3-SiC-C 浇注料的粒度组成（q = 0.21、0.22、0.23、0.24、0.25），通过对不同温度热处理后的显气孔率、透气度、孔径分布和平均孔径等材料气孔参数的表征，揭示了温度和粒度分布（q）对其气孔参数规律及演变过程的作用机理，发现当 q = 0.23、0.24 时，其孔径分布呈现集中连续性分布，温度对气孔参数的影响与加热过程中体系内的物理化学变化有关，且温度效应强于粒度分布。感应抗渣性测试发现：具有集中连续孔径分布的 Al_2O_3-SiC-C 浇注料具有较好的抗渣性。（3）热力学计算及实验研究表明，在 Al_2O_3-SiC-C 浇注料中添加 Cr_2O_3 微粉，即使是在空气气氛下热处理后，也并未发现有六价铬离子形成，而仅可能以 Cr_2O_3 或 Cr_3C_2 的形式存在。在 Al_2O_3-SiC-C 浇注料中引入 1%（质量分数）的 400～500 目❶的 Cr_2O_3 微粉，可以改善其流动性、强度，但是对于体积密度和显气孔率的作用不明显，而其突出的作用在于显著地改善了 Al_2O_3-SiC-C 浇注料的抗渣性。（4）Factsage 热力学软件模拟与实验结果均表明可以通过 TiO_2 在 Al_2O_3-SiC-C 浇注料中的原位反应而合成 TiC 和/或 TiN。在 Al_2O_3-SiC-C 浇注料中直接引入适量的 TiO_2 微粉，其流动性显著改善，体积密度增大，显气孔率降低，改善了材料的力学性能；热学理论计算和实验结果都证明适量的 TiO_2 有利于 Al_2O_3-SiC-C 浇注料抗热震性的提高。另外，TiO_2 在 Al_2O_3-SiC-C 浇注料中的原位反应过程中，将体系中的部分游离碳转变为 TiC 中的结合碳，增强了材料的抗氧化能力，同时原位形成 Ti(C, N) 提高了熔渣与耐火材料界面区域内熔渣的黏度，减缓了熔渣向材料内部的渗透，提高了其抗渣性。（5）选择合适的粒度组成，由试验确定了加入 12%～14%（质量分数）的 JN-40 硅溶胶为结合剂，使 Al_2O_3-SiC-C 浇注料具备了优异的流动性，实现了可泵送性。骨料中复合采用棕刚玉和特级矾土，在基质中引入 Si 粉作为抗氧化剂，蓝晶石作为膨胀剂并引入有效的速强剂，所研制的 Al_2O_3-SiC-C 泵送料的理化性能良好。（6）应用 Andreassen 粒度分布理论，临界粒度采用 8mm，改变 $q = 0.21$、0.22、0.23、0.24、0.25、0.26、0.27，制备了具有不同粒度组成的 Al_2O_3-SiC-C 喷注料。对其流动性、显气孔率和体积密度及常温力学性能的评价结果表明：当 $q = 0.21$ 时，Al_2O_3-SiC-C 喷注料的流动值高达 190mm，且其显气孔率较小，体积密度较大，力学性能较好。在此基础上，通过骨料种类的调整，提高了材料的体积密度，并进一步系统地研究了单质 Si 对于 Al_2O_3-SiC-C 喷注料性能的影响，综合分析认为引入 4%（质量分数）的单质 Si 比较合适。随后，借助 Physical MCR301 流变仪和 zeta 电位仪，研究了 Al_2O_3-SiC-C 喷注料浆体在不同分散剂和促凝剂作用下的流变学特性，确定了该体系比较适宜的促凝剂。现场扩大试验证明所研制的喷注料具有低反弹率，能够快速凝固并强化[32]。

The stable working with long life of blast furnace has great significance for the whole iron making process, as the blast furnace iron craft would be predominant in a long period. Although the life of the blast furnace depends on the initial design, instruments configuration and refractory application et al in some extent, the effect of routine maintenance on the life of blast furnace should not be ignored, after its first running. Specific to the refractory materials of blast furnace in cast house, mainly relates to the quality, application and maintenance technology of refractory castable for main

---

❶ 400 目 = 0.038mm，500 目 = 0.029mm。

through and tap hole mixture. While online maintenance technology at high temperature is the main approach to increase their service time for the refractory products used in main body, thus, to prolong the life of the blast furnace. Colloid bonded **Al_2O_3-SiC-C castable** for blast furnace main through has been developing recent years, but it has not been systematically and theoretically researched, and not widely used in industrial practice yet. Al_2O_3-SiC-C castable with ultra low cement still occupied the main market of the cast house refractory. With the progress of modern metallurgy technology, the blast furnaces become more and more huge and effective, that means the increment of the iron cast frequency and its total amount, namely, there factory castable for main through face a grim challenge for its severe service conditions, such as intensified erosion, corrosion and temperature fluctuations. Both the installation instruments and refractory materials have a remarkable progress after almost 20 years' development, and almost basically meet the fundamental requirement of online maintenance technology at high temperature. However, the domestic refractory materials for online maintenance technology at high temperature still has a great gap compared with developed countries, such as America, England and Japan, especially for the fast drying, rapid hardening. So, in this paper, the performance of traditional Al_2O_3-SiC-C castable for blast furnace main through has been improved through microstructure and compositions optimization. Also, a Al_2O_3-SiC-C pumping castable and a Al_2O_3-SiC-C gunning material using colloid as binder have been developed, both of these two materials have been tested through a small-scale live installation, the derived useful conclusions are as follows: (1) Based on the design and analysis method of, the influences of particle size distribution (q), PAAS, STPP and FDN on the flowability of Al_2O_3-SiC-C castable have been studied. Dates derived by extreme difference analysis indicate that PAAS has the most obvious effect on the system's flowability, STTP takes the second place, while the effect of FDN seems slight. In additional, the above factors also have some effects on the cold crushing strength, modulus of rapture, bulk density and apparent porosity. Based on the measured viscosity, zeta potential, and apparent heat released during curing of Al_2O_3-SiC-C slurry with different dispersants, the action mechanism of each dispersant has been explored, and it is believed that their effect on the properties of Al_2O_3-SiC-C castable realized through affecting the hydration process and products of cement component. The optimal experiment scheme suggested by the orthogonal experiment should have a $q=0.23$, and include 0.05wt.% of PAAS, 0.025wt.% of STPP and 0.1wt.% of FDN. (2) Al_2O_3-SiC-C castables with different particle size distribution were fabricated via the q value vary from 0.21 to 0.25, while the dispersants and their contents were fixed. The mechanism of temperature and particle size distribution acting on the pore characteristics evolution has been revealed, based on the evaluation of apparent porosity, permeability, and pore size distribution of Al_2O_3-SiC-C castables treated at different temperatures. Results indicate that the temperature has greater influence on the pore structure evolution process, because of its effect on the reactions occurred during heating, particle size distribution has a little effect despite of the pore size distribution tend to be focus and continuous with $q = 0.23$ or 0.24. The slag resistance test carried out using inductive furnace proof the better slag resistance of Al_2O_3-SiC-C castable with focus and continuous pore size distribution. (3) Thermal dynamic calculation and experimental results both suggest that the Cr_2O_3 introduced into Al_2O_3-SiC-C castable would not transfer to Cr^{6+}, in spite heat treated in the air atmosphere, but just exists in the form of Cr_3C_2 or original Cr_2O_3. Experiments focused on the effect of $1 wt.\%$ Cr_2O_3 micro-powder (about 400-500 meshes) indicate its addition could improve the flowability, mechanical strength of Al_2O_3-SiC-C castable, but has no obvious effect on the bulk density and the apparent porosity, while it plays a prominent role in improving the slag resistance of castable. (4) The possibility to synthesize Ti(C, N) by in-situ reactions between TiO_2 and components of Al_2O_3-SiC-C castable has been confirmed by thermal dynamic simulation carried out by Factsage, as well as the experimental results. The direct introduction of TiO_2 could significantly improve the flowability of Al_2O_3-SiC-C castable, thus increase its packing density and decrease the porosity, thereby, enhance the mechanical performance. Thermal theoretic calculations and tested thermal shock resistance results show suitable amount of TiO_2 would be benefit to the thermal shock resistance improvement. In addition, the oxidation resistance has been also enhanced attributed to the carbon existence form transformed from free carbon into chemical bonded carbon through *in situ* reaction. Meanwhile, *in situ* synthesized Ti(C, N) has made an increase of the slag viscosity in the local area between the refractory and slag, thus slow down the penetration of slag into the deep structure of refractory, and finally, the slag resistance of Al_2O_3-SiC-C castable has been improved. (5) Selecting suitable particle size distribution and adding $12 wt.\%$-$14 wt.\%$ of silica colloid as binder, excellent flowability of Al_2O_3-SiC-C castable has been derived, thereby, the pumping capacity has been realized. Using brown fused corundum and high-grade bauxite as aggregate, silicon powder as antioxidant and cyanite as expansion agent, and meanwhile, introducing effective quick-setting additive, the performance of developed Al_2O_3-SiC-C pumping castable is good. (6) Al_2O_3-SiC-C gunning mixtures with different particle size distribution have been fabricated applying Andreessen model in which the critical grain size is selected as 8mm and q values vary from 0. 21 to 0. 27 with an interval of 0. 01. Evaluation of their flowability, bulk density, apparent porosity and mechanical properties normal temperature show that 0. 21 might be the best q value than the others, as with $q = 0. 21$, Al_2O_3-SiC-C gunning mixture gains a flowability of 190mm, a bigger density, lower porosity and better mechanical strength. Based on the above results, the density of material has been improved by adjusting the aggregate type, and 4wt.% of silicon powder is believed to be suitable for this system to have better oxidation resistance. Successively, an effective quick-setting additive has been selected with the assist of Physical MCR301 reometer and zeta potential meter, which give some useful rheological characteristics dates of the sully of Al_2O_3-SiC-C gunning mixture with different dispersants and quick-setting additives. Live experiment has confirmed the developed gunning mixture has a low return rate, the capacity of quick setting and to be dried fast[32].

参考文献

[1] 朱天彬, 李亚伟, 桑绍柏. 低碳化镁碳耐火材料的研究进展 [J]. 中国材料进展, 2020, 39 (7/8): 609-617.

[2] Behera S, Sarkar R. Nano carbon containing low carbon magnesia carbon refractory: an overview [J]. Protection of Metals and Physical Chemistry of Surfaces, 2016, 52 (3): 467-474.

[3] Zhou Zhenyu, Tang Ping, Hou Zibing, et al. Melting and crystallization behaviors of modified vanadium slag for maintenance of MgO-C refractory lining in BOF [J]. ISIJ International, 2019, 59 (4): 709-714.

[4] Wei Xingwen, Yehorov A, Storti E, et al. Phenomenon of whiskers formation in Al_2O_3-C refractories [J]. Advanced Engineering Materials, 2022, 24 (2): 2100718.

[5] Yu Chao, Dong Bo, Deng Chengji, et al. In-situ formation of plate-like Al_4O_4C and MWCNTs in Al_2O_3-C refractories with Al_4SiC_4 additives [J]. Materials Chemistry and Physics, 2021, 263: 124363.

[6] Lv Lihua, Xiao Guoqing, Ding Donghai. Improved thermal shock resistance of low-carbon Al_2O_3-C refractories fabricated with $C/MgAl_2O_4$ composite powders [J]. Ceramics International, 2021, 47 (14): 20169-20177.

[7] 陈海军, 徐恩霞, 李森, 等. 热处理温度对 ZrO_2 纤维复合 ZrO_2-C 材料性能的影响 [J]. 硅酸盐通报, 2021, 40 (10): 3219-3226.

[8] Musante L, Muñoz V, Labadie M, et al. High temperature mechanical behavior of Al_2O_3-MgO-C refractories for steelmaking use [J]. Ceramics International, 2011, 37 (5): 1473-1483.

[9] Sarath Chandra K, Sarkar D. Structural properties of Al_2O_3-MgO-C refractory composites improved with YAG nanoparticle hybridized expandable graphite [J]. Materials Science and Engineering: A, 2021, 803: 140502.

［10］ 任桢. Al/Zn 复合低碳 MgO-Al$_2$O$_3$-C 材料组成、结构与性能的研究［D］. 郑州：郑州大学, 2015.

［11］ 何龙, 王玺堂, 王周福, 等. SiC 对 MgO-CaO-C 耐火材料抗氧化性能的影响［J］. 人工晶体学报, 2013, 42（5）：985-989.

［12］ 孙铭成. 加入 Al、Al/Zn 对超低碳 MgO-CaO-C 材料结构与性能的影响［D］. 郑州：郑州大学, 2010.

［13］ Wang Zihao, Su Kai, Gao Jinxing, et al. Preparation, microstructure and properties of Al$_2$O$_3$-ZrO$_2$-C slide plate material in presence of nanoscale oxides［J］. Ceramics International, 2022, 48（7）：10126-10135.

［14］ Ban Jinjin, Zhou Chaojie, Feng Long, et al. Preparation and application of ZrB$_2$-SiC$_w$ composite powder for corrosion resistance improvement in Al$_2$O$_3$-ZrO$_2$-C slide plate materials［J］. Ceramics International, 2020, 46（7）：9817-9825.

［15］ Fan Haibing, Li Yawei, Huang Yupeng, et al. Microstructures and mechanical properties of Al$_2$O$_3$-ZrO$_2$-C refractories using silicon, microsilica or their combination as additive［J］. Materials Science and Engineering：A, 2012, 545：148-154.

［16］ Preisker T, Gehre P, Schmidt G, et al. Kinetics of the formation of protective slag layers on MgO-MgAl$_2$O$_4$-C ladle bricks determined in laboratory［J］. Ceramics International, 2020, 46（1）：452-459.

［17］ Ren Xinming, Ma Beiyue, Liu Hao, et al. Designing low-carbon MgO-Al$_2$O$_3$-La$_2$O$_3$-C refractories with balanced performance for ladle furnaces［J］. Journal of the European Ceramic Society, 2022, 42（9）：3986-3995.

［18］ 姚华柏. Al$_4$SiC$_4$ 在镁碳体系中的高温行为及低碳镁碳砖的研制［D］. 北京：北京科技大学, 2021.

［19］ Wang Shijie, Zhou Pingyi, Liu Xin, et al. Effect of modified coal tar pitch addition on the microstructure and properties of Al$_2$O$_3$-SiC-C castables for solid waste incinerators［J］. Ceramics International, 2022, 48（14）：20778-20790.

［20］ Li Tianqing, Chen Junfeng, Xiao Junli, et al. Formation of liquid-phase isolation layer on the corroded interface of MgO/Al$_2$O$_3$-SiC-C refractory and molten steel：Role of SiC［J］. Journal of the American Ceramic Society, 2021, 104（5）：2366-2377.

［21］ Ju Maoqi, Cai Manfei, Nie Jianhua, et al. Advanced Al$_2$O$_3$-SiC-SiO$_2$-C refractories with B$_2$O$_3$ addition［J］. Ceramics International, 2021, 47（20）：29525-29531.

［22］ Wu Muhan, Huang Ao, Yang Shuang, et al. Corrosion mechanism of Al$_2$O$_3$-SiC-C refractory by SiO$_2$-MgO-based slag［J］. Ceramics International, 2020, 46（18）：28262-28267.

［23］ Chen Yang, Deng Chengji, Wang Xing, et al. Effect of Si powder-supported catalyst on the microstructure and properties of Si$_3$N$_4$-MgO-C refractories［J］. Construction and Building Materials, 2020, 240：117964.

［24］ Yao Huabai, Xing Xinming, Wang Enhui, et al. Oxidation behavior and mechanism of Al$_4$SiC$_4$ in MgO-C-Al$_4$SiC$_4$ system［J］. Coatings, 2017, 7（7）：85.

［25］ Wu Jun, Bu Naijing, Li Hongbo, et al. Effect of B_4C on the properties and microstructure of Al_2O_3-SiC-C based trough castable refractories ［J］. Ceramics International, 2017, 43（1）: 1402-1409.

［26］ Lian Jianwei, Zhu Boquan, Li Xiangcheng, et al. Growth mechanism of in situ diamond-shaped mullite platelets and their effect on the properties of Al_2O_3-SiC-C refractories ［J］. Ceramics International, 2017, 43（15）: 12427-12434.

［27］ Zuo Haibin, Wang Cong. Oxidation behavior and kinetics of Al_2O_3-SiC-SiO_2-C refractories in CO_2 atmosphere ［J］. Ceramics International, 2016, 42（13）: 14765-14773.

［28］ Chen Junfeng, Li Nan, Yan Wen, et al. Influence of Ti_3AlC_2 on microstructure and thermal mechanical properties of Al_2O_3-Ti_3AlC_2-C refractories ［J］. Ceramics International, 2016, 42（12）: 14126-14134.

［29］ Yang Dexin, Liu Yangai, Fang Minghao, et al. Study on the slag corrosion resistance of unfired Al_2O_3-SiC/β-Sialon/Ti(C,N)-C refractories ［J］. Ceramics International, 2014, 40（1）: 1593-1598.

［30］ Luz A P, Miglioli M M, Souza T M, et al. Effect of Al_4SiC_4 on the Al_2O_3-SiC-SiO_2-C refractory castables performance ［J］. Ceramics International, 2012, 38（5）: 3791-3800.

［31］ Luz A P, Pandolfelli V C. Thermodynamic evaluation of SiC oxidation in Al_2O_3-$MgAl_2O_4$-SiC-C refractory castables ［J］. Ceramics International, 2010, 36（6）: 1863-1869.

［32］ 尹玉成. 高炉维护用 Al_2O_3-SiC-C 不定形耐火材料性能研究 ［D］. 武汉: 武汉科技大学, 2014.

7 非氧化物系特种耐火材料原料

7.1 碳化硅/氮化硅质耐火材料原料

7.1.1 术语词

无烟煤 anthracite
焦炭 coke

硅石 quartzite
木屑（锯末）sawdust

7.1.2 术语词组

无烟煤 anthracite coal
黑色碳化硅 black carborundum, black silicon carbide
结晶碳化硅 crystalline silicon carbide
立方碳化硅 cubic silicon carbide
绿碳化硅 green silicon carbide
工业硅粉 industrial silicon powder
金属硅粉 metallic silicon powder
硅微粉 silica fume
石油焦 petroleum coke
石英砂 quartz sand

硅石粉 quartzite powder
石英矿物 quartz mineral
反应烧结碳化硅 reaction sintered silicon carbide
反应烧结氮化硅 reaction sintered silicon nitride
二次碳化硅 secondary silicon carbide
自结合碳化硅 self-bonded silicon carbide
硅砂 silica sand
碳化硅粉 silicon carbide powder
氮化硅 silicon nitride
木屑 wood shaving

7.1.3 术语例句

煤炭产品主要是**无烟煤**与有烟煤等。
The major products are **anthracite coal** and bituminous coal, and so there.

无烟煤当然也可以用于烧制水泥。
Anthracite, of course, also can be used for firing cement.

无烟煤焦分支孔贫乏，比表面积很小。
Anthracite char has poor pore structure and small specific surface area.

石油焦的燃烧特性处于烟煤和**无烟煤**之间。
The combustion characters of petroleum coke were between those of soft coal and **anthracite**.

石煤的燃烧特性与**无烟煤**相近，但比无烟煤差。
The combustion characteristics of stone coal are similar to **anthracite coal** and worse than anthracite.

采用**无烟煤**和焦炭对国产普通沥青进行了改性研究。
The modification of the asphalt with **anthracite coal** and coke were studied in this paper.

无烟煤是一种最有价值的煤，它燃烧慢，烟很少但火焰却很大。
The **anthracite** is the most valuable type of coal. It burns slowly, with little smoke and a very hot flame.

地质学家估计中国北部的**无烟煤**和烟煤煤层形成于侏罗纪时期之前。
Geologists estimate the seams of **anthracite** and bituminous coal in northern China, for instance, were formed from the Jurassic period onward.

对**无烟煤**的配合比和改性焦油洗油残渣用量等主要影响因素进行了探讨。
Right mix of **anthracite** and modification of tar washing oil residue usage and other major factors are discussed.

整个住宅的地板都是由**无烟煤**、矿物涂层覆盖的，建造痕迹被有意保存。
The floor throughout the entire house is covered in an **anthracite**, mineral coating in which the production traces are deliberately conserved.

杂质用**焦炭**烧掉，产生高质量的炼铁。

Impurities were burnt away with the use of **coke**, producing a high-quality refined iron.

焦炭是一种很经济的燃料。
Coke is an economical fuel.

没有**焦炭**就不能炼铁。
Without **coke** iron cannot be smelted.

用测定焦炭电阻率的方法来计算**焦炭**的反应性。
Coke reactivity is calculated with the method of coke resistivity measurement.

焦炉厂中生成的**焦炭**在高炉中还原铁矿石（铁的氧化物）成铁。
As a simple overview, the blast furnace needs **coke** from the coke plant to reduce the iron ore(iron oxide)to iron.

使用**焦炭**，一方面煤产量增加，同时也增加了现代社会大气中二氧化碳的含量。
The rise of coal, from which **coke** is produced, began, and so did the modern rise of carbon dioxide in the atmosphere.

重点研究**焦炭**在高炉块状带内的抗压强度与温度和反应程度的关系。
The relationship between temperature, carbon loss rate and compression strength of **coke** in the lumpish section of blast furnace was investigated.

其次，他们正在用工业垃圾和废弃轮胎代替目前正在使用的煤炭和**焦炭**等矿物燃料。
Second, they are replacing fossil fuels such as coal and **coke** with alternatives such as farm waste or used tyres.

木头的碎渣和**锯末**可以压成木板。
Wood cuttings and **sawdust** can be compressed into boards.

另外还有泥炭、**锯末**、棉籽壳、淤泥等。
Some of the others are peat, **sawdust**, cotton seed hulls, and sludge.

适用于以**锯末**、秸秆为主要原料加工燃料颗粒。

Suitable to process fuel pellets which raw materials are **sawdust** and straw.

模型对于**锯末**、纤维素和木质素的催化裂解适用比较准确。

The proposed model fits well with the calculated data got from pyrolysis tests of wood **sawdust**, lignin and cellulose.

锯末的比重很轻，直上直下，锯末进入后由压轮旋转甩到四周，均匀压制颗粒。

The proportion of **sawdust** is very light, straight up and down, after entering the sawdust composed of a pressing wheel spin around, uniform particle suppression.

砂浆的主要成分是切割液和**碳化硅微粉**及其他杂质。

Mortar has the main components of cutting fluids and **silicon carbide powder** and other impurities.

以**炭黑**和二氧化硅微粉为原料，对双重加热法合成碳化硅晶须进行了研究。

The synthesis of silicon carbide whiskers by double-heating method using **carbon black** and SiO$_2$ powder as raw materials is studied.

硬度和脆性较**黑碳化硅**高、磨粒锋利、导热性好。

Harder and more brittle than **black silicon carbide**, very sharp for grinding and of good heat conductivity.

浸渍强化处理是提高**反应烧结氮化硅**性能的有效措施。

Infiltration strengthening is an effective means to improve the mechanical properties of **reaction bonded silicon nitride**.

氮化硅可以通过适当掺杂引入杂质能级，用于制造量子阱获得蓝光激光。

Silicon nitride can be used to grow quantum well for obtaining blue laser by dopanting.

7.1.4　术语例段

超高性能混凝土（UHPC）的特点是致密的微观结构，可产生超高的强度和耐久性能。**石英砂**（QS）最大粒径为 600μm，代表 UHPC 中的粗颗粒。通过破碎粗砂或岩石，可获得具有最佳级配曲线的 QS，但这是一个耗时、昂贵且污染严重的过程。本文报道了一项研究，以确定生产和使用玻璃砂（GS）部分或全部替代 UHPC 中 QS 的可能性。研究结果表明，平均粒径（d_{50}）为 275μm 的 GS 可作为替代 QS 颗粒的最佳 PSD。用 GS 替代 50% 和 100% QS 时，热养护 2d 后的抗压强度分别为 196MPa 和 182MPa，而含 100% QS 的 UHPC 的抗压强度为 204MPa。因此，提高 GS 替代量可以改善 UHPC 的流动性和致密微观结构，从而缓解骨料的碱-硅反应。

Ultra-high-performance concrete（UHPC）is characterized by a dense microstructure that yields ultra-high strength and durability properties. **Quartz sand**（QS）with maximum particle sizes of 600μm represents the coarse particles in UHPC. The QS with optimum grading curve is obtained from crushing coarse sand or rocks, however this is a time-consuming, costly, and polluting process. This paper reports on a study to determine the possibility of producing and using glass sand（GS）for partial or total replacement of QS in UHPC. The results show that GS with a mean particle size（d_{50}）of 275μm could be recommended as an optimal PSD to replace QS particles. The results demonstrate that compressive strength values of about 196MPa and 182MPa after two days of hot curing can be achieved when replacing 50% and 100% of QS with GS, respectively, compared to 204MPa for reference UHPC containing 100% QS. Incorporating higher replacement rates of GS was shown to produce UHPC of accepted flowability and dense microstructure that mitigated the aggregate alkali-silica reaction[1].

硅砂颗粒的典型表面纹理富含粗糙和矿物碎屑。因此，名义上的晶间接触由许多"接触点"组成。与接触加载过程相关的是纹理特征的微破裂，导致接触演变。即使在恒定载荷下，微压裂过程仍在继续，尽管频率降低。这种在恒定负载下接触（或接触成熟）的静态疲劳被认为是硅砂老化的主要原因。使用扫描电子显微镜和原子力显微镜展示了硅砂颗粒丰富的表面微观形态。晶粒尺度测试表明在恒定载荷下随时间变化的变形，归因于晶粒接触处纹理特征的延迟断裂。该过程很大程度上取决于表面的初始粗糙度。初始粗糙度大的晶粒（晶粒表面高程的均方根（RMS）= 621nm）在加载 2.4N 时的偏转速率被发现为 17.6nm/h。加载后的第一天，而对于具有大约一半初始粗糙度（RMS=321nm）的晶粒来说，它只有大约 2nm/h。颗粒组合中接触成熟过程的结果是宏观刚度随时间增加，并导致约束砂中应力状态的改变。这在对承受恒定载荷的试样进行的软环测得仪测试中得到证实。测径试样中径向应力增加的时间尺度与单晶粒收敛测试中的时间尺度大致相同。水分和压力溶解的存在加速了接触成熟，这在用孔隙流体饱和的试样的测试中得到了体现。颗粒测试和砂试样测试的结果都与假设一致，表明接触熟化是**硅砂**中时间效应的主要因素。

A typical surface texture of silica sand grains is rich in asperities and mineral debris. Consequently, a nominal inter-granular contact is composed of many "contact points". Associated with the contact loading process is micro-fracturing of the textural features, causing the contact to evolve. Even under a constant load, the process of micro-fracturing continues, although with a decreasing frequency. This static fatigue at contacts (or contact maturing) under a constant load is considered to be a major contributor to ageing of silica sand. The rich surface micro-morphology of silica sand grains is demonstrated using scanning electron microscopy and atomic force microscopy. Grain-scale tests indicate time-dependent deflection under constant load, attributed to delayed fracturing of textural features at grain contacts. The process is greatly dependent on the initial roughness of the surfaces. The rate of the deflection for a grain with large initial roughness(root mean square of grain surface elevation (RMS) = 621nm), loaded with 2.4N, was found to be 17.6nm/h at the end of the first day after loading, whereas for a grain with approximately half the initial roughness(RMS=321nm) it was only about 2nm/h. A consequence of the contact maturing process in grain assemblies is a time-dependent increase in macroscopic stiffness, and a resulting alteration of the stress state in a confined sand. This was confirmed in soft-ring oedometer tests on specimens subjected to constant load. The temporal scale of the radial stress increase in the pedometric samples was about the same as that in convergence tests on single grains. Contact maturing is accelerated by the presence of moisture and pressure dissolution, and this was manifested in tests on samples saturated with pore fluid. The results of both grain-scale and sand specimen tests are consistent with the hypothesis indicating contact maturing as the major contributor to time-dependent effects in **silica sand**[2].

流化床反应器中形状选择性催化剂上的**焦炭**分布对目标产物的选择性至关重要。本研究发展了一个种群平衡模型（PBM），用来描述由于催化剂的反应和循环作用，以及焦炭分布的演化。将 PBM 与多尺度计算流体力学（CFD）耦合，模拟了甲醇制烯烃流化床反应器。因此，可以预测焦炭含量的广泛、不均匀分布。所期望产物乙烯和丙烯的质量分数与实验数据相比，无 PBM 时表现出更好的一致性。而且乙烯与丙烯的选择性比也得到较好的预测。此外，还考虑了新鲜催化剂的弛豫时间来表示诱导期对 MTO 反应的影响。合理的预测表明，将 PBM 与多尺度 CFD 耦合有助于理解 MTO 反应器，并可能在其优化和放大方面得到更多的应用。

The **coke** distribution over shape-selective catalysts in fluidized bed reactors is critical to the selectivity of target products. A population balance model (PBM) is developed in this study to describe the evolution of coke distribution due to both reactions and circulation of catalysts. The PBM is coupled with multiscale computational fluid dynamics (CFD) to simulate a methanol-to-olefins fluidized bed reactor. A wide, uneven distribution of coke content can thus be predicted. The mass fractions of the desired products, ethylene and propylene, show better agreement with the experimental data than those without PBM. And the selectivity ratio of ethylene to propylene is also better predicted. Furthermore, the relaxation time of fresh catalyst is considered to express the effects of the induction period in MTO reactions. The reasonable prediction suggests that coupling PBM with multiscale CFD is helpful to understand MTO reactors and may find more application with respect to its optimization and scale-up[3].

地层热处理一直被认为是煤层气采收率增强的潜在刺激技术。但热处理对煤层气解吸效率的影响尚未阐明。本研究对 25℃、200℃、400℃、600℃ 和 800℃ 处理的柱状无烟煤进行等温吸附实验，并考察了甲烷吸附电位和解吸效率的变化规律。同时利用核磁共振测量和扫描电镜对煤的孔隙结构进行了表征。通过工业分析和傅里叶变换红外光谱监测了煤中有机质和官能团含量的变化。结果表明，孔隙结构和组成是影响煤吸附势的主要因素。随着温度升高到 200~600℃，由于孔隙体积和官能团含量的增加，煤的最大吸附电位提高了 7%。在 800℃ 的热处理温度下，官能团含量降低，进而使最大吸附电位降低 76%。通过对不同煤阶煤等温吸附数据的比较发现，与无烟煤相比，烟煤的朗缪尔参数更容易受到热处理的影响。基于煤层气等温吸附解吸过程的可逆性理论，提出了加权压力，可综合评价不同温度热处理后煤的解吸效率，通过加权压力可确定最佳解吸效率的热处理温度。

Formation heat treatment has been considered as a potential stimulation technology for the enhanced recovery of coalbed methane. However, the influence of heat treatment on the desorption efficiency of coalbed methane has not been clarified. In this study, isothermal adsorption experiments of a cylindrical anthracite coal treated at 25℃, 200℃, 400℃, 600℃, and 800℃ were conducted to investigate the variation in the methane adsorption potential and desorption efficiency. The pore structure of coal was characterized by nuclear magnetic resonance measurements and scanning electron microscopy. Changes in the organic matter and functional group content of coal were monitored by proximate analysis and Fourier transform infrared spectroscopy. Results reveal that pore structure and composition are the main factors that affect the coal adsorption potential. With the increase in the temperature to 200-600℃, the maximum adsorption potential of coal increased by 7% due to the increase in the pore volume and the functional group content. At a heat-treatment temperature of 800℃, the functional group content decreased, which in turn reduced the maximum adsorption potential by 76%. Comparison of isothermal adsorption data of coals with different ranks reveals that the Langmuir parameter of bituminous coals is more easily affected by heat treatment in comparison with that of **anthracite coal**. Based on the reversible theory of coalbed methane isothermal adsorption and desorption process, weighted pressure was proposed, which could comprehensively evaluate the desorption efficiency of coal after heat treatment at different temperatures, and the heat treatment temperature for the optimal desorption efficiency can be determined by the weighted pressure[4].

　　由于高硫**石油焦**的产量逐年增加,其燃烧发电的污染排放严重,而炼油厂又需要大量的氢气,所以石油焦制氢工艺很有前景。因此,本研究建立了石油焦直接化学循环制氢工艺的工艺模型并进行了参数优化。优化后的载氧体(Fe₂O₃)、蒸汽和空气与石油焦的质量比为12.71、2.98和2.18;但当进入蒸汽反应器的还原载氧体比例从100%降低到60%~70%时,即为12.71、1.79~2.08和4.53~5.32,可在空气预热温度为278.54~816.80℃的自加热模式下运行。当比例从60%提高到65.32%和70%时,在CO₂捕获率约为100%的情况下,氢气的产品放能从454.58MW提高到494.92~530.36MW,放能效率因此从59.56%~60.75%提高到62.32%~63.59%和62.65%~63.91%。本文还研究了硫的分布和硫化物的现有形式。最后,本研究发现,比例为65%~70%的化学循环制氢工艺的放能效率明显高于石油焦气化制氢工艺的放能效率。

The petroleum coke-to-hydrogen process is promising since the output of high-sulfur **petroleum coke** is increasing year by year, the pollution emissions of its combustion for power generation are serious, and the refinery needs a lot of hydrogen. Therefore, this study establishes the process model and conduct the parameter optimization of the petroleum coke direct chemical looping hydrogen production process. The optimized mass ratios of oxygen carrier(Fe_2O_3), steam, and air to petroleum coke are 12.71, 2.98, and 2.18; however, that are 12.71, 1.79-2.08, and 4.53-5.32 when the proportion of the reduced oxygen carrier entering steam reactor is reduced from 100% to 60%-70%, which can be operated in a self-heating mode with air preheating temperature of 278.54-816.80℃. When the proportion is increased from 60% to 65.32% and 70%, the product exergy of hydrogen is increased from 454.58MW to 494.92-530.36MW with about 100% CO_2 capture and the exergy efficiency is therefore increased from 59.56%-

60. 75% to 62. 32% -63. 59% and 62. 65% -63. 91%. This paper also studies the sulfur distribution and existing forms of sulfides. Finally, this study found that the exergy efficiency of the chemical looping hydrogen process with the proportion of 65% -70% is significantly higher than that of the hydrogen production process by petroleum coke gasification[5].

含有铜、铅和锌的未经处理的雨水的直接流动对水道中的水生生物有直接影响。迄今为止，大多数生物炭都是在受控的实验室条件下用惰性气体净化的炉子合成的。在这项研究中，使用工业规模的双室下行热解反应器，用稻壳和**锯末**原料合成的生物炭去除铜、铅和锌。通过进行批量吸附实验，评估了热解温度和原料对铜、铅和锌去除率的影响。使用近似分析、零点电荷、扫描电子显微镜、X射线衍射和傅里叶变换红外光谱对合成的吸附剂材料进行了表征。与文献相比，生物炭的产量较低，这是因为热解器的加热速度较高（50℃/min）。当初始pH值最接近，低于重金属的溶解极限时，观察到最大的去除效率。在350~450℃和450~550℃温度范围内合成的稻壳生物炭和锯末生物炭对三种重金属的去除效果最好。化学吸附是去除这三种重金属的主要机制。稻壳生物炭对Cu和Zn的最大吸附量为10. 27mg/g和6. 48mg/g，锯末生物炭对Pb的最大吸附量为17. 57mg/g。表面配合、共沉淀、p-电子相互作用、物理吸附和表面沉淀是去除这三种重金属的主要机制。

Direct flow of untreated stormwater containing Cu, Pb and Zn is of immediate concern to aquatic life in waterways. To date, most biochar used has been synthesized under controlled laboratory conditions using furnaces purged with inert gasses. In this study, the removal of Cu, Pb and Zn using biochar synthesized using paddy husk and **sawdust** feedstocks has used an industrial scale double chamber downdraft pyrolysis reactor. The effect of pyrolysis temperature and the effect of feedstock in the removal of Cu, Pb and Zn was evaluated by conducting batch adsorption experiments. Synthesized adsorbent materials were characterized using proximate analysis, zero-point charge, scanning electron microscopy, X-ray diffraction and Fourier transform infrared spectroscopy. The biochar yield was in a lower range compared with the literature attributed to the higher heating rate (50℃/min) in the pyrolizer. Maximum removal efficiencies were observed when the initial pH was at the value closest, when below the solubility limit for the heavy metals. The paddy husk biochar and sawdust biochar synthesized in the temperature range 350-450℃ and 450-550℃ performed best in the removal of the three heavy metals. Chemisorption was the main mechanism for the removal of the three heavy metals. The maximum adsorption capacities of Cu and Zn were 10. 27mg/g and 6. 48mg/g was achieved with paddy husk biochar and a maximum Pb adsorption capacity of 17. 57mg/g was achieved by sawdust biochar. Surface complexation, coprecipitation, p-electron interactions, physical adsorption and surface precipitation were the main mechanisms of removal of the three heavy metals[6].

针对宁夏无烟煤具有"三低六高"和世界稀缺性资源的特点，太西炭基公司组织开展了无烟煤冶炼绿碳化硅科研试验和工业化生产，成功生产出煤基绿碳化硅，其产品指标达到了石油焦基绿碳化硅产品指标要求，甚至个别指标优于石油焦基产品。煤基绿碳化硅工业化生产的成功，推动了碳化硅冶炼技术进步，同时产品具有明显的成本优势和市场竞争力，在一定程度上，促进了碳化硅行业的技术革新[7]。

For Ningxia **anthracite** with "three low and six high" and the characteristics of the scarcity of resources, Taixi Carbon-based Company organized anthracite smelting tests and industrialized production. Coal-based green silicon carbide was produced successfully. The indexes of green silicon carbide with anthracite were reached as one with petroleum coke as raw material. Even individual indexes are better than that of petroleum coke products as raw material. The success production of green silicon carbide with coal as raw materials well drives the SiC smelting technology progress. At the same time, the product has obvious cost advantages and market competitiveness, to a certain extent, promote the technology of silicon carbide industry innovation[7].

分析了**纳米碳化硅**与氧化铁在真空和氩气中分别于1200℃和1400℃下相互作用形成的超细复合粉体的特性。碳化硅（β-SiC）、硅化铁和碳化物、氧化硅和氧氮化硅是复合粉体的主要成分。在SiC-Fe$_2$O$_3$体系中合成的复合粉体中SiC的晶格参数被确定。在真空中的SiC-Fe$_2$O$_3$粉体混合物的相互作用中，合成的二级SiC的晶格参数与立方β-SiC的标准参数一致。在氩气环境中的相互作用伴随着二级SiC的合成，其晶格参数下降。最小的晶格参数（0.4336nm）比立方β-SiC的标准参数小0.6%。研究了在SiC-Fe$_2$O$_3$体系中合成的复合粉体的形态。粉末复合材料的平均粒径随着二次SiC质量含量的增加而减小。

The characteristics of superfine powder composites formed in the interaction of **nanosized silicon carbide** with iron oxide in vacuum and argon at 1200℃ and 1400℃, respectively, are analyzed. Silicon carbide (β-SiC), iron silicide and carbide, silicon oxide, and silicon oxynitride are main components of the powder composites. The lattice parameter of SiC in the powder composites synthesized in the SiC-Fe$_2$O$_3$ system is determined. In the interaction in the SiC-Fe$_2$O$_3$ powder mixture in vacuum, secondary

SiC is synthesized with a lattice parameter that corresponds to the standard parameter for cubic β-SiC. The interaction in an argon atmosphere is accompanied by the synthesis of secondary SiC with a decreased lattice parameter. The minimum lattice parameter (0.4336nm) is 0.6% smaller than the standard parameter for cubic β-SiC. The morphology of the powder composite synthesized in the $SiC-Fe_2O_3$ system is studied. The average particle size of the powder composite decreases with increasing weight content of secondary SiC[8].

采用 X 射线 K 值法测定工业**硅粉**中二氧化硅的含量。结果表明，使用 X 射线 K 值法对工业硅粉中二氧化硅进行定量相分析，可计算得出工业硅粉中二氧化硅的含量，能对工业硅粉中硅石掺假现象进行监督防治，是一种切实有效的检测方法[9]。

The X-ray K-value method is adopted for the determination of silicon dioxide content in industrial **silicon powder**, the results show, the quantitative phase analysis is processed for the silicon dioxide in industrial silicon powder by X-ray K-value method, therefore, the silicon dioxide content in the industrial silicon powder can be calculated, which can supervise and prevent the silica adulteration in the industrial silicon powder, which is a practical and effective detection way[9].

用 α-氧化铝粉末和纳米大小的碳酸钙，采用直接发泡技术，加入铝酸钙水泥和**硅微粉**，制备了六铝酸钙（CA_6）基多孔陶瓷。微硅石导致了 CA_6 的加速形成。此外，室温下的收缩率、致密性和弹性模量都随着**硅微粉**含量的增加而增强。但如果微硅石含量（质量分数）超过 0.5%，热弹性模量在 1100~1200℃之间出现明显下降。此外，在 1550℃处理后，在有**硅微粉**和无**硅微粉**的试样中出现了 α-Al_2O_3，添加**硅微粉**后，具有高长径比的 CA_6 晶粒的比例增加。

Calcium hexaluminate (CA$_6$)-based porous ceramic was prepared with α-alumina powders and nanometer-sized calcium carbonate using a direct foaming technique with the addition of calcium aluminate cement and **microsilica**. The **microsilica** leads to the acceleration of CA_6 formation. Moreover, the shrinkage, densification, and elastic modulus at room temperature were enhanced with increasing **microsilica** content. But the hot elastic modulus showed obvious reduction between 1100℃ and 1200℃ if the microsilica content was over 0.5wt.%. Furthermore, α-Al_2O_3 appeared in the specimens with and without **microsilica** after treatment at 1550℃, and the proportion of CA_6 grains having a high aspect ratio increased after **microsilica** addition[10].

本课题系统研究了高性能 **Si_3N_4-SiC** 耐火材料的合成原理和应用性能，对比了硅粉混合 SiC 颗粒与纯硅粉、硅粉混合氮化硅铁（Fe_3Si-Si_3N_4）颗粒在流动氮气中烧结时 Si 氮化过程的差异，探索 Si_3N_4-SiC 耐火材料生产工艺的优化方向，研究了 Si_3N_4-SiC 在不同气氛中的高温稳定性，分析了 Si_3N_4-SiC 在不同氧分压环境中长期应用的氧化行为。纯硅粉压制成型的坯体在流动氮气中分别在 1180℃ 和 1420℃ 进行氮化烧结，结果表明，硅粉氮化率随温度升高而升高，但是在直接升温至目标温度的烧结制度下，硅粉难以实现完全氮化。原因是氮化产物（SiO_2、Si_2N_2O、Si_3N_4）在硅粉表面形成致密的保护膜，隔断硅粉内部与氮气的接触。为了实现 Si 的完全氮化，应采用分段保温方式，不断破坏体系中达到的反应平衡，为氮化反应提供持续动力。本课题研究了 Si_3N_4-SiC 耐火材料在氮化梭式窑内的合成原理，结果表明，Si 的氮化有低氧分压下的直接氮化和高氧分压下的间接氮化两种方式。1450℃ 时，Si 只有极低氧分压条件下才能与 N_2 发生直接氮化反应生成柱状 β-Si_3N_4。在纯度 99.999% 的流动氮气中，Si 首先与 O_2 发生氧化反应生成大量中间产物气态 SiO。这一反应消耗流动氮气中的 O_2，降低体系中的氧分压，使 Si 的直接氮化可以发生。SiO 与 N_2 反应生成纤维状 α-Si_3N_4。SiO 在体系内的流动，导致 α-Si_3N_4 在 Si_3N_4-SiC 中的不均匀分布。并且 SiO 从砖中逸出造成物料损失，使 Si_3N_4-SiC 生产中 Si 的氮化率难以提高。通过研究 Si-SiC 生坯在碳管炉内的高温氮化和 Si_3N_4 结合氮化硅铁材料的制备，讨论了 Si_3N_4-SiC 耐火材料生产工艺的优化。在碳管炉内，氧气几乎全部转变为 CO，Si 主要与 N_2 发生直接氮化反应生成 β-Si_3N_4。Fe_3Si-Si_3N_4 原料中的硅铁合金增加了体系内液相量，降低了液相的出现温度，有利于 α-Si_3N_4 向 β-Si_3N_4 的转变和 β-Si_3N_4 的生成。因此降低烧结气氛中的氧分压、提高氮化烧结温度、在原料中加入硅铁合金，有利于增大材料中 β-Si_3N_4/α-Si_3N_4 比例。此外，在 Si_3N_4 结合氮化硅铁材料生产过程中，原料中的 Si_3N_4 先于 Si 发生氧化，在表面生成 Si_2N_2O。Si_3N_4 的氧化降低了体系氧分压，使原料中的硅粉发生直接氮化，极大地减少生产过程中 SiO 的产量，降低了物料损失，提高了 Si 的氮化率。为研究 Si_3N_4 和 SiC 的高温稳定性，将 Si_3N_4-SiC 耐火材料在低氧分压条件下进行了 1600℃ 和 1800℃ 高温重烧，并将 Si-SiC 生坯和 Si_3N_4-SiC 材料浸碳后进行了 1500℃ 和 1800℃ 高温重烧。实验结果表明，在含有碳单质的条件下，Si_3N_4 会向 SiC 发生转变，Si 在流动氮气不会发生氮化，而是与 C 反应生成 SiC。Si_3N_4 向 SiC 转变方式有两种：温度低于 1537℃ 时，Si_3N_4 通过与 CO 反应生成气态 SiO 的间接反应生成 SiC；温度高于 1537℃ 时，Si_3N_4 直接与 C 反应生成 SiC。1800℃ 时，Si_3N_4 不稳定，会发生分解反应，生成 Si 蒸气和氮气，并与 CO 反应生成 SiC。在高温条件下 β-Si_3N_4 稳定性高于 α-Si_3N_4。将用后 Si_3N_4-SiC 匣钵砖进行分层分析，研究长期服役过程中，处于空气–氮气的渐变气氛中的 Si_3N_4 和 SiC 的氧化行为。研究结果表明，在匣钵砖外侧，氧分压高，Si_3N_4 和 SiC 发生被动氧化，氧化产物为 SiO_2 和 Si_2N_2O；在匣钵砖内侧氮气气氛中，Si_3N_4 和 SiC 发生活性氧化生成气态 SiO。使用过程中匣钵内 Si 氮化生成的气态 SiO 渗透进入匣钵砖内侧，破坏活性氧化反应平衡，与氮气反应生成纤维状 α-Si$_3$N$_4$。在长期服役过程中纤维状 α-Si$_3$N$_4$ 转变为颗粒状 β-Si$_3$N$_4$，并填充和封闭部分开口气孔。因此用后 Si$_3$N$_4$-SiC 匣钵砖气孔率降低，体积密度升高，在接触氮气的工作面此现象更明显。匣钵砖外侧生成的大量 SiO$_2$ 玻璃相，会影响材料的高温性能，是材料失效的重要原因[11]。

The thesis researched the synthesis mechanism and application properties of high performance **Si$_3$N$_4$-SiC refractory** by comparing silicon nitridation mechanism of Si-SiC with pure silicon powder and Si-Fe$_3$Si-Si$_3$N$_4$ in flowing nitrogen, discussing the manufacturing technique improvement of Si$_3$N$_4$-SiC, analyzing the high temperature stability and oxidation of Si$_3$N$_4$-SiC at different atmosphere. Bricks pressed and molded with pure silicon powder were sintered at 1180℃ and 1420℃ in flowing nitrogen. Test result showed that the nitridation rate of silicon increased along temperature. However, silicon was hardly fully nitrided by directly heating up to target temperature. The reason is that the nitridation product, such as SiO$_2$, Si$_2$N$_2$O and Si$_3$N$_4$, forms a compact protective film on the surface of silicon powder and blocks the contact of silicon inner part with nitrogen. In order to realize full nitridation of silicon, it is necessary to use a stepped sintering technique to break the reaction equilibrium repeatedly and provide continuous reaction kinetics. The thesis analyzed the synthesis mechanism of Si$_3$N$_4$-SiC refractory in flame-isolation nitridation shuttle kiln. There are two modes of silicon nitridation, one of which is direct nitridation at low oxygen partial pressure and the other is indirect nitridation at high oxygen partial pressure. In flowing nitrogen with a purity of 99.999%, Si reacts with oxygen first and generates great amount of gaseous SiO, which consumes oxygen and decreased the oxygen partial pressure. Only when $p(O_2)$ reaches a rather low level, silicon can react with nitrogen directly and generate column β-Si$_3$N$_4$ at 1450℃. SiO reacts with nitrogen and generates fibrous α-Si$_3$N$_4$. The flowing of SiO is the cause of nonuniform distribution of α-Si$_3$N$_4$ in Si$_3$N$_4$-SiC. The effusion of SiO from the bricks causes the material loss and low nitridation rate of silicon in Si$_3$N$_4$-SiC. By analyzing the nitridation of Si-SiC in graphite vacuum furnace and the preparation of Si$_3$N$_4$ bonded ferrosilicon nitride, the possibilities of Si$_3$N$_4$-SiC manufacturing technique improvement were discussed. In graphite vacuum furnace, Si is mainly nitrided to β-Si$_3$N$_4$ directly with nitrogen, for almost all the oxygen transforms to CO and oxygen partial pressure is low. Ferrosilicon in raw Fe$_3$Si-Si$_3$N$_4$ increases the liquid amount and decreases the temperature of liquid appearance, which favors the α-Si$_3$N$_4$ to β-Si$_3$N$_4$ transformation and β-Si$_3$N$_4$ generation. β-Si$_3$N$_4$/α-Si$_3$N$_4$ ratio can be increased by lowering oxygen partial pressure, increasing sintering temperature or adding ferrosilicon alloys to raw materials. Otherwise, during the production of Si$_3$N$_4$ bonded ferrosilicon nitride, Si$_3$N$_4$ in raw materials reacts with oxygen taking precedence over Si and generates Si_2N_2O on surface, which leads to low oxygen partial pressure and direct nitridation of silicon without generating SiO. As a result, there is no material loss and silicon nitridation rate is increased. In order to analyze the high temperature stability of Si_3N_4 and SiC, Si_3N_4-SiC was re-sintered in low oxygen partial pressure at 1600℃ and 1800℃, and Si-SiC and Si_3N_4-SiC were carburized and sintered at 1500℃ and 1800℃. The results indicated that Si_3N_4 and silicon transformed to SiC in an environment containing carbon. There are two Si_3N_4 to SiC transformation modes. When the temperature is below 1537℃, Si_3N_4 transforms to SiC by reacting with CO and generating gaseous SiO. When the temperature is above 1537℃, Si_3N_4 transforms to SiC by reacting with C directly. When the temperature is up to 1800℃, Si_3N_4 is unstable and resolves into gaseous Si and nitrogen, which reacts with CO and generates SiC. β-Si_3N_4 is more stable than α-Si_3N_4 at high temperature. An Si_3N_4-SiC saggar brick after-use was sliced to 5 layers and the oxidation of Si_3N_4 and SiC in a gradually varied air-nitrogen atmosphere during long-term service was analyzed. The outer layer of saggar brick worked in air with high oxygen partial pressure and Si_3N_4 and SiC were passively oxidized to SiO_2 and Si_2N_2O. The inner layer of saggar brick worked in nitrogen and Si_3N_4 and SiC were actively oxidized to gaseous SiO. SiO generated by silicon oxidation in the saggar infiltrated into the saggar brick breaking the equilibrium of active oxidation, and reacts with nitrogen generating fibrous α-Si_3N_4, which transformed to column β-Si_3N_4 filling and closing some pores of the brick during long-term usage. Therefore, Si_3N_4-SiC saggar bricks after-use had a lower open porosity and higher bulk density, especially the inner layer working in nitrogen. A great amount of SiO_2 glass was generated in the outer layer, affecting its high temperature properties, and might cause material failure[11].

7.2 赛隆质耐火材料原料

7.2.1 术语词

多水高岭土 ablykite
石墨 graphite
石墨的 graphitic

石墨化 graphitization
高温石墨 pyrographite
碳 carbon

木炭 charcoal

焦炭 coke

刚玉 corundum

黑高岭土 hisingerite

高岭石族矿物 kandite

高岭土 kaolin

高岭石 kaolinite

高岭土化 kaolinization

蓝晶石 kyanite,cyanite,disthene,zianite

红柱石 andalusite,apyre

莫来石 mullite

锰红柱石 viridine

绿硅线石 bamlite

硅线石 sillimanite,fibrolite

叶蜡石 pyrophyllite,pyrauxite,roseki

叶蜡石基 pyrophyllite-based

明矾石质叶蜡石 alunite-pyrophyllite

红柱石质叶蜡石 andalusite-pyrophyllite

硬水铝石质叶蜡石 diaspore-pyrophyllite

石英质叶蜡石 quartz-pyrophyllite

煤矸石 coal gangue

粉煤灰 fly ash

7.2.2 术语词组

氮化铝 aluminium nitride

工业硅粉 industrial silicon powder

金属硅粉 metallic silicon powder

偏高岭土 metakaolin

偏高岭石 metakaolinite

胶状高岭土 schroetterite

硅微粉 silica fume

石油焦 petroleum coke

石英砂 quartz sand

硅石粉 quartzite powder

石英矿物 quartz mineral

硅砂 silica sand

氮化硅 silicon nitride

土状石墨 amorphous graphite

人造石墨 artificial graphite,synthetic graphite

隐晶石墨 cryptocrystalline graphite

膨胀石墨 expanded graphite

细分散石墨 finely dispersed graphite

鳞片石墨 flake graphite

石墨碎屑 graphite fragment

石墨粉 graphite powder

高纯石墨 high-purity graphite

低灰分石墨 low-ash graphite

天然鳞片石墨 natural flake graphite

天然石墨 natural graphite

天然石墨粉 natural graphite powder

造粒石墨 pelleting graphite

显晶石墨 phanerocrystalline graphite

粉状石墨 powdered graphite

球状石墨 spheroidal graphite

鳞片石墨微粉 ultrafine flake graphite powder

超纯石墨 ultra-pure graphite

精制石墨 washed graphite

不定型碳 agraphitic carbon,amorphous carbon

炭黑 carbon black

蓝晶石硅线石精矿 disthene-sillimanite concentrate

蓝晶石硅线石火泥 disthene-sillimanite mortar

蓝晶石精矿 kyanite concentrate

蓝晶石族矿物 kyanite minerals

蓝晶石火泥 kyanite mortar

蓝晶石片岩 kyanite schist facies

红柱石精矿 andalusite concentrate

硅线石精矿 sillimanite concentrate

硅线石片麻岩 sillimanite gneiss

硅线石族 sillimanite group

硅线石页岩 sillimanite shale

合成硅线石 synthetic sillimanite

叶蜡石块 pyrophyllite in lumps

叶蜡石粉 pyrophyllite in powder,pyrophyllite powder

叶蜡石矿 pyrophyllite deposit, pyrophyllite ore	高岭土矿物 kaolin mineral
碱性高岭土 alkaline kaolin	高岭土原料 kaolin raw material
煅烧高岭土 calcined kaolin	高岭土泥浆 kaolin slip
煤系高岭土 coal series kaolinite	再沉淀次生高岭土 secondary sedimentary kaolin
粗粒高岭土 coarse-grained kaolin	沉淀高岭土 sedimentary kaolin
沉积高岭土 deposited kaolin	净高岭土 washed kaolin
黏土高岭土熟料 kaolin clay chamotte	煤矸石 coal gangue, coal waste
高岭土熟料 kaolin grog	粉煤灰 coal ash

7.2.3 术语例句

但用高能球磨直接氮化合成**氮化铝**的方法尚未报道。

However, there has not article about direct synthesis of **aluminum nitride** in high-energy ball mill.

氮化铝的热膨胀系数与硅的十分接近，因此不会引入大的热应力。

Since coefficients of thermal expansion between **aluminum nitride** and silicon are near, large thermal stress will not be introduced.

本发明提供即使是像**氮化铝**那样介质损耗因数极小的陶瓷也高效且牢固地接合的方法。

Provided is a method of efficiently and strongly joining ceramics even when they are ceramics of extremely low dielectric loss factor, such as **aluminum nitride**.

煤矸石的环境治理问题已成当务之急。

The environmental management of **coal waste** has become an impending problem.

热解温度是影响**煤矸石**热解反应最主要的因素。

Temperature is the main factor to influence the **coal residue** pyrolysis.

煤矸石保温材料的烧结过程是其制备工艺的关键。

The sintering process is the heart of the preparing process of **coal gangue** thermal insulation material.

采用废弃物粉煤灰和**煤矸石**为主要原料焙烧实心墙体保温砖。

Fly ash and **coal gangue** were used as main raw materials to roast solid insulating brick.

研究了**煤矸石**的种类、掺量以及化学激发剂对水泥水化性能的影响。

The influence of **coal gangue** and the activator on the hydration property of cement is studied in this paper.

大量堆存的**煤矸石**不仅占用大量的土地，而且对环境污染也相当严重。

Such vast amount of **coal gangue** not only occupies the farmland but also causes severe pollution to the surrounding environment.

实践表明，道路建设中利用**煤矸石**在技术上是可行的，并具有明显的经济和社会效益。

It shows that the use of **coal gangue** in road construction is feasible and has significant economic and social benefits.

从强度方面考虑，**粉煤灰**的掺量最好在 40% 以下。

The optimization content of the **fly ash** was less than 40% from the strength consideration.

粉煤灰对高贝利特水泥和混凝土的工作性、强度及水化热的影响。

Effect of **fly ash** on the workability, strength and hydration heat of high belite cement and concrete.

高掺量**粉煤灰**混凝土早期二次水化较为缓慢，随后二次速度逐渐加快。

The secondary hydration of high **fly ash** content concrete is slow at the early period and then the velocity of it gradually increases.

粉煤灰加入量增加，泡沫混凝土砌块的干密度递增，抗折和抗压强度增大。

The dry density, bending strength and compressive strength of foam concrete block increased with the increasing of the amount of **fly ash**.

掺入**粉煤灰**将增大混凝土的黏聚性和可塑性，改善混凝土的和易性，减小混凝土的膨胀性。

Cohesion and plasticity of steel slag concrete is increased, its fluidity is improved, and expansibility is decreased owing to addition of **fly ash**.

以**粉煤灰**和炭黑为原料，采用碳热还原氮化法在 1350 ~ 1550℃ 保温 6h 可成功合成出不同组成的 Sialon 环境材料[12]。

Sialon eco-materials with different compositions were successfully synthesized by a carbothermal reduction-nitridation process with **fly ash** and carbon black as raw materials at 1350-1550℃ for 6h.

7.2.4 术语例段

利用固废**粉煤灰**、活性炭和工业 SiC 粉经碳热还原法在氩气气氛下制备了多孔 SiC-Al$_2$O$_3$ 陶瓷。研究了 SiC 添加量（质量分数为 0 ~ 20%）对制备多孔 SiC-Al$_2$O$_3$ 陶瓷物相组成、显微结构、显气孔率、体积密度、吸水率、孔径分布、耐压强度、抗热震性和热扩散系数等性能的影响。发现最终产物是 β-SiC 和 α-Al$_2$O$_3$；同时，SEM 表明，在多孔 SiC-Al$_2$O$_3$ 陶瓷中的孔分布均匀。通过添加适量的 SiC（质量分数为 10%、15% 和 20%）可显著改善制备多孔 SiC-Al$_2$O$_3$ 陶瓷的性能；然而，进一步增加 SiC 用量导致抗热震性和力学性能下降。添加 10%（质量分数）SiC 在 1600℃ 保温 5h 成功制得多孔 SiC-Al$_2$O$_3$ 陶瓷，其平均孔径为 4.24μm，常温耐压强度为 21.70MPa，显气孔率为 48%，热扩散系数为 0.0194cm^2/s。

Porous SiC-Al$_2$O$_3$ ceramics were prepared from solid waste **coal ash**, activated carbon and commercial SiC powder by carbothermal reduction reaction method under Ar atmosphere. The effects of addition amounts of SiC(0-20wt.%) on the properties of as-prepared porous SiC-Al$_2$O$_3$ ceramics, such as phase composition, microstructure, apparent porosity, bulk density, water absorption, pore size distribution, compressive strength, thermal shock resistance and thermal diffusivity have been investigated. It was found that the final products are β-SiC and α-Al$_2$O$_3$. Meanwhile, the SEM shows the pores distribute uniformly in the porous SiC-Al$_2$O$_3$ ceramics. The properties of as-prepared porous SiC-Al$_2$O$_3$ ceramics were found to be remarkably improved by adding proper amounts of SiC(10wt.%, 15wt.%, and 20wt.%). However, further increasing the amount of SiC leads to a decrease in thermal shock resistance and mechanical properties. The porous SiC-Al$_2$O$_3$ ceramics doped with 10wt.% SiC and sintered at 1600℃ for 5h with the median pore diameter of 4.24μm, cold compressive strength of 21.70MPa, apparent porosity of 48%, thermal diffusivity of 0.0194cm^2/s were successfully obtained[13].

以**粉煤灰**和活性炭为原料，通过碳热还原反应工艺在熔盐介质中、氩气气氛下合成了 Al$_2$O$_3$-SiC 复合粉体。研究了 NaCl-KCl 熔盐量对合成过程的影响。采用 XRD 和 SEM-EDS 表征 Al$_2$O$_3$-SiC 复合粉体物相组成和显微结构的影响。与传统碳热还原反应法相比，发现添加 NaCl-KCl 有力促进了相转化和降低合成温度约 100℃。实验结果表明，当**粉煤灰**与活性炭的质量比为 100∶58，引入 100%（质量分数）NaCl-KCl，且在 1500℃保温 5h，可合成 Al_2O_3-SiC 复合粉体；反应产物具有良好的性能，颗粒分布均匀，无团聚，平均粒径为 0.5~1.5μm。

Al_2O_3-SiC composite powders have been synthesized by carbothermal reduction reaction method in molten salts medium under argon atmosphere, using **coal ash** and activated carbon as raw materials. The effect of NaCl-KCl molten salts quantity on the synthesis process was studied. The phase compositions and microstructures of Al_2O_3-SiC composite powders were characterized by XRD and SEM-EDS. Compared with conventional carbothermal reduction reaction method, it was found that the NaCl-KCl addition strongly promotes phase transformation and reduced synthesis temperature of nearly 100℃. Experimental results showed that Al_2O_3-SiC composite powders can be synthesized with the mass ratio of **coal ash** to activated carbon is 100∶58 and an addition amount of 100wt.% NaCl-KCl(mass fraction) at 1500℃ for 5h. The reaction products have good performance with uniform particle distribution, aggregate-free and average particle size is 0.5-1.5μm[14].

利用**粉煤灰**经碳热还原法（CRR）在氩气气氛下成功制备了 Al_2O_3-SiC 复合粉体，详细研究了原料配比、反应温度以及添加 La_2O_3、Sm_2O_3 对合成过程的影响。采用 XRD 和 SEM 表征 Al_2O_3-SiC 复合粉体物相组成和显微结构。获得了适宜的工艺参数和稀土添加剂，并讨论了 Al_2O_3-SiC 复合粉体的形成过程。

Al_2O_3-SiC composite powders were successfully fabricated from **coal ash** by carbothermal reduction reaction(CRR) method in argon atmosphere. The effects of raw materials ratio, reaction temperature, La_2O_3 and Sm_2O_3 additions on the synthesis process were investigated in detail. The phase compositions and microstructures of Al_2O_3-SiC composite powders were characterized by XRD and SEM. Proper processing parameters and rare earths addition amounts were determined, and the formation process of Al_2O_3-SiC powders was also discussed[15].

利用**粉煤灰**可制备**赛隆**基和 SiC 基非氧化物复合粉体材料、**赛隆**基和莫来石基致密陶瓷材料、SiC 基非氧化物多孔陶瓷材料、莫来石和堇青石氧化物多孔陶瓷材料、沸石分子筛、微晶玻璃及铝基和硅基化工原料等高值材料[16]。

High-value materials such as **Sialon**-based and SiC-based non-oxide composite powder materials, **Sialon**-based and mullite-based dense ceramic materials, SiC-based non-oxide porous ceramic materials, mullite and cordierite oxide porous ceramic materials, zeolite molecular sieve, glass-ceramics, aluminum-based and silicon-based chemical raw materials can be prepared from **coal fly ash**[16].

以**粉煤灰**、锆英石和活性炭为原料，通过原位碳热还原氮化工艺成功制备出 **β-Sialon**/ZrN/ZrON 复合材料。研究了原料配比和保温时间对合成过程的影响，并讨论了复合材料的形成过程。借助 XRD 和 SEM 表征复合材料的物相组成和显微结构。研究发现，增加试样中的碳含量和保温时间能促进 β-Sialon、ZrN 和 ZrON 的形成；合成 β-Sialon/ZrN/ZrON 复合材料的适宜工艺参数为：锆英石∶粉煤灰∶活性炭质量比为 49∶100∶100，合成温度为 1550℃，保温时间为 15h。在 1550℃ 保温 15h 合成的 β-Sialon、ZrN（ZrON）的平均粒径分别约为 2μm 和 1μm。β-Sialon/ZrN/ZrON 复合材料的制备过程包括 β-Sialon 和 ZrO$_2$ 的形成以及 ZrO$_2$ 向 ZrN 和 ZrON 的转化。

β-Sialon/ZrN/ZrON composites were successfully fabricated by an in-situ carbothermal reduction-nitridation process, with **fly ash**, zircon and active carbon as raw materials. The effects of raw materials composition and holding time on synthesis process were investigated, and the formation process of the composites was also discussed. The phase composition and microstructure of the composites were characterized by means of XRD and SEM. It was found that increasing carbon content in a sample and holding time can promote the formation of β-Sialon, ZrN and ZrON. The proper processing parameters to synthesize β-Sialon/ZrN/ZrON composites are mass ratio of zircon to fly ash to active carbon of 49∶100∶100, synthesis temperature of 1550℃ and holding time of 15h. The average grain size of β-Sialon and ZrN（ZrON）synthesized at 1550℃ for 15h reaches about 2μm and 1μm, respectively. The fabrication process of β-Sialon/ZrN/ZrON composites includes the formation of β-Sialon and ZrO$_2$ as well as the conversion of ZrO$_2$ to ZrN and ZrON[17].

用**粉煤灰**和炭黑在氮气气氛下经碳热还原氮化工艺成功制备了 **β-Sialon** 基复合材料。研究了加热温度和原料配比对合成过程的影响，并详细讨论了复合材料的形成过程。采用 X 射线衍射仪和扫描电子显微镜表征复合材料的物相组成和显微结构。结果表明，增加加热温度和炭黑与**粉煤灰**的质量比能促进 **β-Sialon** 的形成；当加热炭黑与粉煤灰质量比为 0.56 的试样，在 1723K 保温 6h 可合成 **β-Sialon** 基复合材料；在复合材料中得到的 **β-Sialon** 呈颗粒状存在，其平均理解为 2~3μm；**β-Sialon** 基复合材料的制备过程包括 **O′-Sialon**、**X-Sialon** 和 **β-Sialon** 的形成过程，以及 **O′-Sialon** 和 **X-Sialon** 向 **β-Sialon** 的转化过程。

β-Sialon based composites were successfully prepared from **fly ash** and carbon black under nitrogen atmosphere by carbothermal reduction-nitridation process. Effects of heating temperature and raw materials ratio on synthesis process were investigated, and the formation process of the composites was also discussed in detail. The phase compositions and microstructures of the composites were characterized by X-ray diffraction and scanning electronic microscope. The results show that increasing heating temperature and mass ratio of carbon black to **fly ash** can promote the formation of **β-Sialon**. The **β-Sialon** based composites can be synthesized at 1723K for 6h while heating the sample with mass ratio of carbon black to fly ash of 0.56. The as-received **β-Sialon** in the composites exists as granular with an average particle size of 2-3μm. The preparation process of **β-Sialon** based composites includes the formation of **O′-Sialon**, **X-Sialon** and **β-Sialon** as well as the conversion process of **O′-Sialon** and **X-Sialon** to **β-Sialon**[18].

研究了利用我国**煤矸石**制备 **O′-Sialon** 的物相组成和显微结构。使用硅粉比活性炭或者活性炭–少量硅的还原氮化更有效。掺杂 40% 硅粉的试样在 1500℃ 氮化，制得 **O′-Sialon** 超过 80%。借助 XRD 和 SEM 对形成的 **O′-Sialon** 进行表征。通过基于主成分分析的计算机模式识别程序优化了 **O′-Sialon** 制备参数，采用此方法揭示了 **O′-Sialon** 相对含量较高的目标参数最优区域。

The phase composition and microstructure of **O′-Sialon** prepared from Chinese **coal gangue** have been studied. The use of Si powder is more effective than that of activated carbon or mainly carbon with a little silicon for reduction-nitridation. For specimens with 40% Si addition, more than 80% of **O′-Sialon** may be obtained when nitrided at 1500℃. The formed **O′-Sialon** was characterized by XRD and SEM. The parameters for **O′-Sialon** preparation are optimized by computer pattern recognition program based on principal component analysis, the target parameter optimum regions with higher relative content of **O′-Sialon** was indicated by this way[19].

采用热重分析、XRD 和 SEM，研究了由矾土合成的 **β-Sialon** 和 **O′-Sialon** 粉的氧化机理和动力学。在空气气氛中进行了等温与非等温氧化实验。**β-Sialon** 在约 1000K 时开始氧化，在 1200K 后氧化速率显著增加，且在低温形成了 SiO_2 和 Al_2O_3，在高温形成莫来石。对比发现，**O′-Sialon** 粉开始氧化温度约为 1250K，且 1300K 后，其氧化速率明显增加。直到 1623K，**O′-Sialon** 相占优势，可根据 XRD 分析得以证实，且氧化产物是 SiO_2 以及少量的莫来石。在较高的温度下，SiO_2 与 **O′-Sialon** 粉体中的杂质熔融形成的液相阻碍了氧化的进程，使 **O′-Sialon** 具有更强的抗氧化性。其中，扩散是这两种氧化过程的关键控制步骤。根据实验数据，建立了一种新的氧化过程预测模型。将该模型应用于两个系统，验证了该模型的有效性。**O′-Sialon** 和 **β-Sialon** 的氧化活化能分别为 224.1kJ/mol 和 175.7kJ/mol。

The oxidation mechanism and kinetics of **β-Sialon** and **O′-Sialon** powders synthesized from bauxite have been investigated using thermogravimetric analysis, XRD, and SEM. Oxidation experiments were carried out in both isothermal and non-isothermal modes in oxygen atmosphere. **β-Sialon** begins to oxidize at above 1000K, and the rate increases significantly after 1200K, forming SiO_2 and Al_2O_3 at low temperature and mullite at high temperature. By comparison, the oxidation of **O′-Sialon** powder starts at about 1250K and after 1300K the rate increases significantly. The **O′-Sialon** phase remains dominant up to 1623K, as demonstrated by XRD analysis, and the oxidation products are SiO_2 as well as a small amount of mullite. At higher temperature, liquid phase resulting from the fusion of SiO_2 and the impurities in **O′-Sialon** powder hinders the progress of oxidation, which causes **O′-Sialon** to have more oxidation resistance. Diffusion is the controlling step during both the oxidation processes. Based on the experimental data, a new model for predicting the oxidation process is developed. The application of this new model to the two systems demonstrates the validity of this new model. The activation energy of the oxidation of **O′-Sialon** and **β-Sialon** has been calculated to be 224.1kJ/mol and 175.7kJ/mol, respectively[20].

为了以较低成本制备较高性能的 **ZrN(ZrON)-Sialon** 复合陶瓷材料，先以粉煤灰、锆英石和活性炭为主要原料，经碳热还原氮化法合成 ZrN-Sialon 复合粉；然后在 ZrN-Sialon 复合粉中添加不同量（质量分数分别为 0、5%、10%）的 Y_2O_3，在 1500℃保温 1h 埋碳烧结制备了 ZrN(ZrON)-Sialon 复合陶瓷材料，并研究了 ZrN(ZrON)-Sialon 复合陶瓷材料的相组成、烧结性能和力学性能等。结果表明：Y_2O_3 可加快 ZrN 和 β-Sialon 等主晶相的形成，同时促进材料的烧结致密化。添加 5%（质量分数）Y_2O_3 所制得的 ZrN(ZrON)-Sialon 复合陶瓷材料的显气孔率为 5.3%，体积密度为 3.5g/cm³，常温耐压强度为 29.4MPa，常温电阻率为 30.7Ω·m，性能较优[21]。

In order to prepare high-performance **ZrN(ZrON)-Sialon** composite ceramics with a low-cost, the ZrN-Sialon composite powders were synthesized by carbothermal reduction nitridation with fly ash, zircon and activated carbon as the main raw materials. The ZrN(ZrON)-Sialon composite ceramics were then prepared at 1500℃ for 1h with Y_2O_3 additive(0,5wt.%,10wt.%). At the same time, the effects of Y_2O_3 on the phase composition, microstructure, sintering characters and mechanical properties of the materials were investigated. The results show that Y_2O_3 can accelerate the formation of main crystal phases such as ZrN and β-Sialon and can increase the driving energy of sintering and promote the sintering densification process. When 5wt.% Y_2O_3 was added, the material obtained the best performance. The apparent porosity was 5.3% and bulk density was 3.5g/cm³. The cold compressive strength was 29.4MPa and specific resistance was 30.7Ω · m[21].

在 Al_2O_3-C 耐火复合材料中原位合成了**柱状 β-Sialon** 键合相，并基于第一性原理计算模拟了它们的生长机理。实验结果表明，Fe_2O_3 作为催化剂的加入加速了 Si_3N_4 向 β-Sialon 的转变，形成了发达的柱状结构。在 Si_3N_4 转变为 β-Sialon 的过程中，（100）晶面是晶体生长的主要基面。根据第一性原理计算，由于（Si,N）键被（Al,O）取代，（100）面的表面能大大降低。该催化剂可以促进气相在 Si_3N_4 的（100）晶面上的吸附，即降低了 SiO 和 Al_2O 的气体吸附能。由于原位合成柱状 β-Sialon 键合相的存在，Al_2O_3-C 耐火复合材料在 5 次热冲击循环后的残余抗压强度提高了 25.1%。

Columnar β-Sialon bonding phases were in situ synthesized in Al_2O_3-C refractory composites and their growth mechanism was simulated based on first-principles calculations. The experimental results indicated that the addition of Fe_2O_3 as a catalyst accelerated the transformation of Si_3N_4 to β-Sialon, resulting in a well-developed columnar structure. The(100)facet was the primary surface for crystal growth during the transformation process of Si_3N_4 into β-Sialon. According to first-principles calculations, the surface energy of the(100)facet decreased greatly due to the substitution of(Si,N) pairs with(Al,O). The catalyst could promote the adsorption of gaseous phases on the (100) facet of Si_3N_4 and decreased the gas adsorption energy of both SiO and Al_2O. Owing to the presence of in situ synthesized columnar β-Sialon bonding phases, the residual crushing strength of Al_2O_3-C refractory composites after 5 thermal shock cycles increased by 25.1%[22].

煤矸石是采煤和洗煤的副产品，随着能源消费的增长趋势，其数量正在迅速增加。积累的煤矸石没有得到适当的利用，造成了资源的浪费、废物处理和环境污染问题。在过去的几十年里，由于文化向可持续发展的转变，加上对处理煤矸石堆积的挑战的需求不断增加，开发煤矸石的利用策略受到了广泛关注。然而，据我们所知，目前还没有关于煤矸石再利用系列的全面和深入的评论。尽管利用煤矸石有一些好处，但值得注意的是，负面的环境问题是不容忽视的，因此，科学利用是控制环境影响所必需的。因此，本文的主要目的是对煤矸石在建材生产、能源生产、土壤改良和其他高附加值应用中的利用进行全面的文献综述，分析世界范围内煤矸石利用的研究动态，确定各种途径的潜在环境风险。评审的重点是发现潜在的问题，从而给出解决问题的建议。此外，基于以往的研究进展，本文还指出了该领域内进一步研究的方向。为了了解煤矸石的理化性质和利用特点以及相应的环境风险，在中国国家知识互联网和 Web of Science 中开展了文献计量分析，并对相关的 237 篇文章进行了系统回顾。结果表明，近年来发表的文章数量有所增加，研究主要来自中国，占所选文章总数的 78.94%。此外，这些研究主要集中在煤矸石的利用方面，而对环境风险缺乏关注。本研究的结果为煤矸石的进一步应用开辟了新的途径，希望能推动今后的相关研究，并为政策制定提供指导。

The amount of **coal gangue**, a by-product of coal mining and washing, is rapidly increasing with the growing trend of energy consumption. The accumulated coal gangue without appropriate utilization has resulted in a squander of resources, waste disposal and environmental pollution issues. Over the past few decades, there has been wide attention in developing strategies for the utilization of coal gangue due to a cultural shift towards sustainable development coupled with increasing demand for disposing the challenge of coal gangue accumulation. However, to our knowledge, there is no thorough and in-depth review on the series address of coal gangue reuse. In spite of some advantages of using coal gangue, it is notable that negative environmental problems cannot be ignored, so the scientific utilization is necessary to control the environmental impacts. Therefore, the main objective of this paper is to provide a comprehensive literature review of coal gangue utilization in building material production, energy generation, soil improvement and other high-added applications, analyze the worldwide dynamics of the studies on coal gangue utilization and identify the potential environmental risks in various pathways. The key focus of the review is on detecting the potential problems and thus giving recommendations for the solution. In addition, based on the progress of previous research, this paper also points the directions for further research within the field. A bibliometric analysis was developed in China National Knowledge Internet and Web of Science and a systematic review was conducted for related 237 articles to understand the physicochemical properties and utilization characteristics of coal gangue as well as corresponding environmental risks. The results indicated that the number of published articles has increased in recent years, and the researches were mainly from China, with the contribution of 78.94% of the total selected publications. Besides, these researches mainly focused on the utilization of coal gangue, while there was a lack of attention to environmental risks. The findings of the present study open up a new gate for the further application in coal gangue, hopefully motivate future relevant studies and guide the policymaking[23].

铝矾土是我国的战略资源，合成矾土基莫来石为我国丰富的铝矾土资源提供了综合利用的有效途径，但目前合成的矾土基均质莫来石应用过程中存在热态结合强度低、抗热震性较差以及不耐侵蚀等弱点。在莫来石材料中引入 SiC，可弥补其不足，但目前采用外加 SiC 或碳热还原原位生成 SiC 的方法，存在机械混合不均匀、烧结难度大或工艺复杂、成本高、不易产业化的缺点。为此，本工作以合成矾土基均质莫来石骨料和细粉为主要原料，以酚醛树脂为结合剂，引入 Si 粉，利用 Si 高温还原气氛下原位生成非氧化物晶须的新方法来制备**莫来石-SiC-O′-Sialon 复相材料**，工艺简单，成本低，可实现工业化大规模生产。研究了 Si 粉加入量、温度以及添加物（Al、Zn 和 SiC 粉）对复相材料组成、结构和性能的影响规律，并探讨了莫来石中杂质对 Si 反应和 SiC 晶须生长的催化机理，SiC 和 O′-Sialon 在复杂体系中的生长机理，以及晶须状 SiC 和 O′-Sialon 对复相材料增强增韧机理等，研究结果如下：在矾土基莫来石体系中引入 Si 粉，在高温埋炭条件下，Si 可与 C、CO、N_2 等反应，生成非氧化物 SiC 和 O′-Sialon 晶须，从而制备莫来石-SiC-O′-Sialon 复相材料。SiC 和 O′-Sialon 晶须填充气孔，并与莫来石直接结合，使复相材料结构致密、强度高，材料具有较好烧结性能。莫来石中的 Fe_2O_3、TiO_2 杂质可促进 Si 反应及催化晶须状 SiC 和 O′-Sialon 生长，不需外加催化剂，可形成材料内的自催化。SiC 和 O′-Sialon 晶须的生长机制为 VS 和 VLS。在引入 Si 粉的基础上，添加适量（1%~2%）的 Al、Zn 和 SiC 粉，有助于 Si 反应生成晶须状 SiC 和 O′-Sialon，并提高复相材料的致密度和强度。其原因在于添加物在高温下可增加试样中气相压力，促进气相传输、反应。而过量添加物使试样中气相压力过大而逸出，致使复相材料结构疏松，降低其烧结性能。矾土基莫来石-SiC-O′-Sialon 复相材料具有较高的高温力学性能和优良的抗热震性，其高温强度和抗热震性分别比莫来石砖提高 10 倍和 2 倍以上，Si 粉的较佳加入量为10%。高温强度和抗热震性提高的原因在于 SiC 和 O′-Sialon 与莫来石基体形成直接结合，起钉扎、锚固作用，增强效果显著；SiC 和 O′-Sialon 晶须形成交叉连锁的网络结构，断裂时 SiC 和 O′-Sialon 晶须桥连、拔出以及裂纹偏转等，消耗大量断裂功，增韧效果显著。在引入 Si 粉的基础上，引入适量 Al 粉，材料中除形成晶须状 SiC 和 O′-Sialon 外，还生成针状 AlN，且引入 Al 有助于非氧化物晶体发育长大，增强作用显著，高温强度提高 80%。引入少量 Zn(<1%) 对复相材料高温力学性能影响不大，过量 Zn 粉会劣化材料高温性能，原因是 Zn 在高温下以气态逸出，破坏材料结构。引入 1%~5% 的 SiC 粉，材料高温力学性能变化不大，抗热震性明显提高，主要由于 SiC 总量增加。矾土基莫来石砖抗碱侵蚀性较

差，引入 Si 粉制备的复相材料具有良好的抗碱侵蚀性能。与莫来石材料相比，复相材料 1100℃、1300℃碱侵蚀质量增重逐渐减小，强度增加。侵蚀层结构致密，试样内部 SiC 和 O′-Sialon 仍然存在，且其形貌与碱侵蚀实验前相同。再引入 Al、Zn、SiC 粉体后，复相材料均表现出良好的耐碱性，引入适量 Al、Zn 粉有助于进一步提高复相材料的抗碱侵蚀性能，而引入 SiC 后复相材料的抗碱侵蚀性能略有降低。碱侵蚀的过程为：在活性较高的碱介质中，莫来石、Sialon 和 SiC 首先与 CO 反应生成刚玉和石英相，然后 K 或 K₂O 再与刚玉、石英或者直接与莫来石反应生成钾霞石、白榴石和高钾玻璃相，进而使复相材料遭到侵蚀。复相材料耐碱性改善的机理是试样中的 SiC 和 O′-Sialon 对结构的增强作用及其体系碱侵蚀后形成的钾玻璃相使材料结构致密化，阻碍了碱进一步进入试样内部。矾土基莫来石-SiC-O′-Sialon 复相材料具有优良的抗氧化性，其氧化特性为保护性氧化，即氧化时复相材料表面的 O′-Sialon 和 SiC 先氧化，氧化产物 SiO₂ 与莫来石中杂质形成玻璃膜，封闭气孔，阻碍氧气进入试样内部；氧化产物 SiO₂ 和 Al₂O₃ 反应可形成含莫来石的致密保护层，可减少氧气进入试样。引入适量的 Al、Zn 和 SiC 粉，可提高复相材料的抗氧化性[24]。

Bauxite is China's strategic resource. Synthesis of high-performance bauxite-based mullite provides an effective way for comprehensive utilization of abundant bauxite resources in China. However, the synthesized homogeneous bauxite-based mullite has the disadvantages of poor toughness, low high temperature strength, poor thermal shock resistance and weak corrosion resistance during applications. The introduction of SiC into mullite materials can make up for shortcomings of bauxite-based mullite materials. However, the present methods by adding SiC or through carbothermal reduction in-situ formation of SiC have some shortcomings, such as uneven mechanical mixing, difficult to sinter or complex process, high cost and difficult to industrialize. Therefore, in this study, a novel method has been adopted to prepare bauxite based **mullite-SiC-O′-Sialon composites** at high temperature in carbon embedded condition using the synthesized homogeneous bauxite-based mullite aggregates and fine powder, silicon powder, and phenolic resin as binder. Using the plasticity of Si at low temperature increases the density of green body, and Si reacts CO, N₂ to in-situ formation whisker-like SiC and O′-Sialon at high temperature to prepare mullite-SiC-O′-Sialon composites. The novel method is simple, low cost and suitable for production mullite-SiC-O′-Sialon composites in large scale. The effects of various amount of Si, temperature and additives (Al, Zn and SiC) on the composition, microstructure and properties of the prepared composites have been studied. The catalytic mechanism of the trace impurities in mullite for Si reaction and growth of SiC whiskers, growth mechanism of SiC and O′-Sialon in complicated systems, and the strengthening and toughening mechanism of whisker-like SiC and O′-Sialon in the composites have been also discussed. The research results are as follows. The incorporated Si in bauxite-based mullite will react with C, CO and N_2 to form non-oxide SiC, O′-Sialon whiskers in carbon embedded condition at high temperature, thus the mullite-SiC-O′-Sialon composites are prepared. The SiC, O′-Sialon whiskers fill in pores and directly bond with mullite leading to the dense in structure, high strength and good sintering properties of the composites. The trace impurities of Fe_2O_3 and TiO_2 in mullite can promote the reactions of Si and catalyze the growth of whisker-like SiC and O′-Sialon without other additional catalyst, thus forming self-catalysis in the material. The growth mechanisms of SiC and O′-Sialon whiskers are governed by VS and VLS. On the basis of introducing Si powder, further adding appropriate amount of Al, Zn and SiC powder(1%-2%) can promote Si reaction to form whisker-like SiC and O′-Sialon and improve the density and strength of the composites. The reason is that Al, Zn and SiC additive can increase the gas pressure in the inner of sample at high temperature and promote gas phase transmission and reaction. But the excessive additives can increase the vapor pressure in the sample and more gaseous substances discharging, resulting in the loosening structure of the composite material and reducing its sintering performance. The bauxite-based mullite-SiC-O′-Sialon composite has high strength at high temperature and excellent thermal shock resistance. The high temperature strength and thermal shock resistance of the composite are more than 10 times and 2 times higher than that of mullite brick. The optimum addition of Si powder is 10%. The high strength of the composite is attributed to the direct bond between SiC, O′-Sialon and mullite, which play the role of pinning and anchoring effects. The excellent thermal shock resistance of the composite is due to SiC and O′-Sialon whiskers form an interlocking network structure, which consumes a lot of fracture work and has a remarkable toughening effect when SiC and O′-Sialon whiskers are bridged, pulled-out and cracks deflected. On the basis of introducing Si powder, proper amount of Al powder was introduced. Except whisker-like SiC and O′-Sialon, needle-like AlN is formed in the material. The introduction of Al contributes to the growth of non-oxide crystals. The enhancement effect is remarkable, and the high temperature strength increases 80%. The introduction of a small amount of Zn(<1%) has little effect on the high temperature mechanical properties of the composites. Excess Zn powder can degrade the high temperature properties of the material, because Zn escapes in the gaseous state and destroys the material structure. With the introduction of 1%-5% SiC powder, the high temperature strength properties of the material have slightly change, and thermal shock resistance of the material has been significantly improved, which is mainly due to the increase of the total SiC content. Bauxite-based mullite brick has poor alkali resistance, and the composite material prepared by introducing Si powder has good alkali resistance. Compared with bauxite-based mullite material, the weight gain and strength of the composite materials decreased after alkali corrosion at 1100℃ and 1300℃. The structure of erosion layer is compact, SiC and O′-Sialon still exist in the sample, and its morphology is the same as before alkali erosion experiment. After adding Al, Zn and SiC powder, the composite materials show good alkali resistance. The introduction of proper amount of Al and Zn powder can further improve the alkali resistance of the composite materials, while the alkali resistance of the composite materials is slightly reduced after the introduction of SiC. The process of alkali erosion is as follows: in the alkali medium with high activity, mullite, Sialon and SiC react with CO to form corundum and quartz phase first, then K or K_2O react with corundum, quartz or directly with mullite to form potassium nepheline, leucite and high potassium glass phase, and then the composite material is eroded. The mechanism of improving the alkali resistance of the composite material is that the SiC and O′-Sialon in the sample strengthen the structure and the potassium glass phase formed after the alkali erosion of the system makes the structure of the material compact, which prevents the alkali from further entering the sample. Bauxite-based mullite-SiC-O′-Sialon composite has excellent oxidation resistance, and the oxidation nature of the composite belong to protective oxidation, i. e. , when oxidized, O′-Sialon and SiC in the composite material are oxidized firstly on the surface of the material, and the oxidized product SiO_2 forms a glass film with impurities in mullite, the pores are sealed by the glass film and prevents oxygen from entering the inner of the sample through pores. When the oxidized products SiO_2 and Al_2O_3 react with each other, mullite-containing glass film can be formed. The oxidation resistance of the composite can be improved by introducing appropriate amount of Al, Zn and SiC powders[24].

参考文献

[1] Soliman N A, Tagnit-Hamou A. Using glass sand as an alternative for quartz sand in UHPC [J]. Construction and Building Materials, 2017, 145: 243-252.

[2] Michalowski R L, Wang Z, Nadukuru S S. Maturing of contacts and ageing of silica sand [J]. Géotechnique, 2018, 68 (2): 133-145.

[3] Zhang C, Lu B, Wang W, et al. CFD simulation of an industrial MTO fluidized bed by coupling a population balance model of coke content [J]. Chemical Engineering Journal, 2022, 446: 136849.

［4］ Liu Jiang, Kang Yili, Chen Mingjun, et al. Effect of high-temperature treatment on the desorption efficiency of gas in coalbed methane reservoirs：Implication for formation heat treatment ［J］. International Journal of Hydrogen Energy, 2022, 47 (19)：10531-10546.

［5］ Xiang D, Li P, Yuan X. Parameter optimization and exergy efficiency analysis of petroleum coke-to-hydrogen by chemical looping water splitting process with CO_2 capture and self-heating ［J］. Fuel, 2022, 324：124475.

［6］ Wijeyawardana P, Nanayakkara N, Gunasekara C, et al. Removal of Cu, Pb and Zn from stormwater using an industrially manufactured sawdust and paddy husk derived biochar ［J］. Environmental Technology and Innovation, 2022, 28：102640.

［7］ 张建刚, 杨忠福, 唐晓舟. 无烟煤工业化冶炼绿碳化硅工艺研究 ［J］. 矿产综合利用, 2017 (6)：36-38, 30.

［8］ Tymoshenko Y G, Gadzyra M P, Davydchuk N K. Secondary silicon carbide formed in the interaction of nanosized silicon carbide with iron oxide ［J］. Powder Metallurgy and Metal Ceramics, 2020, 58 (9)：523-528.

［9］ 高珺, 胡耀东. K值法测定工业硅粉中二氧化硅的含量 ［J］. 云南冶金, 2017, 46 (6)：48-51.

［10］ Li Yuanbing, Xiang Ruofei, Xu Nana, et al. Fabrication of calcium hexaluminate-based porous ceramic with microsilica addition ［J］. International Journal of Applied Ceramic Technology, 2018, 15 (4)：1054-1059.

［11］ 龙梦龙. 高性能 Si_3N_4-SiC 耐火材料合成原理及其应用性能 ［D］. 北京：北京科技大学, 2019.

［12］ 马北越, 厉英, 翟玉春. 用粉煤灰合成不同组成的 Sialon 环境材料 ［J］. 东北大学学报 (自然科学版), 2011, 32 (9)：1282-1285.

［13］ Yin Yue, Ma Beiyue, Hu Chuanbo, et al. Preparation and properties of porous SiC-Al_2O_3 ceramics using coal ash ［J］. International Journal of Applied Ceramic Technology, 2019, 16 (1)：23-31.

［14］ Yin Yue, Ma Beiyue, Li Shiming, et al. Synthesis of Al_2O_3-SiC composite powders from coal ash in NaCl-KCl molten salts medium ［J］. Ceramics International, 2016, 42 (16)：19225-19230.

［15］ Ma Beiyue, Ren Xinming, Yin Yue, et al. Effects of processing parameters and rare earths additions on preparation of Al_2O_3-SiC composite powders from coal ash ［J］. Ceramics International, 2017, 43 (15)：11830-11837.

［16］ 高陟, 马北越. 利用粉煤灰制备高附加值材料的新进展 ［J］. 耐火与石灰, 2021, 46 (1)：18-23.

［17］ Ma Beiyue, Sun Minggang, Ding Yushi, et al. Fabrication of β-Sialon/ZrN/ZrON composites using fly ash and zircon ［J］. Transactions of Nonferrous Metals Society of China, 2013, 23 (9)：2638-2643.

［18］ Ma Beiyue, Li Ying, Yan Chen, et al. Effects of synthesis temperature and raw materials composition on preparation of β-Sialon based composites from fly ash ［J］. Transactions of

Nonferrous Metals Society of China, 2012, 22 (1): 129-133.

[19] Zhang Haijun. Preparation and pattern recognition of O'-Sialon by reduction-nitridation from coal gangue [J]. Materials Science and Engineering: A, 2004, 385 (1-2): 325-331.

[20] Hou Xinmei, Zhang Guohua, Chou Kuochih, et al. A comparison of oxidation kinetics of O'-Sialon and β-Sialon powders synthesized from bauxite [J]. International Journal of Applied Ceramic Technology, 2008, 5 (5): 529-536.

[21] 任鑫明, 马北越, 高陟. 添加 Y_2O_3 对 ZrN(ZrON)-Sialon 性能的影响 [J]. 耐火材料, 2020, 54 (2): 134-136.

[22] Zhang Jie, Li Xiangcheng, Gong Wei, et al. First-principles simulation of the growth of in situ synthesised β-Sialon and its effects on the thermo-mechanical properties of Al_2O_3-C refractory composites [J]. Journal of the European Ceramic Society, 2019, 39 (8): 2739-2747.

[23] Li Jiayan, Wang Jinman. Comprehensive utilization and environmental risks of coal gangue: A review [J]. Journal of Cleaner Production, 2019, 239: 117946.

[24] 安建成. 矾土基均质莫来石-SiC-O'-Sialon 复相材料的组成、结构及其性能研究 [D]. 郑州: 郑州大学, 2020.

8 隔热耐火材料原料

8.1 多孔类隔热耐火材料原料

8.1.1 术语词

硅藻土 diatomite,tripolite,bergmeal,kieselguhr | 蛭石 vermiculite
珍珠岩 perlite | 漂珠 cenosphere

8.1.2 术语词组

硅藻土 diatomaceous earth,fossil flour,terra cariosa | 泡沫珍珠岩 foamed pearlite
硅藻土胶泥 diatomaceous mortar,tripolite mortar | 珍珠岩砂 perlite sand
硅藻土熟料粉 diatomite grog powders | 膨胀蛭石 expanded vermiculite,bloating vermiculite
膨胀珍珠岩 expanded pearlite,swell pearlite | 氧化铝空心球 alumina bubble,alumina hollow ball

8.1.3 术语例句

硅藻土是重要的非金属矿物材料。
Diatomite is an important nonmetallic mineral raw material.

对含**黏土硅藻土**和优质硅藻土进行了提纯研究。
The purification of **clay-bearing diatomite** and high-grade ores is investigated in this article.

这种**硅藻土**是由破碎、磨细并筛分的烧过硅藻土生产的。

It is produced by crushing, grinding and screening of burnt **diatomite** products or their breakage.

探讨了作为载体的**硅藻土**的性能对硫酸生产钒催化剂质量的影响。

The effects of quality of **diatomite** as carrier on vanadium catalyst in sulphuric acid production are discussed.

介绍了国外**硅藻土**助滤剂在过滤水中的应用和近年国内研究工作概况。

This article recommends the application of **diatomite** filter aid to water filtration abroad and its recent domestic research.

为了高效廉价地处理电镀废水，对**天然硅藻土**进行处理制备成改性硅藻土。

Natural diatomite was modified to treat electroplating wastewater with high efficiency and low cost.

为提高**硅藻土**矿利用价值，进行了微波作用下硅藻土矿的稳态酸浸提纯研究。

To improve the utilization value of **diatomite** ore, research on isothermal stable state acid leaching purification by microwave was made.

以**硅藻土**悬浮液为絮凝对象，考察了改性反应中影响 APAM 絮凝效果的因素。

Using **diatomite** suspension as flocculating object, the modification factors affecting the flocculation properties of APAM were investigated.

如果**珍珠岩**是重复使用，消毒可能是必要的。

If **perlite** is reused, sterilization may be necessary.

表面光滑、多孔状的**膨胀珍珠岩**独一无二，非常新颖。

The smooth-surfaced vesicular **expanded perlite** is unique and novel.

膨胀珍珠岩是一种传统的建筑保温材料，应用非常广泛。

Expansion perlite is a kind of traditional architectural heat preservation material, which is widely applied.

根据**膨胀珍珠岩**颗粒和水力学特性，可用作水处理过滤池滤料。

Swelled perlite can use as materials of wastewater disposal with its characteristic of granule and waterpower.

在移栽试验中发现，移栽基质以**蛭石**比珍珠岩更好，炼苗成活率达到88%。

Plantlets from test tube grew better on the growth medium **vermiculite** than on perlite in acclimatization stage, the survival rate reached 88%.

试验研究表明，以**膨胀珍珠岩**作葡萄扦插基质生根率为最高，达96.67%。

Through a series of experiments the survival rate of grape cutting with **swell pearlite** as base material is the highest, reaches at 96.67%.

阐述了**膨胀珍珠岩**的绝热机理，介绍了膨胀珍珠岩的性能指标，以及现场膨胀、自动充填的优点。

The paper presents the insulation mechanism of **expanded perlite** and its performance data. The advantages of expansion on site and automatic filling are described.

蛭石在短期内就会分解。

Vermiculite tends to collapse in a short time.

利用红外光谱研究了**蛭石**矿物材料层间水的赋存状态。

The existing status of interlayer water in **vermiculite** was investigated with infrared spectrum.

蛭石对实际废水的处理结果表明**蛭石**可以用于实际废水的处理。

Vermiculite was also applied to treat the electroplating wastewater and the result shows that vermiculite can be used to industry wastewater treatment.

选择的多孔材料有普通纸面石膏板、膨胀珍珠岩、**膨胀蛭石**和膨润土。

There are plaster board, expanded perlite, expanded **vermiculite** and montmorillonite as supporting materials.

用**膨胀蛭石**制成的各类建筑材料不仅耐燃、隔热、隔声，而且生产工艺简单，成本低廉。

All building materials made from **expanded vermiculite** is burn-resisting, heat insulation, soundproof. It is also process simple with low cost.

8.1.4 术语例段

为了替代现有的用于中低温（1000℃以下）保温领域的**硅藻土砖**，同时实现废弃资源的综合利用，本研究用废旧花岗岩通过直接发泡法制备了一种新型发泡保温材料。其中，废旧花岗岩的比例达到88%（质量分数）。实验结果表明，在废旧花岗岩球磨时间为7h、发泡剂含量（质量分数）为8%、煅烧温度为1050℃条件下，所制材料获得最佳综合性能：体积密度为0.7g/cm³，线性收缩率为3.71%，总孔隙率为73.59%，抗压强度为8.0MPa，900℃下8h复烧的线性收缩率为0.2%，300℃下的热导率为0.108W/(m·K)。同时，系统研究了废旧花岗岩的粒径、发泡剂的含量对宏观和微观性能的影响，然后用MIAPS软件得到了孔隙结构参数。最后，基于图像法和分形几何学分析了不同变量下试样的孔隙结构和性能之间的关系。研究结果表明，以废旧花岗岩作为主要原料，完全有可能取代天然硅藻土砖的使用。

To replace the existing **diatomite** brick used in the field of medium and low temperature(below 1000℃) thermal insulation and realize the comprehensive utilization of waste resources, a novel foamed insulation material was prepared with waste granite via a direct foaming method. In particular, the proportion of waste granite reached 88wt.%. The optimum parameter for the specimens were obtained at 1050℃ with 8wt.% foam and milling time of waste granite was 7h. Meanwhile, the experimental results showed that the bulk density was 0.7g/cm³, the linear shrinkage was 3.71%, the total porosity was 73.59%, the compressive strength was 8.0MPa, the linear shrinkage of re-sintering at 900℃ for 8h was 0.2%, and the thermal conductivity was 0.108W/(m·K) at 300℃. The effects of particle size of waste granite, content of foam on the macro and micro properties were systematically investigated, then the MIAPS software was used to obtain the pore structure parameters. Finally, the relationship between the pore structure and properties of specimens under different variables was analyzed based on image method and fractal geometry. The results show the potential use of waste granite as the main raw material to replace the use of natural diatomite brick[1].

不同类型的生物遗骸，从硅藻到碳酸盐沉淀物，都积聚在湖泊和其他水生生态系统的沉积物中。形成硅质沉积物被称为硅藻的单细胞藻类，是水生环境中生态和生物地球化学的重要生物群，通常保存在湖泊或海洋沉积物中。当硅藻大量积聚在沉积物中时，化石残骸可以形成硅藻土。在沉积学文献中，"**硅藻土**"被定义为一种易碎、浅色、硅藻含量至少为 50% 的沉积岩，然而，在第四纪科学文献中，硅藻土通常被用作描述含有"大量"硅藻壳的沉积物类型，而没有对硅藻丰度进行准确描述。在此，我们提出问题：什么是硅藻土？多少硅藻将沉积物定义为硅藻土？它是未压实的沉积物还是压实的沉积物？我们简要概述了先前的做法，并建议若沉积物超过 50% 由硅藻 SiO_2 组成，具有高孔隙率（>70%），并且是未压实的，就是硅藻泥，而压实的是硅藻土。更大的埋藏深度和更高的温度会导致孔隙度下降，并重新结晶为瓷土、燧石和纯石英。

Different types of biogenic remains, ranging from siliceous algae to carbonate precipitates, accumulate in the sediments of lakes and other aquatic ecosystems. Unicellular algae called diatoms, which form a siliceous test or frustule, are an ecologically and biogeochemically important group of organisms in aquatic environments and are often preserved in lake or marine sediments. When diatoms accumulate in large numbers in sediments, the fossilized remains can form diatomite. In sedimentological literature, "**diatomite**" is defined as a friable, light-colored, sedimentary rock with a diatom content of at least 50%, however, in the Quaternary science literature diatomite is commonly used as a description of a sediment type that contains a "large" quantity of diatom frustules without a precise description of diatom abundance. Here we pose the question: What is diatomite? What quantity of diatoms define a sediment as diatomite? Is it an uncompacted sediment or a compacted sediment? We provide a short overview of prior practices and suggest that sediment with more than 50% of sediment weight comprised of diatom SiO_2 and having high (>70%) porosity is diatomaceous ooze if unconsolidated and diatomite if consolidated. Greater burial depth and higher temperatures result in porosity loss and recrystallization into porcelanite, chert, and pure quartz[2].

原始**硅藻土**首先通过酸处理纯化，然后用 γ-甲基丙烯酰氧基丙基三甲氧基硅烷分子（KH570）改性，在酸处理的硅藻土表面引入疏水链。傅里叶变换红外光谱和热重分析（TGA）表明，硅烷偶联剂（KH570）通过共价键成功接枝在硅藻土上。数码照片显示，硅烷化过程改变了硅藻土的表面性质。聚氯乙烯（PVC）/原始硅藻土和 PVC/改性硅藻土复合材料通过双辊磨制备。通过热重分析、力学性能测试和动态力学分析研究了 PVC 复合材料的热稳定性和力学性能。结果表明，复合材料的热稳定性得到改善，含有 1 份改性硅藻土的 PVC 复合材料的最大失重温度（T_{max}）比不含硅藻土的 PVC 复合材料高约 20℃。PVC/改性硅藻土复合材料由于 PVC 基体与改性硅藻土之间更强的界面相互作用而表现出更好的力学性能。但当硅藻土添加量超过 1 份时，冲击强度急剧下降。产生这种现象的原因是硅藻土在 PVC 中起到了缺陷的作用，它不利于吸收冲击强度能量。扫描电镜结果证明了这一推论。

Pristine **diatomite** was first purified by acid treatment and then modified with γ-methacryloxy propyl trimethoxysilane molecule(KH570)to introduce hydrophobic chains on the surface of acid-treated diatomite. Fourier-transform infrared spectroscopy and thermogravimetric analysis(TGA)indicated that the silane coupling agent(KH570)was successfully grafted on the diatomite through covalent bonding. The digital photos showed that the silanization process changed the surface property of the diatomite. The poly (vinyl chloride)(PVC)/pristine diatomite and PVC/modified diatomite composites were prepared via two-roll mill. The thermal stability and mechanical properties of PVC composites were investigated by TGA, mechanical properties tests, and dynamic mechanical analysis. The results showed that the thermal stability of the composites improved and maximum weight loss temperature(T_{max})of the PVC composite with 1 phr modified diatomite was about 20℃ higher than that of PVC composite without diatomite. The PVC/modified diatomite composites exhibited better mechanical properties owing to the stronger interfacial interaction between PVC matrix and modified diatomite. But the impact strength reduced sharply when the addition of diatomite was more than 1 phr. The reason of the phenomenon is that the diatomite plays the role of defects in PVC, and it works against the absorption of impact strength energy. It was proved by the results of scanning electron microscopy[3].

在许多行业中，基于二氧化碳的超临界流体萃取（CO_2 SFE）已成为溶剂萃取的绝佳替代品。在这项研究中，重质原油被吸附到**硅藻土**上，并用于比较不同的提取工艺。以二氯甲烷索氏提取的油提取物为基准。对另一批试样进行连续的短时间和长时间 CO_2 SFE，以及随后的二氯甲烷索氏萃取，得到三种不同的提取物。通过标准的饱和物/芳烃/树脂/沥青质（SARA）方法对所有这四种提取物进行分级分离。然后通过高温气相色谱模拟蒸馏（SIMDIS）、气相色谱-质谱（GC-MS）和傅里叶变换离子回旋共振质谱（FT-ICR MS）分析每个 SARA 馏分。结果发现 CO_2 SFE 选择性地提取具有低分子缩合度的相对低分子量化合物。剩余的未萃取组分主要存在于极性树脂和沥青质馏分中，含有碳数大、分子凝聚度高的化学物质。这些组分中的大多数是低挥发性的，沸点高于 500℃，超过了普通 GC 色谱柱的温度上限。如果分析的目标是通过 GC-MS 确定石油生物标志物或成分分析，CO_2 SFE 可以作为索氏溶剂萃取的环保替代品。此外，CO_2 SFE 仅用于萃取小分子量的挥发性化合物，其优点是可以留下对 GC 系统有害的大多数非挥发性成分。

Carbon dioxide-based supercritical fluid extraction (CO_2 SFE) has been an excellent alternative to solvent extraction in many industries. In this study, a heavy crude oil was adsorbed to **kieselguhr** and performed for comparison of different extraction processes. The oil extract by dichloromethane Soxhlet extraction was used as a benchmark. Another batch of the sample was subjected to sequential short-time and longer-time CO_2 SFE, and subsequent dichloromethane Soxhlet extraction, yielding three distinct extracts. All of these four extracts were then fractionated by a standard saturates/aromatics/resin/asphaltene (SARA) method. Each SARA fraction was then analyzed by high temperature gas chromatography simulated distillation (SIMDIS), gas chromatography-mass spectrometry (GC-MS) and Fourier-transform ion cyclotron resonance mass spectrometry (FT-ICR MS). It was found that CO_2 SFE selectively extracts relatively low molecular weight compounds with low degrees of molecular condensation. The remaining un-extracted components are mainly in polar resin and asphaltene fractions, containing the chemical species with large carbon numbers and high degrees of molecular condensation. Most of these components are in low volatility with boiling points higher than 500℃, beyond the upper temperature limit of common GC columns. CO_2 SFE can serve as an environmental-friendly alternative to Soxhlet solvent extraction if the goal of the analysis is to determine petroleum biomarker or compositional analysis by GC-MS. In addition, CO_2 SFE for extracts only volatile compounds of small molecular weight has an advantage of leaving most non-volatile components that would be detrimental to the GC systems[4].

珍珠岩是一种丰富的矿物，在用作原料或膨胀珍珠岩之前需要进行最少的加工，从而形成一种低成本的天然多孔材料。目前，从液体流出物和受污染的水中去除放射性铯的材料的应用引起了人们的极大兴趣。珍珠岩在过去几年中被评估为可吸附多种金属，但之前没有研究过它可以从受污染的水中去除 Cs^+。目前的工作研究了使用阿根廷萨尔塔矿床中的珍珠岩去除水溶液中的 Cs^+。通过粉末 X 射线衍射、热分析、比面积分析和扫描电子显微镜对矿物进行了表征。通过批量实验研究了溶液 pH 值、伴随离子的存在、接触时间、Cs^+ 初始浓度、珍珠岩剂量以及吸附剂的碱性或酸性处理的影响。在高 pH 值和用 NaOH 处理后，去除率增加。珍珠岩对 Cs^+ 的吸附在接触的前 80min 内迅速上升。对于每升 30g 珍珠岩的剂量和 10mg/L 的初始阳离子浓度，所选材料（来自 Pava 矿）在用 NaOH 处理之前和之后分别产生 84% 和 89% 的去除效率。我们的研究结果表明，珍珠岩是一种能够从水溶液中去除 Cs^+ 的材料，即使在低剂量使用时也是如此。这些发现与从核流出物中去除放射性 Cs 同位素以及环境水污染的情况相关。

Perlite is an abundant mineral that requires minimum processing before use either as raw or expanded perlite, resulting in a low-cost, natural porous material. The application of materials for the removal of radioactive cesium from liquid effluents and contaminated waters is currently of great interest. Perlite has been evaluated in the last years for the sorption of a variety of metals, but it had not been investigated before for removal of Cs^+ from contaminated waters. The present work examines the use of perlites from a deposit in Salta, Argentina, for removal of Cs^+ from aqueous solutions. The mineral was characterized by means of powder X-ray diffraction, thermal analysis, analysis of specific area, and scanning electron microscopy. The effect of solution pH, presence of concomitant ions, contact time, Cs^+ initial concentration, perlite dose, and basic or acidic treatment of the sorbent were studied by batch experiments. Removal increased at high pHs and after treatment with NaOH. Sorption of Cs^+ by perlite presented a rapid rise in the first 80min of contact. The selected material (from Pava mine) yielded removal efficiencies of 84% and 89% before and after treatment with NaOH, respectively, for a dose of 30g perlite per litre and initial cation concentration of 10mg/L. Our results demonstrate that perlite is a material capable of removing Cs^+ from aqueous solutions, even when applied at low doses. These findings are relevant in the context of removal of radioactive Cs isotopes from nuclear effluents and in case of contamination of environmental waters[5].

聚合物水泥复合材料（PCCs）是其中聚合物和矿物黏合剂形成互穿网络并相互配合的材料，从而显著提高了材料的性能。另外，需要利用废料是可持续建筑的需求。各种矿物粉末，例如粉煤灰或高炉矿渣，已成功用于水泥和混凝土的生产。本文涉及**珍珠岩**粉末的使用，珍珠岩粉末是 PCCs 的组成部分，它是来自原始珍珠岩热膨胀过程中的繁重废物。复合材料的力学性能测试结果和一些微观观察结果表明有可能合理有效地利用废弃珍珠岩粉末作为 PCCs 的一种成分使用。这属于一种新型的建筑材料，可满足可持续建筑的需要。

Polymer-cement composites (PCCs) are materials in which the polymer and mineral binder create an interpenetrating network and co-operate, significantly improving the performance of the material. On the other hand, the need for the utilization of waste materials is a demand of sustainable construction. Various mineral powders, such as fly ash or blast-furnace slag, are successfully used for the production of cement and concrete. This paper deals with the use of perlite powder, which is a burdensome waste from the process of thermal expansion of the raw **perlite**, as a component of PCCs. The results of the testing of the mechanical properties of the composite and some microscopic observations are presented, indicating that there is a possibility to rationally and efficiently utilize waste perlite powder as a component of the PCCs. This would lead to creating a new type of building material that successfully meets the requirements of sustainable construction[6].

研究了轻质碱黏结材料的热和力学性能，这些性能与微结构和孔隙率相关。该材料是使用**膨胀珍珠岩**的细颗粒作为原料粉末，并使用二硅酸钾水溶液作为碱性活化剂来生产的，而不是通常使用的硅酸钠，以提高热性能。固结后，对该材料进行微观结构分析以及热物理和机械测试。密度为（467±40）kg/m³，总孔隙率（体积分数）为80%，平均孔径为0.2μm，抗压强度为（1.5±0.5）MPa。观察到，除了二硅酸钾作为碱结合剂的作用外，膨胀的珍珠岩颗粒在表面上反应形成地质聚合物纳米沉淀。激光闪光法被用来评价的温度范围25~550℃内的整体热性能：25℃的热导率为0.084W/（m·K），550℃的热导率为0.121W/（m·K）。在高温下，热导率与商业珍珠岩（密度在180~260kg/m³范围内的硅酸钠材料）相比，甚至更低。结果表明，通过利用硅酸钾和克服了有机黏合剂耐高温性差的问题，可以生产出具有改善的热性能的轻质碱结合膨胀珍珠岩材料。

The thermal and mechanical properties of a lightweight alkali bonded material were investigated in correlation with the microstructure and porosity. The material was produced using fine granules of **expanded perlite** as raw powder and potassium di-silicate aqueous solution as alkaline activator, instead of the commonly used sodium silicate, to improve thermal performances. After consolidation, the material was subjected to microstructural analyses, as well as to thermo-physical and mechanical tests. Density was (467 ± 40) kg/m³, the total porosity was 80vol. %, the average pore size was 0.2μm and compressive strength was (1.5 ± 0.5) MPa. It was observed that, besides the action of potassium di-silicate as an alkali binder, expanded perlite granules reacted on the surface forming geopolymer nano-precipitates. Laser Flash Method was employed to evaluate the whole thermal behavior within the temperature range 25-550℃: thermal conductivity was 0.084W/（m·K）at 25℃ and 0.121W/（m·K）at 550℃. At high temperature, the thermal conductivity resulted similar or even lower when compared with commercial perlite-sodium silicate materials with lower densities in the range 180-260kg/m³. Results show that lightweight alkali bonded expanded perlite materials can be produced with improved thermal properties by exploiting potassium silicate and overcoming the poor temperature resistance of organic binders[7].

用表面活性剂对**蛭石**进行改性以实现蛭石层的插层。由于蛭石在其表面带有负电荷，预计阳离子表面活性剂比阴离子表面活性剂更能扩展黏土矿物层。然而，阴离子表面活性剂带负电的部分与蛭石带正电的边缘相互作用并导致超晶格结构，因此出乎意料的是，层的膨胀被确定为蛭石的完全塌陷相。在阴离子、阳离子和非离子表面活性剂存在下检测蛭石分散体的胶体和结构特性。结果表明，阳离子表面活性剂用第二层覆盖蛭石表面，但与阴离子表面活性剂相比，黏土矿物层的膨胀有限。阴离子表面活性剂与蛭石带正电的边缘产生静电相互作用，充分扩展了蛭石的层状结构。

Vermiculite was modified with surfactants in order to enable intercalation of vermiculite layers. Since vermiculite has negative charges on its surfaces, it was expected that cationic surfactant would expand the clay mineral layers more than an anionic surfactant. Nevertheless, negative parts of the anionic surfactants interacted with the positively charged edges of vermiculite and caused to super lattice structure so, unexpectedly the expansion of the layers was determined to be fully collapsed phase of vermiculite. Colloidal and structural properties of vermiculite dispersions were examined in presence of anionic, cationic and nonionic surfactants. The results showed that cationic surfactant covered the surface of the vermiculite with a second layer, but the expansion of the clay mineral layer was limited compared to the anionic surfactant. The anionic surfactant produced electrostatic interaction with the positively charged edges of vermiculite and fully expanded the layer structure of the vermiculite[8].

通过传统的加热技术，用不同数量的氧化钙浸渍**蛭石**试样，并在热分析设备中进行 CO_2 捕获实验。当分别用纯氧化钙和氧化钙浸渍的蛭石（1:1）进行实验时，氧化钙捕获的 CO_2 量从每摩尔 13g 增加到每摩尔 16.8g。采用 Kissinger-Akahira-Sunose(KAS) 和 Osawa-Flynn-Wall(OFW) 的整体等值转换方法研究该过程的动力学，获得了良好的相关系数。表观活化能值显示，对于低转化率（$\alpha <$ 0.3），该过程的控制步骤是一个混合步骤，其中化学反应和试剂扩散到蛭石中的速率是同一数量级的（$20kJ < E_a < 40kJ$）。而对于较高的转化率（$\alpha > 0.3$），表观活化能值表明，化学反应是控制步骤的关键（$E_a > 40kJ$）。

Vermiculite samples were impregnated with different amounts of calcium oxide by the conventional thermal heating technique and subject to CO_2 capture experiments in thermal analysis equipment. The amount of CO_2 captured by calcium oxide increased from 13g of CO_2 per mol of CaO to 16.8g of CO_2 per mol of CaO when the experiments were carried out with pure calcium oxide and vermiculite impregnated with CaO(1:1), respectively. Integral isoconversional methods of Kissinger-Akahira-Sunose (KAS) and Osawa-Flynn-Wall (OFW) were used for the kinetic study of the process and good correlation coefficients were achieved. The apparent activation energy values showed that for low conversions($\alpha<0.3$) the controlling step of the process is a mixed step where the chemical reaction and the diffusion of the reagents into the vermiculite have rates of the same order of magnitude($20kJ<E_a<40kJ$). For higher conversions values($\alpha>0.3$) the apparent activation energy values suggest that the slow step is a chemical step($E_a>40kJ$)[9].

从**蛭石**中得到甲醇和乙醇脱水反应的两个系列催化剂。用硝酸溶液处理生黏土得到了第一类催化剂。蛭石的这种改性使其比表面积显著增大，同时也使黏土层中的 Al^{3+} 和 Fe^{3+} 离子发生了部分淋溶。蛭石的酸处理导致其在甲醇和乙醇脱水过程中活化。将蛭石与氧化铝柱插层得到第二类催化剂。在柱撑工艺前，先用硝酸溶液处理生蛭石，然后配合 Al^{3+} 和 Fe^{3+} 阳离子，从黏土层中浸出，用草酸或柠檬酸处理。氧化铝柱撑蛭石具有比表面积增大、孔隙率增大、酸性增强等特点，是一种非常有效的醇脱水催化剂。改性蛭石的催化性能主要与试样中存在的酸位的浓度和强度有关。

Two series of catalysts for the reactions of methanol and ethanol dehydration were obtained from **vermiculite**. The first series of the catalysts were obtained by treatment of raw clay with nitric acid solution. Such modification of vermiculite resulted in a significant increase of its specific surface area and also partial leaching of Al^{3+} and Fe^{3+} cations from clay layers. Acid treatment of vermiculite resulted in its activation in the processes of methanol and ethanol dehydration. The second series of the catalysts were obtained by intercalation of vermiculite with alumina pillars. Prior to the pillaring process, raw vermiculite was treated with nitric acid solution followed by complexation Al^{3+} and Fe^{3+} cations, leached from clay layers, with oxalic acid or citric acid. Alumina pillared vermiculites, characterized by increased specific surface area, porosity and acidity, were found to be very active catalysts of alcohols dehydration. Catalytic performance of modified vermiculites was related mainly to the concentration and strength of acid sites present in the samples[10].

在火力发电厂中，煤粉燃烧产生复杂的人为材料组成，例如飞灰（煤）。如果将这些材料丢弃到垃圾填埋场和河流，则会对环境（空气和水等）污染构成重大威胁。在过去二十年，人们一直在努力进行研究，以减少从煤飞灰（如漂珠）中生产和衍生具有潜在价值的材料。**漂珠**是一种低密度、化学惰性的球形材料，充满空气/惰性气体（氮气或二氧化碳）。漂珠被认为是飞灰中最重要的部分，因为它具有高可加工性、耐热性、抗压强度和低导电性、体积密度等优越的特性，因此被用于不同的行业。本综述讨论了从飞灰中提取漂珠、其表征（物理和化学）以及在不同行业中的应用。

In thermal power plants, pulverized coal combusts to give an intricate composition of anthropogenic materials such as fly ash (coal). These materials are a major threat to environmental (air and water, etc.) pollution if dispose of to landfill sites and rivers. Since the last two decades, research and efforts are going on to reduce production and derivation of potentially valuable materials from coal fly ash such as cenosphere. **Cenosphere** is a low density, chemically inert and spherical material filled with air/inert gas (either nitrogen or carbon dioxide). Cenosphere is considered to be the most important fraction of fly ash as it is being used in different industries due to its condescending properties such as high workability, thermal resistance, compressive strength and low conductivity, bulk density. This review discusses the extraction of cenosphere from fly ash, its characterization (physical and chemical) and applications in different industries[11].

本研究探索了一种新型的内部固化剂，即穿孔**漂珠**。漂珠是燃煤发电厂产生的空心飞灰颗粒。漂珠的外壳本质上是多孔的，由一层薄薄的玻璃结晶薄膜密封。通过化学蚀刻去除这层薄膜，可以暴露外壳上的孔隙，从而使漂珠穿孔并为水传播到漂珠内部体积中提供路径。发现穿孔的漂珠具有高达180%（质量分数）的吸水率。在高相对湿度（95%）下，负载的水可以很容易地从漂珠中释放出来。当将饱和漂珠加入水泥砂浆中进行内部养护时，砂浆的自收缩几乎被消除。内部养护还提高了水泥砂浆的抗压强度。所有这些结果表明，穿孔的漂珠可用作HPC的有效内部固化剂。

This study explores a novel internal curing agent, perforated **cenospheres**. Cenospheres are hollow fly ash particles produced from coal burning power plants. The shell of the cenospheres is inherently porous that is sealed by a thin layer of glass-crystalline film. By removing this film through chemical etching, the pores on the shell can be exposed, perforating the cenospheres and providing paths for water propagating into the internal volume of cenospheres. The perforated cenosphere were found to have water absorption as high as 180wt.%. The loaded water can be readily released from the cenospheres under high relative humidity (95%). When incorporating saturated cenospheres into cement mortar for internal curing, the autogenous shrinkage of the mortar was almost eliminated. The internal curing also improved the compressive strength of the cement mortar. All these results suggest that perforated cenospheres can be used as an efficient internal curing agent for HPC[12].

注枪是铁水预处理的关键设备。本工作通过添加**氧化铝空心球**获得了具有高力学性能和抗热震性的喷枪浇注料，该浇注料具有球形和中空结构。研究发现，氧化铝空心球的加入提高了浇注料的流动值。添加4%（质量分数）氧化铝空心球（AB4）的试样的体积密度是所有实验浇注料中最好的。添加氧化铝空心球能有效提高莫来石浇注料的热膨胀、导热系数，并能抵抗1300℃以上热膨胀的急剧下降。添加氧化铝空心球可以提高莫来石浇注料的机械强度。在1400℃加热后，气泡和结合相之间形成强化学键。由于在氧化铝空心球/基体界面处的适当结合，AB4浇注料表现出相对更好的抗热震性。

Injecting lance is a key equipment for hot metal pretreatment. In this work, the lance castables with high mechanical properties and thermal shock resistance were obtained by adding **alumina bubbles**, due to its spherical shape and hollow structure. It is found that the addition of alumina bubble improves the flow values of castables. The bulk density values of samples with the addition of 4wt.% alumina bubbles(AB4) were the best of all the experimental castables. Alumina bubble addition is effective in heightening the thermal expansion, thermal conductivity of mullite castables and can resist the sharp decrease in thermal expansion above 1300℃. The addition of alumina bubble can improve the mechanical strength of mullite castables. After heating at 1400℃, a strong chemical bond forms between the bubbles and the binding phases. Due to the appropriate bonding at the alumina bubble/matrix interface, AB4 castable exhibits a relatively better thermal shock resistance[13].

莫来石浇注料广泛用于铁水预处理加工设备。本工作研究了硅溶胶**改性氧化铝空心球**对莫来石浇注料性能的影响。氧化铝空心球经硅溶胶改性后，试样的容重略有增加，显气孔率有所降低。实验浇注料的常温抗折强度在110℃干燥和1400℃加热3h后显著增加。试样的热膨胀明显增加，而试样的热导率显著降低。氧化铝空心球改性后的氧化铝空心球/基体界面形成更多柱状莫来石晶体，形成连续的互锁网络结构，有助于提高浇注料的抗热震性。

Mullite castables are widely used in the hot metal pretreatment processing equipment. In this work, the effects of **modification of alumina bubbles** with silica sol on the properties of mullite castables were studied. After the alumina bubbles were modified by silica sol, the bulk density of the samples was slightly increased, and the apparent porosity was decreased. The cold modulus of rupture values of experimental castables increased remarkably after drying at 110℃ and heating at 1400℃ for 3h. The thermal expansion of the samples was obviously increased, while the thermal conductivity of the samples was decreased significantly. More columnar mullite crystals were formed at the alumina bubble/matrix interface after the alumina bubble modification and form a continuous inter-locking network structure, which was helpful to improve castables' thermal shock resistance[14].

　　莫来石浇注料广泛用于铁水预处理加工设备。本文研究了**氧化铝空心球**粒径对莫来石浇注料显微组织和物理性能的影响。灰色关联分析是分析多变量案例中模式的有用方法。本文采用灰色关联分析方法估计试样性质与氧化铝空心球结构之间的关系。采用灰色分析法，确定氧化铝气泡的比表面积对显气孔率、冷力学性能和抗热震性能影响最大。氧化铝空心球的空心半径与试样的体积密度、高温抗折强度、流动值和热膨胀系数的关系最大。最后，通过添加粒径在 0.5~1mm 之间的氧化铝空心球，获得了具有高断裂模量和抗热震性的莫来石浇注料。高断裂模量被证明是由氧化铝空心球球形、中空结构和氧化铝空心球/基体界面处的适当结合引起的。

　　Mullite castables are widely used in hot metal pretreatment processing equipment. In this work, the effects of **alumina bubble** particle size on the microstructure and physical properties of mullite castables were studied. Grey relational analysis is a useful method for analyzing patterns in multivariate cases. In this paper, the grey relational analysis was used to estimate the relationship between the properties of samples and the structure of alumina bubbles. Using grey analysis, the specific surface area of alumina bubbles was determined to have the greatest influence on the properties of apparent porosity, cold mechanical properties and thermal shock resistance. The hollow radius of alumina bubbles has the greatest relationship to bulk density, hot modulus of rupture, flow values and coefficient of thermal expansion of samples. Finally, the mullite castables with a high modulus of rupture and thermal shock resistance was obtained by adding alumina bubbles with a particle size between 0.5mm and 1mm. The high modulus of rupture was shown to be caused by the alumina bubble spherical shape, hollow structure and the appropriate bonding at the alumina bubble/matrix interface[15].

　　能源短缺是世界上所有国家面临的共同难题，解决能源短缺最有效的途径是节能，而节能的主要措施之一就是发展和应用隔热材料。膨胀蛭石不但本身具有较低的导热系数和体积密度，而且其颗粒表面的鳞片结构还具有反射热辐射的能力，是一种有潜在应用价值的高温隔热材料。本文采用**膨胀蛭石**与镁橄榄石复合制备出了复合隔热材料，利用灰色关联分析法，对膨胀蛭石-镁橄榄石复合材料

的制备工艺进行了优化。优化后制备出的膨胀蛭石-镁橄榄石复合材料抗折强度
为 9.33MPa，耐压强度为 15.74MPa，200℃时，热导率在 0.19W/(m·K) 左右，
300℃时约为 0.22W/(m·K)，600℃时约为 0.26W/(m·K)。基于离子极化理
论和结合系统的固化反应机理，分析了膨胀蛭石复合材料强度的影响因素，考察
了磷酸二氢铝的浓度及用量、促凝剂镁砂的种类和 MgO/P_2O_5 摩尔比对复合材料
的显气孔率、体积密度和力学性能的影响。为进一步提高复合材料的高温隔热性
能，以热传导的基本原理和协同作用原理为基础，选取了合适的外加剂提高复合
材料的隔热性能。$K_2Ti_6O_{13}$ 晶须的添加量（质量分数）为 2%，二氧化钛添加量
（质量分数）为 2%，添加 8%（质量分数）的 BiOCl 的膨胀蛭石-镁橄榄石复合
材料具有最优的隔热性能。300℃时热导率为 0.117W/(m·K)，600℃时热导率
为 0.169W/(m·K)，800℃时热导率为 0.184W/(m·K)，1000℃时热导率为
0.190W/(m·K)。采用原位凝胶法对膨胀蛭石进行了改性，将膨胀蛭石微米级
孔隙转化为网状结构的纳米级孔隙，有效改善了复合材料的微观结构，增强了膨
胀蛭石复合材料的力学性能和隔热性能，并研究了改性膨胀蛭石微观结构的影响
因素。原位凝胶改性的膨胀蛭石的制备工艺为：n（环氧丙烷）：n（Al）：n（甲酰
胺）：n（乙醇）= 5.5：1：0.8：30，非临界干燥。制备的铝凝胶原位改性的膨胀
蛭石的结构孔隙中，构成铝凝胶的骨架由近似球状的氧化铝颗粒相互聚结而成，
颗粒粒度比较均匀，平均粒径约为 40nm，颗粒间形成的孔径约为 45nm，孔径分
布较均匀。经 900℃和 1000℃煅烧 4h 后仍能保持较好的多孔网络结构，没有出
现明显的团聚或孔结构塌陷的现象。改性的膨胀蛭石质量分数越高，膨胀蛭石-
镁橄榄石复合材料的热导系数越低，高温隔热性能越好。当改性膨胀蛭石占蛭石
总量百分比为 50% 时，复合材料的热导率最低，在 300℃，其热导系数为
0.13W/(m·K)，600℃时为 0.157W/(m·K)，800℃时为 0.169W/(m·K)，
900℃时为 0.168W/(m·K)。与未改性的膨胀蛭石复合材料相比，改性的膨胀蛭
石复合材料的热导率在 300℃降低了 20%~30%，600℃下降了 30%~40%，
800℃和 900℃下降了 35%~45%。为拓展蛭石的应用领域，增加其附加价值，采
用离子交换法对蛭石进行无机改性，将聚羟基铝离子插入蛭石层间，利用煅烧后插层
离子在层间留下的微孔和氧化物柱子，进一步改善蛭石的微观结构，并对蛭石柱化的
影响因素和离子交换反应的动力学做了基础理论研究。XRD 分析表明柱化插层后 Al-
柱化蛭石 001 晶面的层间距为 $18.42×10^{-10}$m，煅烧后的 Al-柱化蛭石的层间距为 $17.26×$
10^{-10}m，通过聚羟基铝的插层，蛭石获得了永久的 $8.82×10^{-10}$m 的层间自由空间。采
用 TEM、TG-DSC、XRD、FT-IR 和 N-吸附脱附等方法对柱化前后的蛭石微观结构进
行了表征，与蛭石原矿相比，经过聚羟基铝离子柱化的蛭石在热稳定性和层间微观结
构都有所改善。蛭石与 keggin-Al_{13}^{7+} 离子的离子交换反应中，粒内扩散步骤为反应的速

控步骤，其反应的表观活化能为26.79kJ/mol，动力学方程可表示为：$1 - \frac{2}{3}\alpha - (1 - \alpha)^{\frac{2}{3}} = k_0 r_0^{-2} \exp\left(\frac{-26.79}{RT}\right) t$ [16]。

Energy shortage is an important problem faced by all countries in the world, and the most effective way to solve the energy shortage is energy saving. The most effective way to save energy is development and application of insulation materials. Expanded vermiculite itself has low thermal conductivity and bulk density, moreover, the scale structure of its granule surface has the ability to reflect thermal radiation. It is a potential insulation refractory material with high application value. In this work, the insulation material was prepared from **expanded vermiculite** and forsterite. Based on analysis of grey relational degree, the preparation process of expanded vermiculite-forsterite composites was optimized. Prepared by the optimized preparation process, the insulation material can endure rupture strength of 9.33MPa and compression strength of 15.74MPa. At the temperature of 200℃, its thermal conductivity is about 0.19W/(m · K), at 300℃, about 0.22W/(m · K), at 600℃, about 0.26W/(m · K). Based on the polarization theory and curing mechanism of bond system, the bond system of composites was optimized, and the relationship between the properties of composites and many factors were studied, such as the concentration and dosage of aluminum dihydrogen phosphate, the types of magnesia and MgO/P_2O_5 molar ratio. Based on the synergistic effect theory, we try to improve the thermal insulation property of composites by way of selecting appropriate admixture to improve the intensity and thermal insulation of composites material. The addictive amount of $K_2Ti_6O_{13}$ whisker is 2wt.%, Titanium dioxide of 2wt.%, BiOCl of 8wt.%, which can produce the best heat-resistance of expanded vermiculite and forsterite composites material. At the temperature of 300℃, its thermal conductivity is 0.117W/(m · K), at 600℃ of 0.169W/(m · K), at 800℃ of 0.184W/(m · K), at 1000℃ of 0.190W/(m · K). In order to improve the microstructure of composites material, furthermore, to enhance the mechanical and thermal insulation properties of composites, we adopted method of gel in situ to modify expanded vermiculite and research the conditions for preparation process of modified expanded vermiculite. Through modification of gel in situ, the pore structure of expanded vermiculite varied from micrometers to nanometer. So, the microstructure of composites can be improved, and the mechanical strength and thermal insulation properties of composites increase. The optimized preparation process of gel in situ modification is: n (epoxy propane) : n(Al) : n(methanamide) : n(alcohol) = 5.5 : 1 : 0.8 : 30, ambient drying. As is shown in SEM of expanded vermiculite after being optimized by aluminum gel in situ, the framework of aluminum gel in situ is made up of sphere-like Al_2O_3 particles, the particle size is fairly regular, the mean particle diameter is 40nm, the pore size among particles is 45nm, and the distribution of pore size is fairly regular. After being calcined under the temperature of 900℃ and 1000℃ for 4 hours, it can be still kept in good multi-porous network, without any distinct signs of conglobation or porous structure collapse. By means of in-situ modification to expanded vermiculite, thermal conductivity of composites material has been remarkably reduced. The higher proportion the modified expanded vermiculite takes up in total weight of vermiculite, the lower the thermal conductivity of composites material is, and the better the thermal insulation. When modified expanded vermiculite takes up 50wt.% of total amount of vermiculite, the thermal conductivity of composites material is the lowest. At the temperature of 300℃, its coefficient of thermal conductivity is 0.13W/(m · K), at 600℃ of 0.157W/(m · K), at 800℃ of 0.169W/(m · K), at 900℃ of 0.168W/(m · K). Compared with non-modified expanded vermiculite composites material, the modified expanded vermiculite composites material can reduce thermal conductivity by 20%-30% at 300℃, 30%-40% at 600℃, 35%-45% at 800℃ or 900℃. In order to expand the application of vermiculite and increase its added value, we adopted ion exchange method to modify vermiculite by inorganic salt, insert polynuclear Al into the layers of vermiculite, and take advantage of the micro holes and oxidized pillars in inter-layers of inserting ions after calcination. In order to furthermore improve microstructure of vermiculite, we also discuss the preparation process of pillared vermiculite, and research the dynamics of ion exchange reaction. The analysis of XRD indicates that after being pillared, the interlayer spacing of Al pillared vermiculite crystal face is $18.42×10^{-10}$m, the interlayer spacing of calcinated Al pillared vermiculite is $17.26×10^{-10}$m. Through inserting layer of polynuclear Al, vermiculite gains perpetual $8.82×10^{-10}$m interlayer free space. From the characterization of vermiculite microstructures before and after pillarization by means of TEM, TG-DSC, XRD, FT-IR and nitrogen sorption isotherms, we find out that polynuclear Al pillared vermiculite is much better than raw vermiculite in thermal stability and interlayer microstructure. During the ion exchange of vermiculite and keggin-Al_{13}^{7+}, intra-particle diffusion process is controlling step, the apparent activation energy of its reaction is 26.79kJ/mol, and its kinetic equation can be denoted as follows:$1 - \frac{2}{3}\alpha - (1 - \alpha)^{\frac{2}{3}} = k_0 r_0^{-2} \exp\left(\frac{-26.79}{RT}\right) t$ [16].

8.2 纤维类隔热耐火材料原料

8.2.1 术语词

石棉 amiantus, amianthus, asbestus

铁石棉 amosite

滑石棉 asbecasite

纤维 fiber

8.2.2 术语词组

石棉 earth flax, mountain cork

石棉纤维 asbestos fiber

石棉填料 asbestos packing

硅酸铝纤维 alumina silicate fibre, aluminium silicate fiber

碳酸铝纤维棉 aluminum carbonate fiber cotton

陶瓷纤维 ceramic fiber

普通硅酸铝纤维 common aluminum silicate fiber

连续陶瓷纤维 continuous ceramic fiber

晶体耐火纤维 crystal refractory fiber

纤维棉 fiber wool

莫来石纤维 mullite fiber

高纯硅酸铝纤维 purer aluminium silicate fibre

耐火纤维 refractory fiber

氧化铝空心球 alumina bubble

刚玉空心球 corundum hollow granule

耐火空心球 refractory hollow ball

电熔空心球 fused hollow ball

8.2.3 术语例句

蓝色的**石棉**在建筑中要少见得多，幸好是这样，因为它比白色石棉更危险。

Blue **asbestos** is far less common in buildings, which is just as well because it's more dangerous than white asbestos.

石棉用于阻燃剂，可能导致癌症。

Asbestos, used as a flame retardant, can cause cancer.

石棉，常见的绝缘材料，可引起肺癌。

Asbestos, which was once widely used in insulation materials, is a notorious cause of lung cancer.

受到破坏的**石棉**可能会向空气中释放微细纤维。

Disturbing **asbestos** can release microscopic fibers to the air.

石棉的主要形式是温石棉（白石棉）及青石棉（蓝石棉）。

The principal forms of **asbestos** are chrysotile（white asbestos）and crocidolite（blue asbestos）.

石棉纤维会破坏肺组织，留下损伤，削弱器官的氧气处理机能，有时还会引起肺癌。

Asbestos fibers can bruise the lung tissue, leaving scars that cripple the organ's ability to process oxygen and sometimes cause lung cancer.

概要介绍了**耐火纤维**的发展趋势。

The development trend of **refractory fibers** was briefed as well.

介绍高发射率涂料对**耐火纤维**的强化作用。

The strengthening effect of high emissivity coating on **refractory fibre** is described in this paper.

在钢铁工业中应用**耐火纤维**取得了明显的节能效果。

It has shown remarkable energy-saving results since the **refractory fiber** was used in steel industry.

耐火纤维材料是一种具有复杂微空间结构的多孔介质材料。

Refractory fiber is a kind of porous medium with complex microstructure.

密封圈由隔热套、**耐火纤维**层和耐热橡胶层组成，制成封闭环状密封圈。

The seal ring consists of a heat insulation sleeve, a **refractory fiber** layer and a heat-resistant rubber layer and is made into a sealed circular shape.

耐火纤维制品作为隔热耐火材料的一种，目前在国内还未得到充分的利用。

Refractory fibre product is a kind of insulating refractory. It is not fully utilized at home recently.

根据实验数据，通过回归分析得到了普通**硅酸铝耐火纤维**毡导热系数的经验公式。

On the basis of experimental data, an equation to determine the thermal conductivity of ordinary **alumino-silicate refractory fiber** felts has been developed by regression method.

结果表明：用**耐火纤维**制品替代传统的耐火材料，节约了能源，提高了经济效益。

The results show that replacing the traditional refractories with **refractory fibre** products in the boiler has saved energy and increased economic benefit.

通过**耐火纤维**喷涂在常压炉衬里上的成功应用，为加热炉衬里改造开拓了一条新路。

A new method was found for heating up furnace lining by painting **refractory fiber** on constant pressure furnace lining.

介绍**耐火纤维**及其制品的特性、分类及几种施工工艺，并着重介绍耐火纤维喷涂技术。

To introduce the characteristics, classification and some kinds of construction technologies of refractory fibre and its products. **Refractory fibre** spraying technology is emphatically introduced.

8.2.4　术语例段

六种商业**石棉矿物**（温石棉、纤维阳起石、青石棉、铁石棉、纤维透闪石和纤维直闪石）被 IARC 列为对人类致癌的物质。目前有几项研究涉及石棉矿物的惰性化，其中干磨过程受到了极大的关注，而在偏心振动磨中干磨对透闪石石棉和直闪石石棉的影响还没有被研究。本研究的目的是沿着石棉机械处理的研究路线，通过改变研磨时间（30s、5min 和 10min），评估在偏心振动磨中干磨对 Val d'Ala（意大利）的透闪石石棉和 Paakkila 矿（芬兰）的 UICC 标准直闪石石棉的结构、温度稳定性和纤维尺寸的影响。研究结果表明，在研磨 30s 至 10min 后，透闪石石棉和直闪石石棉由于晶格应变的增加和结晶度的降低而导致脱羟和分解温度的降低。此外，在研磨到 10min 后，透闪石和直闪石纤维都低于世界卫生组织定义的可计数纤维的限度。

The six commercial **asbestos minerals** (chrysotile, fibrous actinolite, crocidolite, amosite, fibrous tremolite, and fibrous anthophyllite) are classified by the IARC as carcinogenic to humans. There are currently several lines of research dealing with the inertisation of asbestos minerals among which the dry grinding process has received considerable interest. The effects of dry grinding on tremolite asbestos and anthophyllite asbestos in eccentric vibration mills have not yet been investigated. Along the research line of the mechanical treatment of asbestos, the aim of this study was to evaluate the effects of dry grinding in eccentric vibration mills on the structure, temperature stability, and fibre dimensions of tremolite asbestos from Val d'Ala, (Italy) and UICC standard anthophyllite asbestos from Paakkila mine(Finland) by varying the grinding time(30s, 5min, and 10min). After grinding for 30s to 10min, tremolite asbestos and anthophyllite asbestos showed a decrease in dehydroxylation and breakdown temperatures due to the increase in lattice strain and the decrease in crystallinity. Moreover, after grinding up to 10min, tremolite and anthophyllite fibers were all below the limits defining a countable fibre according to WHO[17].

阳起石是属于**石棉矿物**组的六种矿物之一。人们越来越关注接触天然石棉和含石棉材料的潜在健康风险。纤维状石棉矿物的正确区分不仅从科学的角度来看，而且从立法的角度来看都非常重要。石棉阳起石是目前唯一尚未从热学角度充分表征的石棉矿物。为了弥补科学文献中的这一空白，本文使用热重和差示扫描量热法讨论了阳起石石棉的热行为。采用 X 射线粉末衍射、扫描和透射电子显微镜结合能量色散光谱仪对阳起石纤维在 1000℃ 和 1200℃ 加热前后的特性进行表征，以确定其对高温变化的抵抗力和产物热重结晶。阳起石石棉在大约 1030℃ 时分解。阳起石石棉的热分解过程包括两个不同的事件，然后再结晶成新的稳定晶相，保持原始纤维形态（称为假形态）。热分析可能被证明对阳起石的识别和鉴别有用，特别是在石棉透闪石-阳起石闪石相互混合的天然大块试样的情况下。此外，对这种石棉矿物的热行为的深入了解可以为我们提供相关数据，以了解石棉通过热惰性化处理的晶体化学转变。

Actinolite is one of the six minerals belonging to the group of **asbestos minerals**. There is increasing concern regarding the potential health risks from exposure to naturally occurring asbestos and asbestos-containing materials. The correct distinction of the fibrous asbestos minerals is very important not only from a scientific point of view, but also from a legislative perspective. Asbestos actinolite is currently the only asbestos mineral that has not been fully characterized from the thermal point of view. In order to compensate for this gap in scientific literature, this paper discusses the thermal behaviour of actinolite asbestos using thermogravimetric and differential scanning calorimetry. X-ray powder diffraction, scanning and transmission electron microscopy combined with energy-dispersive spectrometry were used for the characterization of actinolite fibres before and after heating at 1000℃ and 1200℃ in order to determine their resistance to high-temperature changes and the products of thermal recrystallization. Actinolite asbestos breaks down at approximately 1030℃. The thermal decomposition process of actinolite asbestos consists of two distinct events followed by recrystallization into new stable crystalline phases which preserved the original fibrous morphology (known as pseudomorphosis). The thermal analysis may prove to be useful for actinolite identification and discrimination, particularly in the case of natural massive samples where asbestos tremolite-actinolite amphiboles are mutually intermixed. Furthermore, profound knowledge of the thermal behaviour of this asbestos mineral may provide us with the relevant data for understanding the crystal-chemical transformations of asbestos through thermal inertization treatment[18].

Al_2O_3-Cr_2O_3 耐火砖用作各种高温炉的高耐腐蚀性内衬，如废物焚烧炉、气化炉、**玻璃纤维炉**等。由于 Cr_2O_3 在熔渣/玻璃中的溶解度极低，所以表现出高的耐腐蚀性。此外，在 Al_2O_3-Cr_2O_3 中添加 CaO，使其作为浇注料的应用更加广泛，更便于炉衬的修复和安装。但是 Cr_2O_3 在高温和/或氧化气氛下，甚至在 CaO、Na_2O、K_2O 等氧化物的存在下更容易形成不同的高价（Ⅳ和/或Ⅵ）Cr 化合物。此外，粉煤灰、石棉灰、煤渣、炼铁渣等固体工业废物以及城市垃圾中最常见的主要成分为 SiO_2、CaO、Al_2O_3 和 Fe_2O_3，而 MgO、Na_2O 和 K_2O 为次要成分。因此，含 Cr_2O_3 材料的主要问题是形成有毒和致癌的 Cr(Ⅵ) 化合物。此外，Cr(Ⅵ) 化合物易溶于水，因此很容易进入食物链。

Al_2O_3-Cr_2O_3 refractory bricks are used as highly corrosion resistance lining of the various high temperature furnaces such as waste incinerators, gasifiers, **fiber glass furnaces** etc. Extremely low solubility of Cr_2O_3 into molten slags/glass is responsible for these corrosion resistance properties. Moreover, addition of CaO to Al_2O_3-Cr_2O_3 unfolds its refractory application as castables which is more convenient for repairing and installation of the furnace linings. But Cr_2O_3 forms different higher-valent(Ⅳ and/or Ⅵ)Cr-compounds at high temperatures and/or oxidizing atmosphere and even more readily in presence of oxides such as CaO, Na_2O, K_2O etc. Moreover, the solid industrial wastes such as fly ash, asbestos ash, coal slag, iron smelting slag as well as municipal wastes most commonly contain SiO_2, CaO, Al_2O_3 and Fe_2O_3 as major constituents, while MgO, Na_2O and K_2O as minor constituents. Thus, the major concern of Cr_2O_3 containing materials is the formation of Cr(Ⅵ) compounds which are toxic and carcinogenic. Besides, Cr(Ⅵ) compounds are readily water soluble, consequently making its easy entrance to the food chain[19].

　　本文提出了一种以粉煤灰和**石棉尾矿**为原料，通过高温成孔制备多孔微晶玻璃的新方法。系统讨论了石棉尾矿含量和烧结温度对多孔微晶玻璃的相组成、微观结构和性能的影响，并阐述了成孔机制。与 T0 相比，T1、T2 和 T3 的多孔微晶玻璃由于添加了石棉尾矿，在烧结过程中经历了更剧烈的自膨胀。T3 多孔微晶玻璃的孔隙率为 51%，体积密度为 $1.42g/cm^3$，抗弯强度为 19MPa，主晶相为印度钙长石，以及钙长石、顽火辉石、镁橄榄石等几种次生相。由于强度高，该材料有望用作具有承重功能的多孔建筑材料。该工作为利用粉煤灰和石棉尾矿在不添加发泡剂的情况下制备多孔微晶玻璃提供了一种方便且有前景的方法。

This paper presented a new method of preparing porous glass-ceramics by high-temperature pore-forming using coal fly ash and **asbestos tailings** as raw materials. The effects of the content of asbestos tailings and sintering temperature on the phase composition, microstructure and properties of the porous glass-ceramics had been systematically discussed, furthermore, the pore formation mechanism was also expounded. Compared with T0, porous glass-ceramics from T1, T2 and T3 experienced more violent self-expansion during the sintering process due to the addition of asbestos tailings. The porosity of porous glass-ceramics from T3 was 51%, the bulk density was $1.42g/cm^3$, the flexure strength was 19MPa, the main crystal phase was indialite, along with several secondary phases such as anorthite, enstatite, and forsterite. Due to the high strength, the material was expected to be used as a porous construction material with load-bearing function. This work provided a convenient and promising method for the utilization of coal fly ash and asbestos tailings to prepare porous glass-ceramics without adding foaming agents[20].

　　本研究旨在优化**石棉**水泥废料（ACW）的热处理，使其适合用作替代黏合剂。使用以温度（600℃和800℃）、时间（1h和3h）和质量（1kg和5kg）为因子的中心点的 2k 因子设计进行优化。处理过的 ACW 中温石棉、贝利特（C_2S）和方解石的百分比是通过热重分析（TGA/DTG）和 X 射线衍射（XRD）获得的主要实验响应。进行了 11 次实验以开发将实验响应与因素及其相互作用相关联的模型。结果表明，温石棉在温度为 700℃ 或更高的处理中经历了完全脱羟基。随着煅烧温度的升高，处理过的 ACW 中贝利特含量增加；然而，在处理过程中，更多的二氧化碳排放到大气中。通过在 800℃ 的炉子中加热 1h 煅烧 5kg 的 ACW 来进行最佳处理。在这些条件下，温石棉被完全去除，处理 1kg 废物二氧化碳排放量估计为175.60g。此外，发现处理后的残留物含有40.42%的贝利特，为材料提供了结合能力。

This study aims to optimize the thermal treatment of **asbestos** cement waste (ACW) to make it suitable for use as an alternative binder. The optimization was performed using a 2k factorial design with the central point having temperature (600℃ and 800℃), time (1h and 3h), and mass (1kg and 5kg) as factors. The percentages of chrysotile, belite (C_2S), and calcite in the treated ACW were the main experimental responses obtained through thermogravimetric analysis (TGA/DTG) and X-ray diffraction (XRD). Eleven experiments were conducted to develop models correlating the experimental responses with the factors and their interactions. The results show that chrysotile underwent complete dehydroxylation in treatments in which the temperature is 700℃ or higher. With an increase in the calcination temperature, the belite content increased in the treated ACW; however, a higher amount of CO_2 was emitted to the atmosphere during treatment. The optimal treatment was performed by calcining 5kg of ACW heated in a furnace for 1h at 800℃. Under these conditions, chrysotile was completely removed with an estimated CO_2 emission of 175.60gCO_2/kg of treated waste. Moreover, the treated residue was found to have 40.42% belite, providing the material with binding capacity[21].

在 SiO_2-MgO-CaO-Fe_2O_3-Na_2O 体系中，以预烧石棉尾矿和废玻璃为原料，采用粉末烧结自膨胀法制备了无发泡剂的微晶玻璃。采用 X 射线衍射（XRD）、热重-差示扫描量热分析（TG-DSC）、扫描电子显微镜（SEM）和物理性能测试，进一步阐述了试样的成孔机制。用67%（质量分数）的废玻璃制备并在1180℃下烧结的试样分别具有 0.96g/cm³ 的体积密度、93%的体积膨胀率、64%的孔隙率和14MPa 的抗压强度。本研究实现了石棉尾矿和废玻璃的综合利用，为在不添加发泡剂的情况下制备发泡微晶玻璃提供了理论和实用价值。

Foamed glass-ceramics without using foaming agents have been fabricated in the system of SiO_2-MgO single bond CaO-Fe_2O_3 single bond Na_2O by self-expansion during the process of powder sintering, using pre-calcined **asbestos tailings** and waste glass as raw materials. The effects of waste glass content and sintering temperature on phase composition, pore structure and physical properties of foamed glass-ceramic samples were systematically investigated by means of X-ray diffraction (XRD), thermogravimetry-differential scanning calorimetry analysis (TG-DSC), scanning electron microscopy (SEM) and physical properties tests, furthermore, the pore forming mechanism of samples was also expounded. The samples prepared with 67wt.% of waste glass and sintered at 1180℃ possessed bulk density of

0.96g/cm^3，volume expansion rate of 93%，porosity of 64% and compressive strength of 14MPa，respectively. This study realized the utilization of asbestos tailings and waste glass and provided theoretical and practical value for the preparation of foamed glass-ceramics without adding foaming agents[22].

CaZrO$_3$（CZO）前体纤维通过溶胶-凝胶法和静电纺丝技术从含有钙和锆离子的水性前体和聚环氧乙烷的溶液中制备。CZO 纤维的结晶是与有机物分解同时发生的过程。通过傅里叶变换红外（FT-IR）和拉曼光谱、热重和差示扫描量热法（TG/DSC）、X 射线衍射（XRD）和扫描电子显微镜（SEM）对演化过程进行了表征。纤维的导热性能和高温稳定性分别通过热导率和加热永久线性变化的测量来表征。纤维在 80℃ 的 NaOH 溶液中处理以表征其耐碱性。结果表明，CZO 纤维的热导率低于其他报道形式的 CZO 材料，在 1100℃ 下具有优异的稳定性，热收缩率小于 1.2%，对碱具有优异的耐腐蚀性。因此，CZO 纤维可以作为一种合适的耐腐蚀耐火材料用于高温绝热。

CaZrO$_3$（CZO）precursor fibers were prepared by sol-gel method and electrospinning technique from solutions which contained aqueous precursors of calcium and zirconium ions and polyethylene oxide. The crystallization of CZO fibers was a concurrent process with the decomposition of organics. The evolution process was characterized by Fourier transform infrared(FT-IR) and Raman spectra, thermogravimetry and differential scanning calorimetry(TG/DSC), X-ray diffraction(XRD) and scanning electron microscopy(SEM). The heat-conducting property and high temperature stability of fibers were characterized by the measurements of thermal conductivity and heating permanent linear change, respectively. The fibers were treated in NaOH solution at 80℃ to characterize the alkali resistance. The results showed that CZO fibers had the lower thermal conductivity than the other reported forms of CZO materials, and they possessed excellent stability up to 1100℃ with thermal shrinkage less than 1.2% and excellent corrosion resistance to alkalis. Hence, CZO fiber could be used as a suitable corrosion resistant refractory material for high-temperature thermal insulation[23].

耐火陶瓷纤维（RCF）板被广泛使用，因为它们的绝缘性能使其能够降低高温工业过程中的能源消耗。但是，由于传统的 RCF 板在 1300℃ 以上的线收缩率超过 3%，因此由 RCF 板制成的顶板、墙壁和底部等炉衬部件很容易变脆。为了抑制 RCF 板在 1300℃ 以上的收缩，RCF 板表面涂有含有几种氧化铝颗粒的硅溶胶。当 RCF 板涂有硅溶胶浆料时将含有 50%（质量分数）氧化铝薄片在 1400℃ 加热 8h，RCF 板的线收缩率从 4.3% 下降到 1.5%。这反过来又将 RCF 板的耐热性提高到了 1400℃。

Refractory ceramic fiber (RCF) boards are widely used because their insulating properties allow them to reduce energy consumption during high-temperature industrial processes. However, since conventional RCF boards undergo a linear shrinkage of more than 3% above 1300℃, furnace lining parts, such as the ceiling, wall, and bottom made of RCF board easily become fragile. In order to suppress the shrinkage of the RCF board above 1300℃, the RCF board surface was coated with a silica sol containing several types of alumina particles. When a RCF board coated with a silica sol slurry containing 50wt.% alumina platelets was heated at 1400℃ for 8h, the linear shrinkage of the RCF board decreased from 4.3% to 1.5%. This, in turn, improved the heat resistance of the RCF board up to 1400℃[24].

四种具有不同比表面积 （80 ~ 360m²/g） 和粒径的不同胶态二氧化硅溶胶在作为**耐火纤维**黏合剂的应用中进行了研究。使用真空成型技术制造纤维材料，并使用阳离子淀粉作为絮凝剂以获得高保留的二氧化硅。淀粉分子还用作纤维产品中的临时黏合剂。无论所用硅溶胶的类型如何，在淀粉添加量与添加二氧化硅的总表面积之间的恒定比率下，都可获得最佳保留，因此获得最均匀的纤维体。纤维制品的抗压强度主要取决于纤维材料的密度。密度随着淀粉含量的增加而增加。因为淀粉是一种有效的黏合剂，生坯强度高于烧结试样的强度。具有最佳综合性能的溶胶具有约 15nm 的粒径和中等的比表面积 （220m²/g）。

Four different colloidal silica sols with different specific surface areas(from 80m²/g up to 360m²/g) and particle sizes have been examined in an application as **refractory fibre** binder. The fibrous materials were fabricated using a vacuum-forming technique, and a cationic starch was used as flocculent to obtain high retention of silica. The starch molecule also works as a temporary binder in the fibre product. Optimum retention and, hence, the most homogeneous fibre bodies were obtained at a constant ratio between the starch addition and the total surface area of the added silica, regardless of the type of silica sol used. The compressive strength of the fibre products depended mainly on the density of the fibrous materials. The density increased with increased starch content. The green strength was higher than the strength of the sintered samples, as starch is an effective binder. The sol with the best overall properties had a particle size of about 15nm and an intermediate specific surface area(220m²/g)[25].

传统**耐火纤维**制备和使用过程中的粉尘很容易被吸入人体，吸入的纤维在人体内不容易降解，危害了人体的健康，并且空气中飘浮的纤维对环境也造成了一定的污染。传统耐火纤维虽然使用温度较高，但是对生物体和环境存在的潜在威胁限制了它的使用，因此人们开始致力于开发研究既能满足使用要求又具有一定降解性的耐火纤维，就是生物可溶性耐火纤维。为了提高钙镁硅系耐火纤维的成纤率以及纤维的使用温度，降低纤维对人体的危害，本文探讨研究了纤维组成中 SiO_2 含量变化、$m(CaO)/m(MgO)$（以下简写为 C/M）变化和不同添加剂的引入对纤维高温熔体黏度、析晶行为以及溶解性的影响。取得的主要研究结果如下：（1）纤维的高温熔体黏度随纤维组成中 SiO_2 含量的增大而增大，析晶温度及溶解性则呈先增大后降低的趋势，其中 SiO_2 含量为 67%时，纤维的析晶温度最高，溶解性最好。另外不同含量的 SiO_2 不会改变纤维的析晶产物。（2）纤维的高温熔体黏度随纤维组成中 C/M 的增大而增大，析晶温度及溶解性则呈先增大后降低的趋势，其中 C/M 为 5 时，纤维的析晶温度最高，溶解性最好。随着 C/M 的增加，纤维析出的晶相中硅灰石数量不断增加，透辉石数量逐渐减少。（3）当纤维组成中 SiO_2 含量为 67%同时 C/M 为 5 时，纤维的高温黏度、析晶行为和溶解性能较优。（4）选取的几种添加剂对纤维的高温黏度、析晶行为及溶解性都产生了一定的影响：Al_2O_3 能提高纤维的析晶峰值温度，降低纤维的高温熔体黏度以及在 Gamble 溶液中的溶解性；La_2O_3 的引入可以降低纤维的高温熔体黏度，大于 1%的 La_2O_3 会降低纤维的析晶峰值温度而溶解性得到提升；氯化物（NaCl、KCl、$MgCl_2$）的引入影响了纤维的高温熔体黏度，其中 NaCl 和 $MgCl_2$ 可以降低纤维的高温黏度，三种氯化物都能提高纤维的析晶温度及溶解性，其中以添加 $MgCl_2$ 的纤维高温黏度、析晶温度以及溶解性最优[26]。

Traditional **refractory fiber** dust is easy to be absorbed into the body in preparation and using process, the inhaled fiber will hurt the health of human because the fiber is not dissolved easily in human body, and the fiber floating in the air will be caused certain environmental pollution. Although the using temperature is higher, but the potential threatening of the traditional refractory fiber to the organism and environment restrict its use, so people begin to devote to research and develop biological soluble refractory fiber which can satisfy the use requirement and have excellent biodegradability. In order to improve the forming probability and use temperature of refractory fiber of CaO-MgO-SiO2 system and reduce the harm to human body, this paper studies the influence of SiO2 content, $m(CaO)/m(MgO)$ (C/M for abbreviation) and the introduction of different additives on high temperature melt viscosity, crystallization behavior and solubility of refractory fiber. The main research results were obtained as follows: (1) The high temperature melt viscosity of fiber will be increased gradually with the SiO2 content increasing, while the crystallization behavior and solubility increased at first, then decreased, among them when SiO_2 content is 67%, the crystallization temperature of fiber is highest and the solubility is best. Besides, different content of SiO_2 will not change the crystal product of fiber. (2) The high temperature melt viscosity of fiber will be increased gradually with the C/M increasing, while the crystallization behavior and solubility increased at first, then decreased, among them when C/M is 5, the crystallization temperature of fiber is highest, and the solubility is best. Besides, along with the increase of C/M, the amount of wollastonite in fiber precipitation of crystalline phase increase gradually and the amount of diopside decrease gradually. (3) When the content of SiO_2 is 67% and meanwhile C/M is 5, the high temperature viscosity, crystal behavior and solubility have a better performance. (4) The several kinds of additives bring certain effect on high viscosity, crystal behavior and solubility of fiber: Al_2O_3 can improve crystallization peak temperature of fiber and reduce the high temperature melt viscosity and the solubility of fiber in Gamble solution; La_2O_3 can reduce the high temperature melt viscosity, more than 1% of La_2O_3 will reduce the crystallization peak temperature of fiber and that will elevate solubility of fiber; Chloride($NaCl$, KCl, $MgCl_2$) can affect the high temperature melt viscosity, among them $NaCl$ and $MgCl_2$ can reduce the high temperature viscosity, the crystallization temperature and solubility of fiber will be improved by the introduction of chloride, of which the fiber adding $MgCl_2$ display the optimal performance in high temperature viscosity, crystal behavior and solubility[26].

传统耐火纤维在人体内不容易降解,可能对人体的健康产生影响,并且空气中由于粉化而产生的纤维粉尘也对环境造成了一定的影响。因此,人们致力于开发既能满足使用要求又在人体内具有一定溶解性的生物可溶性耐火纤维。生物可溶性耐火纤维的使用温度一般比传统耐火纤维的低。因此,在保证生物可溶性耐火纤维溶解性的前提下,为了提高生物可溶性耐火纤维的使用温度,探索研究了添加剂(BaO 和 SrO)的引入对钙硅系生物可溶性耐火纤维析晶和溶解性能的影响。得到的研究结果如下:(1) BaO 和 SrO 的引入在一定程度上抑制钙硅系陶瓷纤维的析晶,但并没有改变析出的晶相种类。钙硅系陶瓷纤维在模拟人体肺液(Gamble 溶液)中表现出良好的溶解性,BaO 和 SrO 的引入对纤维的溶解性影响不大。(2) 较高 BaO 引入量的试样较易析晶;纤维试样在 Gamble 溶液中表现出良好的溶解性,但溶解性变化不大。(3) 随着纤维组成中 SrO 引入量的增加:整体上纤维试样的析晶温度得到提高,但析出的晶相种类不变;试样在模拟人体肺液中均表现出一定的溶解性,且纤维试样的溶解性呈增加趋势。(4) 复合引入试样的析晶开始温度和析晶峰值温度都低于单独引入的试样,析晶时析出的晶相种类相同;纤维试样在模拟人体肺液(Gamble 溶液)中都具有一定的溶解性,复合引入试样的溶解性较单独引入的差[27]。

Traditional refractory fiber will hurt the health of human beings because the fiber is not dissolved easily in human body, and the fiber floating in the air will cause certain environmental pollution. Therefore, people begin to research and develop biological soluble refractory fiber which can satisfy the use demands and have excellent bio-solubility. In general, the applicable temperature of biological soluble refractory fiber is lower than that of the traditional refractory fiber. As a consequence, under the premise of guaranteeing the bio-solubility of biological soluble refractory fiber, in order to improve the applicable temperature of biological soluble refractory fiber, the influence of BaO and SrO on crystallization behavior and solubility of refractory fiber have been investigated in the paper. The main research results were as follows: (1) The addition of BaO and SrO can restrain the crystallization of the CaO-SiO$_2$ ceramic fiber to a certain extent, however, the types of precipitated crystalline phases are not changed. **The fibers of CaO-SiO$_2$ system** show high bio-solubility in the Gamble solution, and the effect of the addition of BaO and SrO on the bio-soluble properties of the fibers is hardly noticeable. (2) The fiber with higher BaO contents is relatively easy to crystallization; The fibers have outstanding bio-solubility in the Gamble solution. (3) With the content of SrO increasing in the composition of fibers, the crystallization temperature of fibers will be increased, while the types of precipitated crystalline phases are not changed; The fibers have still splendid bio-solubility in the Gamble solution, and the solubility of fibers will increase. (4) The initial crystallization temperature and the crystallization peak temperature of fibers with the additions of BaO and SrO are lower than the fibers with a single additive, however, the types of precipitated crystalline phases are the same. The fibers showed excellent bio-solubility in the Gamble solution, the solubility of fibers with the additions of BaO and SrO become worse than the fibers with a single additive[27].

参考文献

[1] Zhu Lin, Li Shujing, Li Yuanbing, et al. Preparation of castable foam with regular micro-spherical pore structure as a substitute for diatomite brick [J]. Ceramics International, 2022, 48 (15): 21630-21640.

[2] Zahajská P, Opfergelt S, Fritz S C, et al. What is diatomite? [J]. Quaternary Research, 2020, 96: 48-52.

[3] Wu Guangfeng, Ma Siyu, Bai Yu, et al. The surface modification of diatomite, thermal, and mechanical properties of poly (vinyl chloride)/diatomite composites [J]. Journal of Vinyl and Additive Technology, 2019, 25 (s2): 39-47.

[4] Ni Hongxing, Samuel Hsu C, Lee P, et al. Supercritical carbon dioxide extraction of petroleum

on kieselguhr [J]. Fuel, 2015, 141: 74-81.

[5] Cabranes M, Leyva A G, Babay P A. Removal of Cs+ from aqueous solutions by perlite [J]. Environmental Science and Pollution Research, 2018, 25 (22): 21982-21992.

[6] Łukowski P. Polymer-cement composites containing waste perlite powder [J]. Materials, 2016, 9 (10): 839.

[7] Papa E, Medri V, Murri A N, et al. Characterization of alkali bonded expanded perlite [J]. Construction and Building Materials, 2018, 191: 1139-1147.

[8] İşçiS. Intercalation of vermiculite in presence of surfactants [J]. Applied Clay Science, 2017, 146: 7-13.

[9] Simplício Pereira M H, dos Santos C G, de Lima G M, et al. Capture of CO_2 by vermiculite impregnated with CaO [J]. Carbon Management, 2022, 13 (1): 117-126.

[10] Marosz M, Kowalczyk A, Chmielarz L. Modified vermiculites as effective catalysts for dehydration of methanol and ethanol [J]. Catalysis Today, 2020, 355: 466-475.

[11] Danish A, Mosaberpanah M A. Formation mechanism and applications of cenospheres: a review [J]. Journal of Materials Science, 2020, 55 (11): 4539-4557.

[12] Liu Fengjuan, Wang Jialai, Qian Xin, et al. Internal curing of high performance concrete using cenospheres [J]. Cement and Concrete Research, 2017, 95: 39-46.

[13] Li Minghui, Li Yuanbing, Ouyang Degang, et al. Effects of alumina bubble addition on the properties of mullite castables [J]. Journal of Alloys and Compounds, 2018, 735: 327-337.

[14] Li Minghui, Ouyang Degang, Li Canhua. Effect of alumina bubble modification on properties of mullite castables [J]. Journal of the Australian Ceramic Society, 2020, 56 (3): 923-930.

[15] Li Minghui, Li Yuanbing, Ouyang Degang, et al. The impact of alumina bubble particle size on the microstructure and physical properties of mullite castables [J]. Ceramics International, 2019, 45 (2): 1928-1939.

[16] 王春风. 蛭石及其复合隔热材料的组成、结构与性能 [D]. 武汉: 武汉理工大学, 2012.

[17] Bloise A, Kusiorowski R, Gualtieri A F. The effect of grinding on tremolite asbestos and anthophyllite asbestos [J]. Minerals, 2018, 8 (7): 274.

[18] Bloise A. Thermal behaviour of actinolite asbestos [J]. Journal of Materials Science, 2019, 54 (18): 11784-11795.

[19] Nath M, Song Shengqiang, Li Yawei, et al. Effect of Cr_2O_3 addition on corrosion mechanism of refractory castables for waste melting furnaces and concurrent formation of hexavalent chromium [J]. Ceramics International, 2018, 44 (2): 2383-2389.

[20] Zeng Li, Sun Hongjuan, Peng Tongjiang, et al. Preparation of porous glass-ceramics from coal fly ash and asbestos tailings by high-temperature pore-forming [J]. Waste Management, 2020, 106: 184-192.

[21] Carneiro G, Santos T, Simonelli G, et al. Thermal treatment optimization of asbestos cement waste (ACW) potentializing its use as alternative binder [J]. Journal of Cleaner Production, 2021, 320: 128801.

［22］ Zheng W M, Sun H J, Peng T J, et al. Novel preparation of foamed glass-ceramics from asbestos tailings and waste glass by self-expansion in high temperature ［J］. Journal of Non-Crystalline Solids, 2020, 529: 119767.

［23］ Shi Shuying, Yuan Kangkang, Xu Chonghe, et al. Electrospun fabrication, excellent high-temperature thermal insulation and alkali resistance performance of calcium zirconate fiber ［J］. Ceramics International, 2018, 44 (12): 14013-14019.

［24］ Takahashi N, Hashimoto S, Daiko Y, et al. High-temperature shrinkage suppression in refractory ceramic fiber board using novel surface coating agent ［J］. Ceramics International, 2018, 44 (14): 16725-16731.

［25］ Lidén E, Karlsson S, Tokarz B. Silica sols as refractory fibre binders ［J］. Journal of the European Ceramic Society, 2001, 21 (6): 795-808.

［26］ 姜广坤. 组成变化对 SiO_2-CaO-MgO 系生物可溶性耐火纤维性能的影响研究 ［D］. 武汉: 武汉科技大学, 2011.

［27］ 魏哲. BaO 及 SrO 对 CaO-SiO_2 系生物可溶性耐火纤维性能的影响研究 ［D］. 武汉: 武汉科技大学, 2013.

9 耐火材料结合剂

9.1 术语词

禾木树脂 accroides

酚醛树脂黏合剂 aerodux

树脂 resin

热固性树脂 resinoid

树脂的 resinous

可熔酚醛树脂 resol-bakelite

黑沥青 abbertite,albertite

碳沥青 anthraxolite

沥青 asphalite,asphalt,pitch,asphaltum,goudron

含沥青的 bitumeniferous

六聚偏磷酸盐 hexametaphosphate

偏磷酸盐 metaphosphate

正磷酸盐 orthophosphate

磷酸盐 phosphate

多磷酸盐 polyphosphate

焦磷酸盐 pyrophosphate

四聚磷酸盐 tetrapolyphosphate

三偏磷酸盐 trimetaphosphate,tripolyphosphate

9.2 术语词组

乙醛树脂 acerous resin

丙酮树脂 acetone resin

酸性树脂 acid resin

酸凝树脂 acid-cure resin

丙烯酸树脂 acrylic resin,acrylic polymer

聚醛树脂 adlehyde resin

碱性离子交换树脂 base exchange resin

环氧树脂类黏合剂 bond master

环氧树脂胶 epoxide-resin glue

环氧树脂 epoxy resin,ethoxyline resin

热固性树脂 heat convertible resin,thermosetting resin

改性树脂 modified resin

酚醛树脂结合剂 phenol-formaldehyde binder

苯酚糠醛树脂 phenol-furfural resin

酚醛树脂 phenolic resin,bakelite resin,phenolic-

246

formaldehyde resin

聚苯乙烯树脂 polystyrene resin

树脂结合 resin bonding

硅有机树脂 silicon organic resin

固体树脂 solid resin

合成树脂（人造树脂）synthetic resin

热塑性酚醛树脂 thermoplastic phenolic resin, novolac resin

尿素甲醛树脂 urea formaldehyde resin

沥青化合物 bitumen sealing compound, bituminous compound

高温沥青 high temperature pitch

中温沥青 mesothermal pitch

改质沥青 modified coal tar pitch

石油沥青 petroleum pitch

沥青结合剂 pitch binder, asphalt binder, pitch tar binder

沥青焦油 pitch tar

球状沥青 spherical asphalt

气硬性水泥 air setting mortar

加气水泥 air-entrapping cement

铝酸盐水泥（波兰特水泥）aluminate cement

矾土水泥结合剂 aluminous cement bond

石棉水泥 asbestos cement

矾土水泥（高铝水泥）bauxite cement

高炉矿渣水泥 blast cement

铝酸钙水泥 calcium aluminate cement

水泥结合剂 cement binder

矿渣硅酸盐水泥 clinker-bearing slag cement

膨胀水泥 expanding cement

快硬水泥 fast hardening cement

粉煤灰水泥 fly ash cement

电熔纯铝酸钙水泥 fused pure calcium aluminate cement

水硬性耐火水泥 hydraulic-refractory cement

低钙铝酸盐耐火水泥 low-calcium aluminate

refractory cement

镁质水泥 magnesia cement

矿渣水泥 metallic cement, slag cement

普通硅酸盐水泥 ordinary Portland cement

方镁石水泥 periclase cement

磷酸盐耐火水泥 phosphate refractory cement

纯铝酸钙水泥 pure calcium aluminate cement

快速高强水泥 rapid hardening and high strength cement

快速快硬氟铝酸盐水泥 rapid setting and hardening fluo-aluminate cement

快速快硬硅酸盐水泥 rapid setting and hardening Portland cement

耐火水泥 refractory cement, fireproof cement, high-temperature cement

半硅质水泥 semi-acid cement

硅质水泥 silica cement

超低水泥结合 ultra low cement bond

锆铝质水泥 zircon-alumina cement

锆质水泥 zirconia cement

中性水玻璃 neutral water glass

氟硅酸钠 sodium fluorsilicate

偏硅酸钠 sodium metasilicate

正硅酸钠 sodium orthosilicate

硅酸钠（泡花碱）sodium silicate

速溶硅酸钠 soluble sodium silicate

水玻璃 water glass

碱性磷酸盐 alkaline phosphates

铝铬磷酸盐结合剂 aluminochrome phosphate binder

黏土磷酸盐结合剂 clay phosphate binder

含水磷酸盐 hydrous phosphate

磷酸盐结合剂 phopsphate binding agent

钠铬磷酸盐结合剂 sodium chromium phosphate binder

硫酸铝 aluminium sulphate

9.3 术语例句

通过添加**酚醛树脂**二氯化锡固化剂可以获得性能更好的共混物。

A blend with better properties could be obtained by adding **phenolic resin** tin bichloride curative.

利用乳化法制备了**酚醛树脂**微球。

Phenolic resin microspheres were prepared by emulsion method.

介绍了铸造用**碱性酚醛树脂**的合成原理及优点。

The synthesis principle and advantages of **alkaline phenolic resin** for foundry were introduced.

研究了**热塑性酚醛树脂**和原砂对覆膜砂性能的影响。

The effect of **thermoplastic phenolic resin** and sand on performance of resin coated sand was researched.

本文讨论了砂带用**酚醛树脂**黏合剂的合成工艺和条件。

The synthetic technology and conditions of **phenolic resin** as adhesive of abrasive belt were discussed.

热塑性酚醛树脂是人类最早用化学反应方法合成出的树脂黏结剂之一。

Novolak resin is one of the earliest resin binders that were made by human being with chemical reaction.

常规热塑性**酚醛树脂**生产的产率较低，且产生大量高浓度含酚废水，污染环境。

Conventional thermoplastic **phenolic resin** production yield is low, and produces a large number of high concentration phenol wastewater, pollute the environment.

改性树脂具有良好的耐热性和耐水性。

The **modified resin** demonstrates excellent water and heat resistant properties.

讨论了**改性树脂**、反应温度和反应时间等因素对树脂水溶性和涂层防锈性能的影响。

The effect of **modified resin**, reaction temperature, reaction time on the solubility of resin and antirust property of the film have been discussed.

选用合适的复合酰化体系，通过电位滴定，测定了工业碱木质素及其**改性树脂**的羟值。

Hydroxyl values of straw alkali lignin and **modified lignin resin** were determined by potentiometric titration in suitable composite acylation system.

以**改性树脂**和混合树脂为粘料，铜粉为导电性填料，并配以适当添加剂、固化剂制备铜导电胶。

Copper conductive adhesive is prepared by using **modified resins** and mixed resins as cohesive materials and Copper powders as conductive fillers.

目前国内性能较好的水性涂料基料树脂主要是醇酸树脂、环氧树脂、聚氨酯树脂、丙烯酸树脂及它们的**改性树脂**品种等。

At present, alkyd, epoxy resin, polyurethane resin, acrylate and varieties of their **modified resins** are better domestic resins for waterborne coatings.

环氧树脂是一种综合性能优良的热固性树脂，但其韧性不足。

Epoxy resin is a kind of thermosetting resin with excellent general performances, but its toughness is not enough.

该理论可望推广至用固化剂固化的其他类型的**热固性树脂**体系。

It is expected that this theory can be extended to other **thermosetting resin** systems cured by hardener.

聚酯树脂是交联的不饱和聚合物，用过氧化物类催化剂可制成热固性树脂。

Polyester resins are unsaturated, cross-linked polymers which are made thermosetting by use of a peroxide-type catalyst.

相反，有些**热固性树脂**一般不熔化或者只是轻度熔化，如酚醛树脂和环氧树脂。

On the other hand, some **thermosetting resins** generally did not melt or melted slightly, such as phenolic resin and epoxy resin.

热固性树脂中经常采用的原料是电木粉，镶嵌的过程需要使用到金相试样镶嵌机。

The important raw materials usually used in such **thermosetting resin** are bakelite, and metallographic specimen inlaying machine will be used in the same process.

本文介绍了介电分析法在**热固性树脂**及以其为基体的涂料和纤维增强复合材料固化研究中的应用。

The paper reviewed the use of dielectric analysis to study the curing process of **thermosetting matrix** system including organic coatings and composites.

常规**热塑性酚醛树脂**生产的产率较低，且产生大量高浓度含酚废水，污染环境。

Conventional **thermoplastic phenolic resin** production yield is low and produces a large number of high concentration phenol wastewater, pollute the environment.

从**石油沥青**和煤沥青中制取沥青中间相物质。

A **pitch mesophase** can be prepared from petroleum pitch and coal tar pitch.

硫铝酸盐水泥比硅酸盐水泥具有早强、高强、抗渗、耐侵蚀、抗冻、微膨胀等性能。

Sulphoaluminate cement has better properties than Portland cement, high strength, fast hardening, impermeability, corrosive resistance, frost resistance, expansive property and so on.

以激发反应原理研制的超早强熟料，并用以激发粉煤灰的潜在活性制成**粉煤灰硅铝酸盐水泥**。

Super high early strength clinker has been developed and **fly ash aluminosilicate cement** has been produced on the principle of excitation reaction.

本文从界面出发对以石英砂、硅酸盐水泥熟料和**高铝水泥熟料**作集料的砂浆的抗蚀性进行了对比实验。

The chemical corrosion resistances of the mortars of quartz, Portland cement clinker and **aluminous cement clinker** were studied respectively for comparison.

本文还采用**水玻璃**、沥青为固化剂对铬渣的处理进行了研究。

Water glass and bitumen as solidification agents to treat with chromium waste are also researched.

采用先进的**水玻璃**砂造型技术，将有效地改善水玻璃砂铸造落砂性能。

The application of advanced molding technique for **water glass** sand will effectively improve sand shakeout and cleaning.

水玻璃类浆液作为化学灌浆材料，其黏度在随时间变化，因此浆液在砂土中的渗流过程较复杂。

When **water glass** seriflux as chemistry injection material, its kinetic viscosity is varying with time, so the process of seriflux permeating in the sand is complicated.

9.4 术语例段

热固性树脂（TR）是形成刚性固体聚合物的有机化合物，在加热下会分解，不会出现塑性状态或熔化。这样的方面对于它们在耐火系统中的应用是有用的。尽管具有技术重要性，但很少有作品系统地讨论 TR 成分和分子结构、固化条件和添加剂的存在如何影响含碳耐火材料在热解、石墨化和烧结后的性能过程。本文回顾了应用 TR 生产含碳耐火材料的优缺点。第一部分介绍了 TR 合成和化学修饰的一般方面。讨论了加工条件、原料和添加剂对有机材料热解、碳产率和相变的影响。此外，还指出了几种 TR 的毒性和安全问题的重要方面，从可持续原材料中获得的不同等级的使用，以及 TR 黏合耐火材料生产的未来挑战。

Thermosetting resins（TR）are organic compounds that form rigid solid polymers that under heating, decompose with no plastic state or melting. Such an aspect is useful for their application in refractory systems. Despite their technological importance, few works have systemically discussed how TR composition and molecular architecture, the curing conditions and the presence of additives influence the properties of carbon-containing refractories after pyrolysis, graphitization and sintering processes. The present paper reviews the benefits and drawbacks of applying TR to produce carbon-containing refractories. The first sections present general aspects of TR synthesis and chemical modification. The influence of the processing conditions, raw materials and additives on the pyrolysis of organic material, carbon yield generation and phase transformations are discussed. Additionally, important aspects about toxicity and safety issues of several types of TR, the use of different grades attained from sustainable raw materials, and future challenges for TR-bonded refractories' production are pointed out[1].

有机物质在耐火砖制造中用作黏合剂。在这项工作中，首先将目前用于炼钢行业的 MgO-C 砖中的两种商业黏合剂体系与耐火材料的其他成分分开分析。通过常规技术（XRD、FTIR、拉曼光谱、DTA/TGA 和密度、孔隙率、机械和抗氧化性的测量）研究了单独的**甲阶酚醛树脂**和与**焦油沥青**处理产生的生态黏合剂混合的效果，以及在接近制造砖（温度低于 300℃）和使用期间（在 600～1400℃ 范围内）的条件热处理后的效果。所得结果的比较分析表明，生态黏合剂在其原始状态下具有有序碳域，尽管树脂在低温（<300℃）处理后形成了更强的三维结构。当两种黏合剂混合并且温度从 600℃ 开始升高时，实验数据表明它们在没有或几乎没有相互作用的情况下演变；当存在生态黏合剂时，产生的碳具有更多和更大的结晶区域。最后简要分析了结合剂的特性对掺入它们的 MgO-C 砖部分性能的影响。

Organic substances are used in refractory bricks manufacture as binders. In this work, two commercial binder systems currently employed in MgO-C bricks for steelmaking industry were first analyzed separately from the other components of the refractory. A **resol phenolic resin** alone and mixed with an eco-binder produced by treatment of **tar pitch** were studied by several techniques (XRD, FTIR, Raman spectroscopy, DTA/TGA and measurements of density, porosity, and mechanical and oxidation resistances) in their as-received condition, as well as after thermal treatments in conditions near to those used in the manufacture of the bricks (temperatures lower than 300℃) and also during their service (in the range 600℃ to 1400℃). The comparative analysis of the obtained results indicates that the eco-binder has domains of ordered carbon in its original condition, although the resin forms a stronger three-dimensional structure after treatments at low temperature (<300℃). When both binders are mixed and temperature rises from 600℃ onwards, experimental data show that they evolve with no or little interaction; The produced carbon has more and larger crystalline regions when the eco-binder is present. Finally, the effect of the binders' characteristics on some properties of MgO-C bricks which have them incorporated were briefly analyzed[2].

本研究介绍了含有葡萄糖、柠檬酸和金属铝的镁质耐火材料及其对结构、机械强度和抗热震性的影响。结合热力学计算、微观结构和性能分析，结果表明，引入的葡萄糖和柠檬酸形成葡萄糖酸盐和柠檬酸盐，增强了原料的结合力和试样的固化强度。葡萄糖酸盐和柠檬酸盐的热解以及金属 Al 的氧化和氮化对高温处理后镁质耐火材料的组织演变有显著影响，主要是 AlN 和 $MgAl_2O_4$ 孔和颗粒间隙中的晶须。晶须的形成和发展提高了试样的韧性和抗热震性。将金属 Al 含量（质量分数）从 0 增加到 2% 可使 1400℃ 热处理试样的断裂表面能从 82.22kJ/m² 提高到 119.75kJ/m²。同时，经过三个热冲击循环后，剩余强度热处理试样的比例从 52.8% 提高到 67.9%。在不同温度下相变对试样结构和性能的影响的基础上，考虑到冶金领域用镁质耐火材料的结构致密性和承受外应力的能力，建议在使用**葡萄糖**和**柠檬酸**作为结合剂时，金属 Al 的添加量（质量分数）应小于 3%。

A study of magnesia refractories containing glucose, citric acid, and metallic Al and their influences on structure, mechanical strength, and thermal shock resistance, is presented. Combined with thermodynamic calculation, microstructure, and properties analysis, the results show that formation of gluconates and citrates from the introduced glucose and citric acid enhanced the bonding of raw materials and the curing strength of samples. The pyrolysis of gluconates and citrates, and the oxidation and nitridation of metallic Al had significant effects on the microstructural evolution of magnesia refractories after high temperature treatments, mainly AlN and $MgAl_2O_4$ whiskers in pores and particle gaps. The formation and development of whiskers give rise to samples improved toughness and thermal shock resistance. Increasing the metallic Al content from 0 to 2wt.% enhanced the fracture surface energy of samples heat treated at 1400℃ from 82.22kJ/m² to 119.75kJ/m². Meanwhile, after three thermal-shock cycles, the residual strength ratios of heat-treated samples were elevated from 52.8% to 67.9%. On the basis of influences of phase transformation on the structure and properties of samples at different temperatures, considering the structural compactness and ability to withstand external stress for magnesia refractories used in metallurgical field, it is suggested that when **glucose** and **citric acid** are used as binders, the addition of metallic Al should be less than 3wt.% [3].

在 600～1200℃ 的还原气氛中，研究了在不同数量硝酸镍存在下**酚醛树脂**的石墨化过程，处理时间为 3h。作为比较，研究了催化**酚醛树脂**与非催化酚醛树脂在 800～1400℃ 范围内对 Al_2O_3-C 耐火试样的微观结构和力学性能的影响。分别使用 X 射线衍射和 SEM 研究了相组成和微观结构演变。采用拉曼光谱法测定酚醛树脂的石墨化程度。研究结果表明，结晶碳（石墨）仅由镍催化的酚醛树脂在高温下获得（约 1200℃），而非催化树脂导致无定形碳的形成。由于催化酚醛树脂的存在，Al_2O_3-C 试样的力学性能得到改善。SEM 结果表明，这种增强与含 Ni 催化酚醛树脂试样微观结构中晶须状 SiC 相的原位形成有关。

The catalytic graphitization of the **phenolic resin** was investigated in the presence of various amounts of nickel nitrate at 600-1200℃ in a reducing atmosphere for 3h. For comparison, the effect of catalyzed **phenolic resin** versus none-catalyzed one was explored on the microstructure and mechanical properties of the Al_2O_3-C refractory specimens within a range of 800-1400℃. The phase composition and microstructure evolution were investigated using X-ray diffraction and SEM, respectively. Raman spectroscopy was used to measure the graphitization level of phenolic resin. Results indicated that crystalline carbon (graphite) was achieved only from Ni-catalyzed phenolic resin at a high temperature (about 1200℃), while non-catalyzed resin led to the formation of amorphous carbon. Mechanical properties of the Al_2O_3-C specimens were improved due to the presence of catalyzed phenolic resin. SEM results illustrated that the enhancement was associated with the in-situ formation of the SiC phase in whisker shape in the microstructure of specimens containing Ni-catalyzed phenolic resin[4].

在开发、生产和使用通过**酚醛树脂**作为黏合剂制备的 MgO-C 耐火材料期间释放出来的有害成分，例如甲醛和酚醛化合物，可能会对环境和人类健康造成严重伤害。在本文中，制备了用蔗糖作为黏合剂和金属铝制粉末作为抗氧化剂的 MgO-C 耐火材料。不同的炉渣温度和时间因素在很大程度上影响了准备试样的炉渣腐蚀性。通过分析试样的微结构和组成，发现在恒定时间的条件下，温度越高，炉渣腐蚀现象越明显。在等温条件下，时间越长，炉渣腐蚀现象就越明显。同时，据显示，通过添加蔗糖作为黏合剂制备的 MgO-C 耐火材料的主要腐蚀过程是碳氧化和炉渣腐蚀，这为优化 MgO-C 耐火材料的抗渣性提供了理论基础。

The harmful components such as formaldehyde and phenolic compounds released during the development, production and use of MgO-C refractories prepared by using **phenolic resin** as a binder can cause serious harm to the environment and human health. In this paper, MgO-C refractory material with sucrose as binder and metal aluminium powder as antioxidant is prepared. Different slag temperature and time factors have a largely influence on the slag corrosion resistance of as-prepared samples. By analyzing the microstructure and composition of the samples, it is found that under the condition of constant time, the higher the temperature, the more obvious the slag corrosion phenomenon, under isothermal conditions, the longer the time, the more obvious the slag corrosion phenomenon. At the same time, it is revealed that the main corrosion process of MgO-C refractories prepared by adding sucrose as a binder is carbon oxidation and slag corrosion, which provides a theoretical basis for optimizing the slag resistance of MgO-C refractories[5].

中空多壁碳纳米管（MWCNTs）通过添加纳米二氧化锰（nano-MnO$_2$）作为催化剂的**酚醛树脂**催化裂化原位形成。研究了催化剂含量、焦化温度和加热速率对碳纳米管（CNTs）原位生长的影响。通过X射线衍射、场发射扫描电子显微镜、透射电子显微镜和激光-拉曼光谱对其组成和微观结构进行了表征。结果表明，纳米MnO$_2$可以催化酚醛树脂在材料内部部分形成多壁碳纳米管，从而改变热解碳结构，提高碳化后材料的石墨化程度。随着催化剂含量的增加，碳纳米管的长度减小，最佳添加量（质量分数）为1.0%。碳化温度越高，催化剂颗粒的活性越大。碳纳米管的最佳生长温度为1200℃。在较高的加热速率下，碳氢化合物分子难以沉积在催化剂颗粒表面形成碳纳米管。以2℃/min的加热速率获得的碳纳米管具有最高的石墨化程度。碳纳米管的生长符合顶部生长机制。

Hollow multi-walled carbon nanotubes(MWCNTs) were formed in situ by catalytic cracking of **phenolic resin** with addition of nano-manganese dioxide(nano-MnO$_2$) as a catalyst. The effects of catalyst content, coking temperature and heating rate on in situ growth of carbon nanotubes (CNTs) were investigated. The composition and microstructure were characterized by X-ray diffraction, field emission scanning electron microscopy, transmission electron microscopy and laser-raman spectrum. The results showed that nano-MnO$_2$ could catalyze phenolic resin to partly form MWCNTs inside of the materials, which transformed pyrolytic carbon structure and improved graphitization degree of the material after carbonization. With the increase of catalyst content, the length of CNTs decreased and the optimum addition amount was at 1.0wt.%. The higher the carbonization temperature, the bigger the activity of catalyst particles. The optimum growth temperature of CNTs was at 1200℃. At a higher heating rate, hydrocarbon molecules were hard to deposited on the surface of the catalyst particles to form CNTs. CNTs obtained at a heating rate of 2℃/min possessed the highest degree of graphitization. The growth of carbon nanotubes was accordance with the mechanism of top growth[6].

以四水柠檬酸钙和铝为原料，采用碳床烧结法合成原位含碳铝酸钙水泥（CCAC）作为原材料。通过X射线衍射、场发射扫描电子显微镜、高分辨率透射电子显微镜、拉曼光谱和红外碳硫分析对合成产物进行了表征。研究结果表明，在1500℃烧结4h后，产品的相组成接近商品水泥Secar71。产物中的原位碳具有部分石墨化域和多孔结构，均匀嵌入铝酸钙中，产物碳含量为 1.45%。CCAC 的漂浮率和氧化率低于碳背/Secar71（S71CB）复合粉末，表明水分散性和抗氧化性 CCAC 的改进。此外，分别比较了 CCAC 和 S71CB 结合刚玉基浇注料的冷压强度（CCS）和常温抗折强度（CMOR）。与 CCAC 结合的浇注料在 1100℃烧制 3h 后的 CCS 和 CMOR 值分别比与 S71CB 结合的浇注料高 20% 和 21%，表明 CCAC 可以作为一种很有应用前景的**耐火浇注料的黏合剂**。

In-situ carbon-containing calcium aluminate cement（CCAC）was synthesized through carbon-bed sintering with calcium citrate tetrahydrate and Al as raw materials. The synthesized product was characterized by X-ray diffraction, field-emission scanning electron microscopy, high-resolution transmission electron microscopy, Raman spectroscopy, and infrared carbon-sulfur analysis. The results show that after sintering at 1500℃ for 4h, the phase compositions of the product approached that of the commercial cement Secar71. The in-situ carbons in the product had partially graphitized domains and porous structures, were uniformly embedded in calcium aluminate, and the carbon content of the product was 1.45%. The floating ratios and oxidation ratios of the CCAC were lower than those of carbon back/Secar71（S71CB）composite powders, implying that the water dispersion and oxidation resistance of CCAC were improved. Furthermore, the cold crushing strength（CCS）, and cold modulus of rupture（CMOR）of the corundum-based castables bonded with CCAC, and S71CB, respectively, were compared. The CCS and CMOR values of the castables bonded with CCAC after being fired at 1100℃ for 3h are higher by 20% and 21%, respectively, than those of the castables bonded with S71CB, suggesting that CCAC can be applied as a promising **binder for the refractory castables**[7].

这项工作通过添加原位纳米铁颗粒作为催化剂，介绍了**酚醛树脂**（PR's）的催化石墨化过程。在还原气氛下在 600~1200℃下对制备的包含不同含量的纳米铁颗粒的组合物进行热解处理 3h，并通过不同的技术评估石墨化过程，如 X 射线衍射（XRD）、场发射扫描电子显微镜（FESEM）、高分辨率透射电子显微镜（HRTEM）、同步热分析（STA）和拉曼光谱，主要用于识别相和微观结构分析、抗氧化性和石墨化碳形成的扩展。研究结果表明，在还原气氛下，在 800℃下烧制试样 3h 后，已经观察到原位石墨碳发展，提高温度和纳米 Fe 的量导致更有效的石墨化水平。此外，在含纳米 Fe 试样的石墨化过程中，原位鉴定了洋葱和竹状等不同纳米晶碳形状和碳纳米管（CNT）。有人提出，这些不同纳米碳结构的形成与纳米铁催化剂行为和碳壳生长有关。

This work presents the catalytic graphitization process of **phenolic resins**（PR's）by addition of in situ nano-Fe particles as catalyst. Pyrolysis treatments of prepared compositions including various contents of nano-Fe particles were carried out at 600-1200℃ for 3h under reducing atmosphere and graphitization process were evaluated by different techniques such as X-ray diffraction(XRD), field emission scanning electron microscopy(FESEM), high resolution transmission electron microscopy(HRTEM), simultaneous thermal analysis(STA) and Raman spectroscopy that mainly performed to identify the phase and microstructural analysis, oxidation resistance and extend of graphitized carbon formation. Results indicate that, in situ graphitic carbon development were already observed after firing the samples at 800℃ for 3h under reducing atmosphere, increasing temperature and amount of nano-Fe led to a more effective graphitization level. In addition, the different nano crystalline carbon shapes such as onion and bamboo like and carbon nanotubes(CNTs) were in situ identified during graphitization process of nano-Fe containing samples. It was suggested that formation of these different nano carbon structures related to nano-Fe catalyst behavior and the carbon shell growth[8].

含碳耐火材料的耐化学性和热力学性能高度依赖结构类似于石墨的碳质相的存在。然而，大多数热固性树脂（用作黏合剂）被归类为非石墨化碳源。因此，人们对寻找替代途径以在与使用中的耐火材料相似的温度和条件下诱导此类聚合物组分的有效石墨化具有高度的兴趣。本研究评估了不同催化剂（二茂铁、赤铁矿和纳米 Fe_2O_3 粉末）和一些关于**酚醛清漆树脂**石墨化过程的加工参数（即温度、停留时间、加热速率、交联添加剂的量）。采用 X 射线衍射、SEM/EDS 和热重分析仪研究所制备试样的相组成和微观结构演变。研究结果表明，在1000℃和1400℃的还原气氛下处理5h 的试样观察到了树脂碳的石墨化。添加二茂铁（质量分数>2%）有利于在1400℃时产生更多的氧化铁颗粒，从而导致更有效的石墨化，而在最初试样的混料过程中，由于纳米 Fe_2O_3 粉末的团聚，抑制了高温下树脂碳的结晶。此外，对工艺参数的影响研究发现，当热处理温度为1400℃时，加热速率为3℃/min，保温时间为5h，添加剂为酚醛清漆树脂+3%（质量分数）二茂铁时，试样中的石墨碳生成量最大，为33.1%。

The chemical resistance and thermo-mechanical properties of carbon-containing refractories are highly dependent on the presence of carbonaceous phases with structures similar to graphite ones. However, most of the thermosetting resins(used as binders) are classified as non-graphitizing carbon sources. Consequently, there is a high degree interest in finding alternative routes to induce an effective graphitization of such polymeric components at temperatures and conditions similar to the ones that refractories are submitted to in service. This research evaluates the effect of different catalytic agents (ferrocene, hematite and nano-Fe$_2$O$_3$ powder) and some processing parameters (i. e. , temperature, dwell time, heating rate, amount of cross-linking additive) on the graphitization process of a **novolak resin**. X-ray diffraction, SEM/EDS and thermogravimetric analyses were carried out to identify the phase and microstructural evolution of the prepared compositions. According to the attained results, carbon graphitization was observed after firing the samples at 1000℃ and 1400℃ for 5h under a reducing atmosphere. The addition of ferrocene (>2wt.%) favored the generation of a higher amount of iron oxide particles in the microstructure at 1400℃ and led to a more effective graphitization, whereas the greater tendency for nano-Fe$_2$O$_3$ powder to agglomerate during the initial samples' mixing step inhibited carbon crystallization. Regarding the effect of the processing parameters, the use of a heating rate of 3℃/min and dwell time of 5h at the maximum temperature of 1400℃, led to the generation of a maximum amount of graphitic carbon, 33.1%, in the composition with novolak resin + 3wt.% ferrocene[9].

以硝酸镍为催化剂前驱体，研究了在 Ar 气氛下在 600~1200℃温度范围内催化热解**酚醛树脂**、大规模制备碳纳米管（CNTs）的低成本方法。采用 X 射线衍射、扫描电镜、透射电镜和能量色散 X 射线光谱对热解树脂的形貌和结构进行了表征。结果表明，碳纳米管的形貌和结构取决于热解温度，碳纳米管的起始生长温度约为 600℃。当热解温度升高时，碳纳米管的长度和结晶度显著增加。在 1000℃下可以获得平均直径为 50~60nm、微米级长度的高纵横比和结晶良好的 CNT。扫描电子显微镜和透射电子显微镜分析表明催化剂颗粒位于每个碳纳米管的顶部，这表明了碳纳米管的尖端生长机制。碳纳米管的生长过程将经历以下几个阶段：碳氢化合物成分分解成碳原子、溶解、扩散和偏析。碳纳米管的生长符合气固（VS）模型。

Using nickel nitrate as catalyst precursor, a cost-effective method for large scale preparation of carbon nanotubes (CNTs) was investigated through catalytic pyrolysis of **phenol resin** in the temperature range of 600-1200℃ under an Ar atmosphere. The morphology and structure of pyrolyzed resin were characterized by X-ray diffraction, scanning electron microscope, transmission electron microscope and energy dispersive X-ray spectroscopy. The results show that the morphology and structure of CNTs depend on pyrolysis temperature, and the starting growth temperature of CNTs is about 600℃. When the pyrolysis temperature is raised, the length and crystallinity of CNTs increase remarkably. High aspect ratio and well crystallised CNTs with average diameter of 50-60nm and micrometer scale length could be obtained at 1000℃. Scanning electron microscopy and transmission electron microscopy analysis reveal that the catalyst particle was located at the top of each CNT, which indicates the tip growth mechanism for CNTs. The growth process of CNTs will go through the following stages: decomposition of hydrocarbon components into the carbon atoms, dissolution, diffusion and segregation. The growth of CNTs agrees with vapour-solid(V-S)model[10].

炭黑（CB）和碳纳米管（CNTs）被用作**酚醛树脂**的添加剂。通过差示扫描量热分析、X 射线衍射、扫描电子显微镜和压汞法研究了添加剂和热处理温度对热解酚醛树脂碳的抗氧化性和结构的影响。据观察，CNTs 和 CB 都提高了热解碳的石墨化程度和抗氧化性。含 CNTs 的碳的石墨化程度高于含 CB 的碳，但由于显微组织中孔隙率较高，前者的抗氧化性较低。热处理温度的升高也导致碳的石墨化程度和抗氧化性的提高。

Carbon black (CB) and carbon nanotubes (CNTs) were used as additives in **phenolic resin**. The effects of the additives and heat treatment temperature on the oxidation resistance and structure of carbon from the pyrolysed phenolic resin were investigated by differential scanning calorimetric analysis, X-ray diffraction, scanning electron microscopy, and mercury intrusion porosimetry. It was observed that both CNTs and CB improved the graphitization degree and oxidation resistance of the pyrolytic carbons. The graphitization degree of carbons containing CNTs was higher than that of those with CB, but the oxidation resistance of the former was lower due to the higher porosity in the microstructure. An increase in the heat treatment temperature also resulted in an improvement in the graphitization degree and oxidation resistance of the carbons[11].

本文详细研究了硝酸铁（FN）对热塑性**酚醛树脂**的催化石墨化反应。通过透射电子显微镜（TEM）、高分辨率透射电子显微镜（HRTEM）、X 射线衍射和拉曼光谱测量研究了包括洋葱状碳纳米颗粒和竹状碳纳米管在内的产物的形貌和结构特征。研究结果表明，随着 FN 负载量和1000℃停留时间的变化，产物呈现出不同的形貌。TEM 图像显示，竹状碳纳米管由数十根竹棍组成，洋葱状碳纳米颗粒由准球形同心封闭碳纳米笼组成。

The catalytic graphitization of thermal plastic **phenolic-formaldehyde resin** with the aid of ferric nitrate(FN) was studied in detail. The morphologies and structural features of the products including onion-like carbon nanoparticles and bamboo-shaped carbon nanotubes were investigated by transmission electron microscopy(TEM), high-resolution transmission electron microscopy(HRTEM), X-ray diffraction and Raman spectroscopy measurements. It was found that with the changes of loading content of FN and residence time at 1000℃, the products exhibited various morphologies. The TEM images showed that bamboo-shaped carbon nanotube consisted of tens of bamboo sticks and onion-like carbon nanoparticle was made up of quasi-spherically concentrically closed carbon nanocages[12].

酚醛树脂已广泛用作含碳耐火材料的黏合剂。然而，酚醛树脂中的碳被称为非石墨化碳，其抗氧化性较差。本文研究了含碳耐火材料中添加酚醛树脂的碳的抗氧化性和结晶性。在 MgO-C 耐火材料中，由于 MgO 和 C 反应生成的 Mg(g) 和 $CO_2(g)$ 的作用，增强了碳从酚醛树脂中的结晶，提高了抗氧化性，而在 Al_2O_3-C 耐火材料，由于 Al(g) 的平衡分压低，碳的结晶和抗氧化性没有变化。

Phenolic resin has been widely used as a binder for carbon-containing refractories. Carbon from phenolic resin, however, is so called non-graphitizing carbon and its oxidation resistance is poor. In this paper, oxidation resistance and crystallization of carbon from phenolic resin added to carbon-containing refractories were investigated. In MgO-C refractory, owing to the effect of Mg(g) and $CO_2(g)$ formed by the reaction between MgO and C, the crystallization of the carbon from phenolic resin was enhanced and the oxidation resistance was improved, whereas in Al_2O_3-C refractory, owing to the low equilibrium partial pressure of Al(g), no changes were found in the crystallization and the oxidation resistance of the carbon[13].

苯基硼酸（PBA）**改性酚醛树脂**（PBPR）由于其良好的加工性能和较高的成炭率，是最重要的酚醛树脂（PR）之一。然而，PR 和 PBPR 的热解机理仍然知之甚少，这限制了这些材料的合成、开发和应用。在这项工作中，通过酚醛树脂（NR）和 PBA 交联 NR（PBNR）在高温下研究了 PBPR 的热解过程，并探讨了 PBPR 高炭收率的原因。研究结果表明，NR 的热解始于分子骨架中共价单键的裂解反应。由 C—C 单键断裂形成的单酚和双酚等挥发性有机化合物（VOCs）的释放是 NR 在高温下炭收率降低的主要原因。同时，提高 NR 的分子量和交联 NR 可以降低 VOCs 的形成概率，有利于获得高产炭率树脂。在 PBPR 中，硼酸酯键使 VOCs 的形成更加困难，并有助于树脂的高炭产率。本工作可为深入了解 PR 和 PBPR 的热解机理以确保它们在高温下的高热稳定性和成炭率提供指导意义。

Phenylboronic acid(PBA) **modified phenolic resin**(PBPR) is one of the most important phenolic resins(PR) due to its good processability and high char yield. However, the pyrolysis mechanisms of PR and PBPR still remain poorly understood, which imposes limits on the synthesis, development and application of these materials. In this work, the pyrolysis process of PBPR has been studied through novolac resin(NR) and PBA crosslinked NR(PBNR) at high temperatures, and the reasons for the high char yield of PBPR were explored, as well. According to the results, the pyrolysis of NR begins with the cleavage reaction of the covalent single bonds in the molecular backbone. The release of volatile organic compounds (VOCs) such as monophenols and bisphenols formed by the cleavage of C—C single bonds is the major cause of the decreased char yield of NR at elevated temperatures. Meanwhile, increasing NR molecular weight and crosslinking NR could reduce the formation probability of VOCs, which is beneficial to obtaining the high char yield resin. In PBPR, the boronic ester linkages make the formation of VOCs more difficult and contribute to the high char yield of resin. The present work provides guiding significance for deeply understanding the pyrolysis mechanism of PR and PBPR to ensure their high thermal stability and char yield at high temperatures[14].

　　酚醛树脂作为含碳耐火材料的结合剂，由于其具有可在常温下混练、结合强度高、残碳率高以及环境污染少等优点，而被广泛应用于耐火材料领域。但是，酚醛树脂中存在酚羟基和亚甲基，其在固化过程中会放出水，容易使碱性耐火材料的组分产生水化，而使其使用性能降低。本文针对热塑性酚醛树脂的结构特性，对树脂进行接枝改性研究，获得如下结论：（1）增大甲醛与苯酚的摩尔比，延长合成反应时间，增加催化剂用量均可增加热塑性酚醛树脂的相对分子质量，提高其软化温度、聚合速度、固含量和残碳率，而流动度和游离酚含量降低。热塑性酚醛树脂的适宜合成条件：甲醛与苯酚摩尔比为 0.85，反应时间为 3.5h，催化剂用量为 1.5%。合成获得的热塑性酚醛树脂的结构以邻位取代和对位取代为主，并有少量 2、4、6 位同时取代。（2）随着氯丙烯用量的增加，烯丙基酚醛树脂的特性黏数增加，相对分子质量增大，产率降低。烯丙基酚醛树脂的固化是通过碳碳双键的加成反应来实现的，其固化温度比普通酚醛树脂高约 100℃，固化窗口较宽。固化工艺参数为：在 180℃下固化 2h，升温到 225℃时再固化 4h。树脂固化后表面平整，为连续相结构，没有孔洞出现。烯丙基酚醛树脂的残碳率为 31.34%，低于热塑性酚醛树脂。烯丙基酚醛树脂含水量低，可减少氧化钙的水化。利用 Kissinger 法和 Ozawa 法分别求出了热塑性酚醛树脂和烯丙基酚醛树脂的固化活化能 E_a，建立了固化动力学模型，模型与实验数据吻合良好。固化初期，烯丙基酚醛树脂的活化能低于酚醛树脂；固化后期，两种树脂的固化反应为扩散控制，活化能随着固化温度的升高而增大。（3）与硼酸改性酚醛树脂相比，利用有机硼改性的酚醛树脂的特性黏数和固含量较高，聚合速度较低，但其软化点相同。硼酸改性酚醛树脂的固化温度较低，固化过程中仍有羟基参与反应。两种树脂的固化工艺参数为：硼酸改性酚醛树脂在 126℃ 下固化 2h，升温到 144℃ 时再固化 4h；有机硼改性酚醛树脂在 135℃ 下固化 2h，升温到 142℃ 时再固化 4h。硼酸改性酚醛树脂固化后，内部有尚未反应的硼酸纤维存在；而利用有机硼改性的酚醛树脂表面平整、光滑、无孔洞和相分离现象。两种硼改性酚醛树脂的残碳率相同，增加硼元素含量，可提高树脂的残碳率。有机硼改性的酚醛树脂放水量低，可减少氧化钙的水化。利用 Kissinger 法和 Ozawa 法分别求出了两种树脂的固化活化能 E_a，其中利用硼酸改性的酚醛树脂的活化能较大，为 361.33kJ/mol，建立出两种树脂的固化动力学模型。（4）正硅酸乙酯改性硅烷偶联剂酚醛树脂（SKPF）固含量为 90.73%，树脂内部呈网络状结构，受热过程中不发生软化。硼酸改性硅烷偶联剂酚醛树脂（BKPF）固含量为 87.04%，相对分子质量高，特性黏数大。BKPF 固化的同时，伴随着缩聚反应发生，固化温度与热塑性酚醛树脂相比差别较小。两种树脂固化后，基体表面无孔洞出现。BKPF 的残碳率高于 SKPF，正硅酸乙酯的用量对 SKPF 的残碳率影响不大，增大硼酸的用量可提高 BKPF 的残碳率。BKPF 分解时放出的水较少，可以减少氧化钙的水化[15]。

Phenolic resin has been widely used for refractories as binder, because of its some excellent properties such as easy mixing at room temperature, high strength and carbon residue as well as less pollution, however, there are a few weakness, especially, the water component could be lost from the phenolic resin in heating process because the phenolic hydroxyl groups and methylene existed in the structure of the phenolic resin, and would react with the component of basic refractories to form hydroxide. It will cause the damage of the basic refractories. Basic above reasons, the phenolic resin was grafted and modified for decreasing the generation of water component in the carbonization of the resin. The obtained results are as follows. (1) The amount of relative molecular mass, softening temperature, polymerization rate, solid content and carbon residue of the phenolic resin could increase with the increasing of the molar ratio of formaldehyde to phenolic and the amount of catalyst, and the prolonging of the synthesis reaction time, however, the fluidity and the free phenol content could decrease. The optimum condition for the synthesis of the phenolic resin was as follows. That was the molar ratio of formaldehyde to phenol was 0.85, the reaction time was 3.5h, and the amount of catalyst is 1.5%. The structure of synthesized phenolic resin was mainly ortho-substituted and para-substituted, and a small amount of 2,4,6-substituted. (2) With the increasing of the amount of allyl chloride, the limiting viscosity number of allyl phenolic resin increased, the relative molecular mass increased, and the yield decreased. The curing of allyl phenolic resin was achieved by the addition reaction of the carbon-carbon double bond, the curing temperature was higher about 100℃ than ordinary phenolic resin, and the curing window was wider. The curing process parameters could be obtained that curing for 2h at 180℃, the temperature was raised to 225℃ and cured 4h. The surface of cured resin was flat, continuous phase structure and nonporous. The carbon residue rate of allyl phenolic resin was 31. 34% , lower than that of the thermoplastic phenolic resin, but it had lower moisture content, and could reduce the hydration of the calcium oxide. The curing activation energy E_a of thermoplastic phenolic resin and allyl phenolic resin was separately calculated by Kissinger and Ozawa methods, and the curing kinetics model was established, the model with experimental data were in good agreement. The activation energy of allyl phenolic resin was lower than phenolic resin at curing initial stage, and the curing reaction of the two resins was controlled by diffusion, the activation energy was increased with the curing temperature increased. (3) Compared with the phenol resin modified by boric acid, the phenol resin modified by organic boron showed lower limiting viscosity, polymerization rate and higher solid content, however the softening point was same as each other. The curing temperature of phenolic resin modified by boric acid was lower, and the hydroxyl groups were still involved in the reaction during curing process. The curing process parameters of two kinds of resins could be obtained that the phenolic resin modified by boric acid, was cured for 2h at 126℃, the temperature was raised to 144℃ and cured 4h, and the phenolic resin modified by organic boron, was cured for 2h at 135℃, the temperature was raised to 142℃ and cured 4h. The phenolic resin modified by boric acid, had unreacted borate fibers after curing, and the surface of phenolic resin modified by organic boron was smooth, nonporous and no phase separation. Two kinds of phenolic resins had the same residual carbon ratio, increasing the boron element content, the residual carbon ratio of the resin could be improved. The water flowage of the phenolic resin modified by organic boron was low, and it could reduce the hydration of the calcium oxide. The curing activation energy E_a of two kinds of resins were separately calculated by Kissinger and Ozawa methods, the curing activation energy of the phenolic resin modified by boric acid was large, it was 361. 33kJ/mol, and the curing kinetics model was established. (4) The solid content of SKPF was 90. 73% , its interior structure looked like a net, and the softened phenomenon would not take place in the heating process. The solid content of BKPF was 87. 04% , and both of the relative molecular mass and the intrinsic viscosity number were high. When the BKPF cured, the polycondensation reaction was processing. Compared with the thermoplastic phenolic resin, the difference on the curing temperature was small. There was no hole on the surface of substrate after the curing of the two resins. The carbon residue of BKPF was higher than that of SKPF. The amount of TEOS made little effect on the carbon residue rate of SKPF. The carbon residue rate of BKPF increased with the increase of the amount of boric acid. The water released during the decomposition of BKPF was less, which would reduce the hydration of calcium oxide[15].

参考文献

[1] Luz A P, Salomão R, Bitencourt C S, et al. Thermosetting resins for carbon-containing refractories: Theoretical basis and novel insights [J]. Open Ceramics, 2020, 3: 100025.

[2] Pássera C L, Manfredi L B, Tomba Martínez A G. Study of the thermal behavior of organic binders used in oxide-carbon refractory bricks [J]. Metallurgical and Materials Transactions B, 2021, 52: 1681-1694.

[3] Liu Hao, Jie Chuang, Zhang Zhongzhuang, et al. Microstructure and mechanical properties of magnesia refractories containing metallic Al by the incorporation of glucose and citric acid as binders [J]. Ceramics International, 2021, 47 (18): 26310-26318.

[4] Darban S, Kakroudi M G, Vandchali M B, et al. Characterization of Ni-doped pyrolyzed phenolic resin and its addition to the Al_2O_3-C refractories [J]. Ceramics International, 2020, 46 (13): 20954-20962.

[5] Shen Xiulin, Zhu Ling, Lv Zhenfei, et al. The behavior of slag resistance of MgO-C refractory prepared by sucrose as binder [J]. IOP Conference Series: Materials Science and Engineering, 2019, 678: 012091.

[6] Li Jinyu, Tang Bingjie, Tu Junbo, et al. Effects of the addition of nano-MnO_2 on in situ catalytic formation of carbon nanotubes during the coking of phenolic resin [J]. Journal of the Ceramic Society of Japan, 2019, 127 (4): 191-198.

[7] Xiao Guoqing, Yang Shoulei, Ding Donghai, et al. One-step synthesis of in-situ carbon-containing calcium aluminate cement as binders for refractory castables [J]. Ceramics International, 2018, 44 (13): 15378-15384.

[8] Rastegar H, Bavand-vandchali M, Nemati A, et al. Catalytic graphitization behavior of phenolic resins by addition of in situ formed nano-Fe particles [J]. Physica E: Low-dimensional Systems and Nanostructures, 2018, 101: 50-61.

[9] Bitencourt C S, Luz A P, Pagliosa C, et al. Role of catalytic agents and processing parameters in the graphitization process of a carbon-based refractory binder [J]. Ceramics International, 2015, 41 (10): 13320-13330.

[10] Zhu B Q, Wei G P, Li X C, et al. Preparation and growth mechanism of carbon nanotubes via catalytic pyrolysis of phenol resin [J]. Materials Research Innovations, 2013, 18 (4): 267-272.

[11] Liang Feng, Li Nan, Li Xuanke, et al. Effect of the addition of carbon black and carbon nanotubes on the structure and oxidation resistance of pyrolysed phenolic carbons [J]. New Carbon Materials, 2012, 27 (4): 283-287.

［12］ Zhao Mu, Song Huaihe. Catalytic graphitization of phenolic resin ［J］. Journal of Materials Science and Technology, 2011, 27 (3): 266-270.

［13］ Yu Jingkun, Yamaguchi A. Crystallization and oxidation behavior of carbon from phenolic resin in MgO-C and Al_2O_3-C refractories ［J］. Journal of the Ceramic Society of Japan, 1995, 103 (1195): 274-277.

［14］ Xing Xiaolong, Zhang Ping, Zhao Yuhong, et al. Pyrolysis mechanism of phenylboronic acid modified phenolic resin ［J］. Polymer Degradation and Stability, 2021, 191: 109672.

［15］ 刘洋. 酚醛树脂的接枝改性研究 ［D］. 沈阳：东北大学, 2013.

10 耐火材料原料常用词汇术语查询

10.1 汉英对照

B

白黏土（陶土）argil

白云石（白云岩）dolomite, pearl spar, magnesium limestone

白云石化 dolomitization

半稳定性方石英 meta-cristobalite

变质石英岩 metaquartzite

C

超石英 stishovite

纯橄榄岩 dunite

D

低铁水镁石 ferrobrucite

电熔刚玉 electro-corundum

电熔石英 electro-quartz

多磷酸盐 polyphosphate

多水高岭土 ablykite

多水菱镁矿 lansfordite

E

二次煅烧白云石 magdolite

F

方解石（冰洲石）calcite, calcspar

方石英 cristobalite

方石英化 cristobalization

酚醛树脂黏合剂 aerodux

粉煤灰 flyash

风化黏土 aeroclay

G

钙橄榄石 calcio-olivine

钙镁橄榄石 shannonite

橄榄石 olivine, peridotite

刚玉 corundum

刚玉砂 emery

高硅膨润土 distribond

高岭石 kaolinite

高岭石族矿物 kandite

高岭土 kaolin

高岭土化 kaolinization

高锰白云石 mangan-dolomite

高温石墨 pyrographite

锆刚玉 zirconia-corundum

铬铁矿 chromite

硅石 quartzite

硅石（石英岩）silica, quartzite, ganister

硅线石 sillimanite, fibrolite

硅藻土 diatomite, tripolite, bergmeal, kieselguhr

H

含沥青的 bitumeniferous

含铝的（含铝土的）aluminous

含铝土的 aluminous

禾木树脂 accroides

黑高岭土 hisingerite

黑沥青 abbertite, albertite

红柱石 andalusite, apyre

红柱石质叶蜡石 andalusite-pyrophyllite

琥珀蛇纹石 amber

滑石（寿山石）agalmatolite

滑石棉 asbecasite

滑石质镁石 talc-magnesite

J

胶状高岭土 schroetterite

焦磷酸盐 pyrophosphate

焦石英 lechatelierite

焦炭 coke

K

凯石英 keatite

柯石英 corsite

可熔酚醛树脂 resol-bakelite

苦闪橄榄石 olivinite

块滑石 steatite

L

蓝晶石 kyanite, cyanite, disthene, zianite

沥青 asphalite, asphalt, pitch, asphaltum, goudron

磷石英 tridymite

磷酸盐 phosphate

鳞石英化 tridymization

菱镁矿 magnesite, giobertite

六聚偏磷酸盐 hexametaphosphate

铝矾土（铝土矿）bauxite

铝热莫来石 thermit-mullite

铝土化 bauxitization

绿硅线石 bamlite

绿粒橄榄石 glaucochroite

M

煤矸石 gangue

镁白云石 konite, magnesiodolomite, magnesite-dolomite

镁橄榄石 boltonite, forsterite

镁铬尖晶石 picrochromite

镁铬铁矿 magnesiochromite

镁砂 magnesia

镁铁橄榄石 hortonolite

蒙脱石 smectite, askanite

锰橄榄石 tephroite

锰红柱石 viridine

锰铁橄榄石 knebelite

明矾石质叶蜡石 alunite-pyrophyllite

莫来石 mullite

莫来石化 mullitization

木炭 charcoal

木屑（锯末）sawdust

N

黏土 clay

黏土质的 clayish

镍纤蛇纹石 garnierite

P

膨润土（膨土岩）amargosite, bentonite

偏高岭石 metakaolinite

偏高岭土 metakaolin

偏磷酸盐 metaphosphate

片状蛇纹石 plate-serpentine

漂珠 cenosphere

R

热固性树脂 resinoid

人造刚玉 boule

S

三偏磷酸盐 trimetaphosphate, tripolyphosphate
三水菱镁矿 nesquehonite
烧结尖晶石 spinel-sintered
蛇纹石 serpentine
石棉 amiantus, amianthus, asbestus
石墨 graphite
石墨的 graphitic
石墨化 graphitization
石英 silex, quartz
石英砂岩 silicarenite
石英质叶蜡石 quartz-pyrophyllite
树脂 resin
树脂的 resinous
水白云母 hydro-dolomite
水白云石 hydromuscovite
水滑石 hudrotalcite
水菱镁矿 hydromagnesite
水镁铬石 barbertonite
水镁石 brucite
水蛇纹石 deweylite
水纤菱镁矿 artinite
四聚磷酸盐 tetrapolyphosphate

T

钛菱镁矿 titanomignesite
碳 carbon
碳沥青 anthraxolite

铁白云石 ankerite, ferrodolomite
铁橄榄石 fayalite
铁菱镁矿 breunerite
铁石棉 amosite
铜菱镁矿 cupromagnesite

W

无烟煤 anthracite

X

纤滑石 agalite, talcum, talc
纤维 fiber
纤维蛇纹石 chrysotile

Y

亚黏土（砂质黏土）loam
叶蜡石 pyrophyllite, pyrauxite, roseki
叶蜡石基 pyrophyllite-based
叶蛇纹石 antigorite
异纤蛇纹石 asbophite
硬化黏土 clunch
硬黏土 leck
硬水铝石质叶蜡石 diaspore-pyrophyllite

Z

皂石（块滑石）soapstone
珍珠岩 perlite
正磷酸盐 orthophosphate
蛭石 vermiculite

10.2   英汉对照

A

abbertite, albertite 黑沥青
ablykite 多水高岭土

accroides 禾木树脂
aeroclay 风化黏土
aerodux 酚醛树脂黏合剂
agalite, talcum, talc 纤滑石

agalmatolite 滑石（寿山石）

aluminous 含铝的（含铝土的）

alunite-pyrophyllite 明矾石质叶蜡石

amargosite，bentonite 膨润土（膨土岩）

amber 琥珀蛇纹石

amiantus，amianthus，asbestus 石棉

amosite 铁石棉

andalusite，apyre 红柱石

andalusite-pyrophyllite 红柱石质叶蜡石

ankerite，ferrodolomite 铁白云石

anthracite 无烟煤

anthraxolite 碳沥青

antigorite 叶蛇纹石

argil 白黏土（陶土）

artinite 水纤菱镁矿

asbecasite 滑石棉

asbophite 异纤蛇纹石

asphalite，asphalt，pitch，asphaltum，goudron 沥青

B

bamlite 绿硅线石

barbertonite 水镁铬石

bauxite 铝矾土（铝土矿）

bauxitization 铝土化

bitumeniferous 含沥青的

boltonite，forsterite 镁橄榄石

boule 人造刚玉

breunerite 铁菱镁矿

brucite 水镁石

C

calcio-olivine 钙橄榄石

calcite，calcspar 方解石（冰洲石）

carbon 碳

cenosphere 漂珠

charcoal 木炭

chromite 铬铁矿

chrysotile 纤维蛇纹石

clay 黏土

clayish 黏土质的

clunch 硬化黏土

coke 焦炭

corsite 柯石英

corundum 刚玉

cristobalite 方石英

cristobalization 方石英化

cupromagnesite 铜菱镁矿

D

deweylite 水蛇纹石

diaspore-pyrophyllite 硬水铝石质叶蜡石

diatomite，tripolite，bergmeal，kieselguhr 硅藻土

distribond 高硅膨润土

dolomite，pearl spar，magnesium limestone 白云石
（白云岩）

dolomitization 白云石化

dunite 纯橄榄岩

E

electro-corundum 电熔刚玉

electro-quartz 电熔石英

emery 刚玉砂

F

fayalite 铁橄榄石

ferrobrucite 低铁水镁石

fiber 纤维

flyash 粉煤灰

G

gangue 煤矸石

garnierite 镍纤蛇纹石

glaucochroite 绿粒橄榄石

graphite 石墨

graphitic 石墨的

graphitization 石墨化

H

hexametaphosphate 六聚偏磷酸盐

269

hisingerite 黑高岭土

hortonolite 镁铁橄榄石

hudrotalcite 水滑石

hydro-dolomite 水白云母

hydromagnesite 水菱镁矿

hydromuscovite 水白云石

K

kandite 高岭石族矿物

kaolin 高岭土

kaolinite 高岭石

kaolinization 高岭土化

keatite 凯石英

knebelite 锰铁橄榄石

konite, magnesiodolomite, magnesite-dolomite 镁白云石

kyanite, cyanite, disthene, zianite 蓝晶石

L

lansfordite 多水菱镁矿

lechatelierite 焦石英

leck 硬黏土

loam 亚黏土（砂质黏土）

M

magdolite 二次煅烧白云石

magnesia 镁砂

magnesiochromite 镁铬铁矿

magnesite, giobertite 菱镁矿

mangan-dolomite 高锰白云石

meta-cristobalite 半稳定性方石英

metakaolin 偏高岭土

metakaolinite 偏高岭石

metaphosphate 偏磷酸盐

metaquartzite 变质石英岩

mullite 莫来石

mullitization 莫来石化

N

nesquehonite 三水菱镁矿

O

olivine, peridotite 橄榄石

olivinite 苦闪橄榄石

orthophosphate 正磷酸盐

P

perlite 珍珠岩

phosphate 磷酸盐

picrochromite 镁铬尖晶石

plate-serpentine 片状蛇纹石

polyphosphate 多磷酸盐

pyrographite 高温石墨

pyrophosphate 焦磷酸盐

pyrophyllite, pyrauxite, roseki 叶蜡石

pyrophyllite-based 叶蜡石基

Q

quartzite 硅石

quartz-pyrophyllite 石英质叶蜡石

R

resin 树脂

resinoid 热固性树脂

resinous 树脂的

resol-bakelite 可熔酚醛树脂

S

sawdust 木屑（锯末）

schroetterite 胶状高岭土

serpentine 蛇纹石

shannonite 钙镁橄榄石

silex, quartz 石英

silica, quartzite, ganister 硅石（石英岩）

silicarenite 石英砂岩

sillimanite, fibrolite 硅线石

smectite, askanite 蒙脱石

soapstone 皂石（块滑石）

spinel-sintered 烧结尖晶石